Workman

Developments in Environmental Modelling, 10

Agricultural Nonpoint Source Pollution: Model Selection and Application

Developments in Environmental Modelling

Series Editor: S.E. Jørgensen
Langkaer Vaenge 9,
3500 Vaerløse,
Copenhagen,
Denmark

1.   ENERGY AND ECOLOGICAL MODELLING
edited by W.J. Mitsch, R.W. Bosserman and J.M. Klopatek
1981 839 pp.

2.   WATER MANAGEMENT MODELS IN PRACTICE:
A CASE STUDY OF THE ASWAN HIGH DAM
by D. Whittington and G. Guariso
1983 xxii + 246 pp.

3.   NUMERICAL ECOLOGY
by L. Legendre and P. Legendre
1983 xvi + 419 pp.

4A.   APPLICATION OF ECOLOGICAL MODELLING IN
ENVIRONMENTAL MANAGEMENT, PART A
edited by S.E. Jørgensen
1983 viii + 735 pp.

4B.   APPLICATION OF ECOLOGICAL MODELLING IN
ENVIRONMENTAL MANAGEMENT, PART B
edited by S.E. Jørgensen and W.J. Mitsch
1983 viii + 438 pp.

5.   ANALYSIS OF ECOLOGICAL SYSTEMS: STATE-OF-THE-ART
IN ECOLOGICAL MODELLING
edited by W.K. Lauenroth, G.V. Skogerboe and M. Flug
1983 992 pp.

6.   MODELLING THE FATE AND EFFECT OF TOXIC SUBSTANCES
IN THE ENVIRONMENT
edited by S.E. Jørgensen
1984 viii + 342 pp.

7.   MATHEMATICAL MODELS IN BIOLOGICAL WASTE WATER
TREATMENT
edited by S.E. Jørgensen and M.J. Gromiec
1985 vi + 802 pp.

8.   FRESHWATER ECOSYSTEMS: MODELLING AND SIMULATION
by M. Straškraba and A. Gnauck
1985 309 pp.

9.   FUNDAMENTALS OF ECOLOGICAL MODELLING
by S.E. Jørgensen
In preparation

Developments in Environmental Modelling, 10

Agricultural Nonpoint Source Pollution: Model Selection and Application

Edited by

ALDO GIORGINI
School of Civil Engineering, Purdue University, West Lafayette, IN 47907 (U.S.A.)

FRANCO ZINGALES
Cattedra di Chimica, Facoltà di Ingegneria, Università di Padova (Italy)

Coedited by

ALESSANDRO MARANI
Facoltà di Chimica Industriale, Università di Venezia (Italy)

JACQUES W. DELLEUR
School of Civil Engineering, Purdue University, West Lafayette, IN 47907 (U.S.A.)

Contributions to a Workshop held in June 1984 in Venice, Italy

Sponsored by
National Science Foundation, U.S.A.
Consiglio Nazionale delle Ricerche, Italy

ELSEVIER
Amsterdam — Oxford — New York — Tokyo   1986

Published jointly by

INTERNATIONAL SOCIETY FOR ECOLOGICAL MODELLING (ISEM)
Langkaer Vaenge 9, 3500 Vaerløse, Copenhagen, Denmark

and

ELSEVIER SCIENCE PUBLISHERS B.V.
Sara Burgerhartstraat 25
P.O. Box 211, 1000 AE Amsterdam, The Netherlands

Distributors for the United States and Canada:

ELSEVIER SCIENCE PUBLISHING COMPANY INC.
52 Vanderbilt Avenue
New York, NY 10017

ISBN 0-444-99505-6 (Vol. 10)
ISBN 0-444-41948-9 (Series)

Printed in Denmark by Fair-Print AS, Roskilde

CONTENTS

PREFACE

This volume contains most of the scientific contributions to the workshop "Prediction of Agricultural Nonpoint Source Pollution: Model Selection and Application" held in Venice, in the historic Ca' Vendramin Calergi, in June 1984. Other contributions of specialists who had not been able to attend the workshop have been included in an attempt to make the work more complete.

It is hoped that this collection be useful to planners who operate in the field of agricultural diffuse source pollution, since several contributions are state-of-the-art presentations and others are specialized studies by American and European researchers.

We wish to thank all the people who have contributed to this volume, starting with Giuseppe Bendoricchio and Andrea Rinaldo, in Italy, and with David B. Beasley and Larry F.Huggins, in the USA, for their collaboration to the venture which has led to the workshop and to the publication of this volume.

A special acknowledgment is offered to Timothy R. Ginn for his invaluable help in serving as English language editor.

We end with expressions of sincere gratitude to all the sponsors of the workshop, in particular the National Science Foundation in the U.S.A. and the Consiglio Nazionale delle Ricerche in Italy, for their contribution in financing the workshop and the publication of this volume; to the academic sponsors, Purdue University (in the persons of Steven C. Beering, President; John F. McLaughlin, Acting Dean of the Schools of Engineering; and Harold L. Michael, Head of the Schools of Civil Engineering) and the Università degli Studi di Padova (in the persons of Luciano Merigliano, Rettore, and Lucio Susmel, Facoltà di Agraria); and finally to Sven E.Jorgensen whose enthusiastic support has been essential to the publication of these contributions to the field of nonpoint source pollution modeling.

Aldo Giorgini   Franco Zingales
Purdue University   Università Padova

A REVIEW OF HYDROLOGIC AND WATER QUALITY MODELS USED FOR SIMULATION OF AGRICULTURAL POLLUTION

Vladimir Novotny
Professor, Department of Civil Engineering
Marquette University
Milwaukee, WI 53233, U.S.A.

ABSTRACT

Nonpoint source simulation programs are part of a category of loading models which describe primarily formation of runoff and generation of pollutants from a source area. They can be divided into continuous simulation models or event oriented models. They also can be based on the distribution parameter or lumped parameter concept. In scope, they range from small field size application models to mostly deterministic, process-oriented, large watershed models. The available models range from simple application of the long term form of the Universal Soil Loss Equation, to medium complexity models that use a simple hydrological component, to multiple parameter-multicomponent models requiring a large computer memory and considerable amount of field data for calibration. Models have been developed to simulate hydrology, erosion and sediment process, nutrient (fertilizer) losses, and transport of organic chemicals from agricultural watersheds.

From a large number of models that have been developed in the United States in the last ten years, this paper focuses on those that are in practical use and/or have been used by practitioners for managing nonpoint pollution from agricultural operations.

INTRODUCTION AND HISTORICAL PERSPECTIVE

Throughout history, the term mathematical model has meant a mathematical description of a process or of a phenomena, usually in a form of an input-output relationship. Today the term mathematical model implies a computer software type program that has evolved from a known mathematical description of a process.

Historically, development of mathematical models that have been used for modeling nonpoint pollution can be divided into four periods. The first period -- the precomputer age -- covered approximately 60 years and lasted from 1900 to the end of the 1950's. In this period the scientific fundamentals of the present mathematical models were formulated. For example, in 1911 Green and Ampt (1) developed their model of infiltration that is now very popular. Their work was followed by Horton (2) and Philip (3).

In the 1950's, scientists from the U.S. Department of Agriculture-Soil Conservation Service studied rainfall and its losses and soil loss phenomena. The former effort resulted in the development of a simple model for estimating surface runoff from a daily rainfall - the SCS Curve Number Model (4). In the latter effort many years of experimental plot data were analyzed to develop the Universal Soil Loss Equation for estimating gross erosion by water (5). Many other processes and phenomena that were later incorporated into mathematical models were studied and formulated in this period.

Introduction of digital computers initiated a second period of model development -- creation -- that lasted for about ten years, through out the 1960's. During this period only large universities and research centers possessed computers. Few researchers could run and understand the models and the general technical public was unaware of them. However, scientists could formulate and create more complex models and the basis for development of complex mathematical models was laid out. In 1966 Crawford and Linsley published one of the first hydrologic models -- the Stanford Watershed Model that was to become the basis for numerous studies (6). Components of the Stanford Watershed Models have been incorporated into several large models in present use.

The passage of the Federal Water Pollution Control Act Amendments by the U.S. Congress in 1972 initiated the third period of modeling activities -- generation. Consequently, by introducing time-sharing and remote terminal features to the computer hardware, the computer has become more available to scientists and engineers. Financial resources were devoted by U.S. federal agencies, namely the U.S. Environmental Protection Agency (EPA), U.S. Department of Agriculture (U.S.D.A.), and the National Science Foundation (NSF), to the development of mathematical hydrologic and nonpoint pollution models.

Most of the field applications of these models in the 1970's were of
an experimental nature.   Only a few models were extensively tested
and occasionally comparisons of measured field data vs. simulated
outputs were not satisfactory.   Therefore, in the 1970's several
supportive field studies were   carried out   throughout the U.S. to
provide the data base necessary for successful application and
verification of models. The U.S. EPA created a data storage bank -
STORET - from which field data on quantity and quality could be
accessed via a remote terminal.

The fourth period -- implementation -- is closely related to the
rapid advance of commercially available inexpensive minicomputers in
the 1980's.   The minicomputers, in spite of their small   size and
relatively low cost, are now reaching or even surpassing the storage
and computing capacity of medium size computers of fifteen to twenty
years ago. The practical user -- an agricultural or civil engineer
or local soil conservation specialist -- is now capable of using a
model for his routine evaluations.   During the same period a new
generation of engineers and specialists educated in the use of computers has begun to appear. In this period no major model development comparable to the period of the 1970's is underway in the U.S.
However, a number of commerical software companies and consultants
as well as the original creators are now adapting the models for the
minicomputers and for a wide practical use.

TYPES OF MODELS

The nonpoint pollution models fall generally into two categories:
the screening (unit loads) planning models and hydrological
assessment models (7, 8).

Screening Models

Screening models are usually simple tools which identify problem
areas in a large basin.   These models usually rely on assignment of
unit loads of pollution to the   various lands within the watershed.
A unit loading is a simple value or function expressing pollution
generation per unit area and unit time for each   typical land use.
The loads are typically expressed in kilograms per hectare-year.

Despite its questionable accuracy, the concept of relating pollution
loading to land use categories has found wide application in area-
wide pollution abatement efforts and planning.   One reason explains
this popularity: the concept provides a simple mechanism and quick
answers to pollution problems of large areas where more complicated
efforts would fail because of the enormous amounts of information
required.   The land use/pollutant loading is also compatible with so
called "overview modeling", whereby unit loadings are combined with

information on land use, soil distribution, and other characteristics to yield watershed loadings, or to identify areas producing or causing the highest amount of nonpoint pollution (Figure 1).

The magnitudes of unit loadings can be obtained for example from a publication by McElroy et al. (9). The modeling methodologies have been published by Johnson et al. (10), Sonzogni et al. (11) and Haith and Tubbs (12).

Loading functions for agricultural areas are commonly based on the Universal Soil Loss Equation (5). Use of the unit load concept presumes that an adequate inventory of land data is available from maps, aerial and terrestrial surveys, remote surveys, and local information. The loading concept is applicable - in most cases - to long term estimates such as average annual loading figures.

Hydrological Models

V.T. Chow divided hydrological models into eight categories (Figure 2). Analog and scale models are of little significance to nonpoint pollution modeling although they are used for studying the fundamental individual processes (e.g., infiltration, soil/water movement, and adsorption - desorption of chemicals on soil). A simulation model reproduces the behavior of a hydrologic phenomenon in detail but does not reproduce the phenomenon itself. Chow (13) classified the Stanford Watershed Model as a simulation model however, it could be also classified as abstract lumped-nonlinear model. Abstract models attempt to represent the prototype theoretically in a mathematical form. These models replace the relevant features of the system by a set of mathematical relationships. According to certainty or uncertainty of such relationships, the models can be further divided into deterministic or indeterministic types. Most of the nonpoint pollution models are in the category of physical-simulation or abstract-deterministic groups.

There are basically two approaches to modeling nonpoint pollution. The most common are the lumped parameter models while some more complex models developed recently are based on the distributed parameter concept.

The lumped parameter models treat the watershed or a large portion of it as one unit. The various characteristics of the unit are then lumped together, often with the use of an empirical equation, and the final form and magnitude of the parameters are simplified to represent the model unit as a uniform homogenous system. For such systems the input-output relationship may be present as:

$$Y = \phi X \qquad\qquad (1)$$

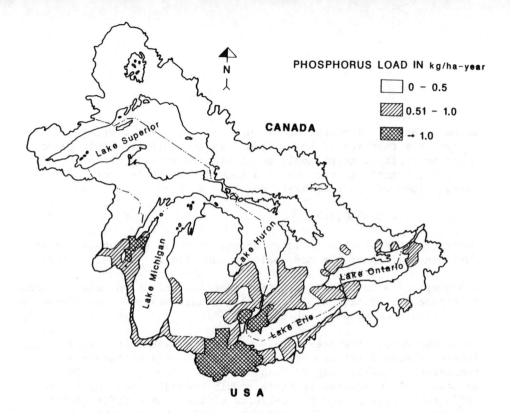

Fig. 1 Agricultural Sources of Phosphorus to the Great Lakes
Estimated by a Screening Model (10)

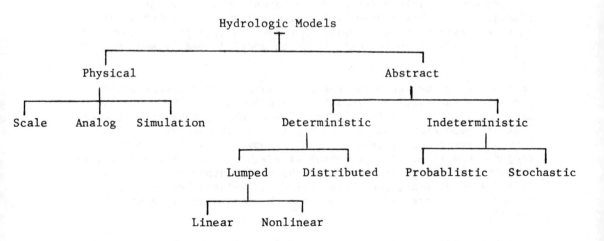

Fig. 2 Classification of Hydrologic Models (13)

where Y, and X are output and input vectors, respectively, and ϕ is
a transform function of the system. An Instantaneous Unit Hydrograph or Sediment graph could be considered as examples of a
watershed transform function in a lumped hydrologic model. A
storage form of the continuity equation such as:

$$X - Y - \text{losses} = \frac{dS}{dt} \tag{2}$$

where S is the storage of water or pollution in the system, is the
fundamental form of the mass balance equation for lumped systems.

A concept of a lumped hydrological nonpoint pollution model is shown
in Figure 3. Note that flow from one unit may overflow either to
the drainage or an adjacent unit.

The distributed parameter models divide the system into very small
finite elements (Figure 4). Each element has uniform system
parameters, soils, imperviousness, crop, slope, etc. The
mathematical foundation of these models commonly uses the finite
difference (element) representation of the basic differential
equation governing the flow and mass continuity and motion in one,
two or even three dimensions.

Theoretically, the lumped parameter model can provide only one
output location, while outputs can be obtained throughout the system
from distributed parameter models, that is, from each element. This
feature of distributed parameter model is one of their primary
benefits, since areal loading maps and graphics can be generated by
the computer. Distributed parameter models require large computer
storage and extensive description of the system parameters, which
must be provided for each unit. However, changes in the watershed
and their effect on the output can be modeled easily and more
effectively.

Models can be designed to run on an event or continuous basis.
Discrete event modeling simulates the response of a watershed to a
major rainfall or snowfall event. The principal advantage of event
modeling over continuous simulation is that it requires relatively
little meterological data and can be operated with a shorter
computer run time. The principal disadvantage of event modeling is
that it requires specification of the design storm and antecedent
moisture conditions, thereby assuming equivalence between the
recurrence interval of the storm and the recurrence interval of the
runoff.

Event oriented models are advantageous and proper for comparative
analyses of impacts of various land management and pollution
mitigation practices on water quality for predetermined (extreme or
average) conditions. Such models can not be used for estimation of
long term loadings of pollutants to a receiving water body without

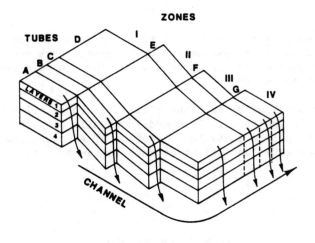

Fig. 3 Lumped Parameter Model Concept

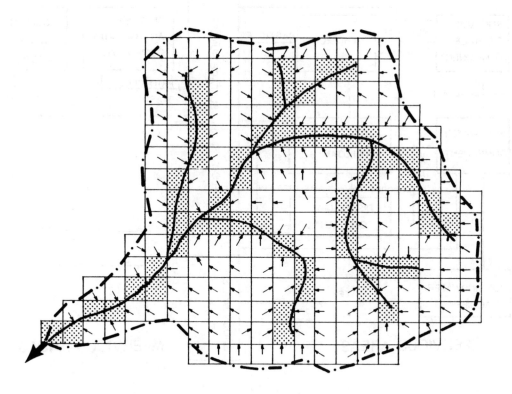

Fig. 4 Distributed Parameter Model Concept

15

difficulty and larger expense. Distributed parameter models due to their discretization of parameters and, hence, large computational time (CPU) requirements can be mostly run for a single event, or at most, a small series thereof.

Continuous process modeling sequentially simulates all processes incorporated in the model. Such models usually operate on a time interval ranging from a day to a fraction of an hour, and continuously balance water and pollutant mass in the system.

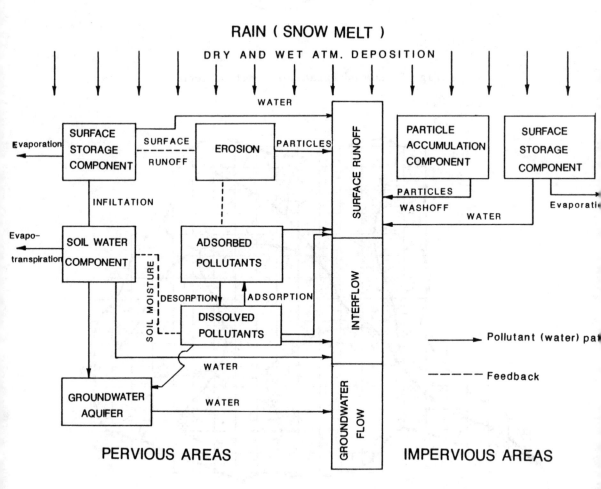

Fig. 5 Components of a Hydrology - Nonpoint Pollution Model

The principal advantage of continuous modeling is that it provides long-term series of water and pollutant loadings that can be analyzed statistically as to their frequency. A principal disadvantage of continuous modeling is that it requires long simulation runs, thus imposing restrictions on the number of alternatives that can be investigated. It also requires historical data on precipitation often in less than hourly intervals, which is not always available.

Indeterministic stochastic and probabilistic nonpoint pollution models are relatively rare and have not left research laboratories yet. Users of deterministic models usually overcome the random or probabilistic nature of the inputs and system parameters by a sensitivity analysis of the model, whereby the magnitudes of the most significant inputs and system parameters are changed within their probabilistic boundaries and the response of the deterministic model to these changes provides an estimate of the ranges of the output.

Bogardi and Duckstein (14) present an example of a stochastic model of P-loadings from nonpoint sources of Lake Balaton in Hungary.

STRUCTURE OF MODELS

Figure 5 shows the typical components of a nonpoint pollution model (8). These are:

1. A surface runoff generation component which describes the transformation of rainfall into runoff and its overland surface flow component. Modeling surface runoff includes the following processes.

    a) Exhaustion of surface storage;
    b) Evapotranspiration;
    c) Snow accumulation and melt.

2. Soil and groundwater components (not common to all models) which describe movement of water through the unsaturated soil zone and into a saturated groundwater zone. These components balances current soil moisture with infiltration rate, evaportranspiration, and water loss into a deep groundwater zone. Since infiltration is a function of soil moisture content, an iterative procecdure is usually employed. If the soil component is not included in the model, infiltration rate of surface runoff are estimated by an empirical equation (for example SCS-Runoff Curve Number Model). The most commonly used infiltration models are those by Green and Ampt (1), Phillip (2), and, Holtan (15).

17

3. Runoff and sediment routing component. In order to obtain runoff flow or pollution histograms, the excess rain and eroded soil must be routed overland to the watershed exit. In lumped parameter models, some form of transform function (e.g., the Nash (16) reservoir routing model) is employed. In distributed parameters modeling, routing is accomplished by a Kinematic wave model in conjunction with the continuity equation.

4. Erosion component estimates soil loss from pervious areas. The most common models are the Universal Soil Loss Equation (5) and Negev's (17) model and their modification.

5. Pollutant accumulation and washoff from impervious areas component estimates the mass of pollutants from impervious urban and suburban areas (8).

6. The soil adsorption/desorption component (not common to all models) determines the distribution of adsorbed and dissolved fractions of pollutants in soils. This component may also include volatilization and decay of such materials as pesticides and nitrogen. In the absence of the soil adsorption segment, modeling of pollutants is accomplished by the use of factors that related pollutant concentration to that of sediments. The adsorbed fraction then moves with the sediment and the dissolved is a part of the soil water movement.

A more detailed mathematical description of the processes involved in modeling nonpoint pollution has been published elsewhere (8, 18) and will be discussed further in the subsequent presentation.

NONPOINT POLLUTION MODELS IN CURRENT USE

This section describes the most prominent models for estimating the nonpoint pollution from agricultural lands and land use. Only models which are documented and currently maintained and have been practically applied with success are presented.

Watershed Models

Areal, Nonpoint Source Watershed Environment Response Simulation – ANSWERS. ANSWERS is a model that simulates behavior of watersheds having agriculture as their primary land use. It is a distributed parameter model, primarily event oriented (19, 20).

In order to use the ANSWERS model, the watershed is divided into square uniform elements as shown on Figure 4. The hydrologic response of each element to water inputs is computed by an explicit

backward solution of the continuity equation (Eq. 2). The water
motion component is provided by the Manning's equation for overland
flow. Overland and tile outflow from an element flows into neighboring elements according to the direction of the element slope
until it reaches a channel element. The overland flow is initiated
when surface detention is exhausted and the residual is greater than
infiltration. The infiltration relationship chosen for ANSWERS was
the one originally developed by Holtan (15).

Sediment detachment is computed by modified version of the Universal
Soil Loss Equation (21, 22). Land use changes, tillage techniques
and management procedures for controlling nonpoint source pollution
are simulated with ANSWERS by using appropriate values At this
moment only water and sediment yields from watersheds can be modeled
by the public version of the model. The ANSWERS model does not
require extensive calibration.

The ANSWERS model has been extensively tested on several watershed
in Indiana and elsewhere (23). The model is maintained and is
provided by the Department of Agricultural Engineering, Purdue
University, West Lafayette, Indiana.

Agricultural Chemical Transport Model - ACTM. This model was
developed by the Agricultural Research Service of the U.S. Department of Agriculture. The model consists of three submodels:
hydrologic, erosion, and chemical transport (24).

The hydrological component is a modified version of the USDAHL - 70
Model of watershed hydrology reported by Holtan and Lopez (25). The
basic areal subunit for the hydrologic model is called a "zone" and
is constructed by grouping together fields of the same physical and
crop management features. Figure 3 represents a schematic breakdown
of a watershed into computational units-zones. The USDAHL and ACTMO
are basically lumped parameter models.

The model continuously accounts for soil moisture by balancing infiltration, evapotranspiration, and seepage into lower soil layers.
Infiltration capacity is expressed by the Holtan (15) equation.
The evapotranspiration rate is estimated from crop stage, pan evaporation, and soil moisture characteristics by a modified evapo-
ration-index equation suggested by Holtan and Lopez (25). Infiltration and rainfall excess are computed for each zone by comparing
rainfall to infiltration capacity for each zone.

Rainfall in excess of infiltration and surface storage is routed
across each zone and cascades if it overflows on adjacent soil
segments. A modified kinematic wave concept is used for routing
excess rainfall.

The erosion – deposition segment predicts soil loss by the Modified Universal Soil Loss Equation, that includes the effect of runoff on detachment and transport of soils particles (26). The model includes rill and interrill contributions and is capable of estimating particle size and distribution by calculating the clay and enrichment ratio.

The chemical submodel traces the movement of a single application of a chemical through and over the watershed. Cultivation is assumed to redistribute the chemical. The model simulates the sorption-desorption process by a linear isotherm and the process is assumed to be instantaneous. The sorbed pollutant fraction moves with the detached soil particles, while surface runoff during ponding mixes with the soil water containing the dissolved phase. The model was tested using data from a 2-year field experiment with the pesticide carbofuron. The model is available from the U.S. Department of Agriculture.

Another version of the USDAHL capable of modeling nitrogen and phosphorus was prepared by the Agricultural Experimental Station of the University of Maryland (27) from where the model is also available. The version of the program is called NONPT. Phosphorus is transported adsorbed to the sediment load. The amount of P adsorbed or dissolved is computed using a concept published by Novotny et al. (28).

The ACTMO and NONPT models do not require extensive calibration.

The Hydrologic Simulation Program HSPF. HSPF is a comprehensive package for simulation of watershed hydrology and water quality developed for the U.S. Environmental Protection Agency (29, 30). The model originally evolved from the Stanford Watershed Model (SWM). It is a large model requiring a considerable effort in preparation of data input and the user should not be limited by the computer storage and time availability.

The concept of a simulation model of this scope grew out of a need to simulate pesticide movement on the land and in the receiving water bodies. The system consists of a set of modules arranged in a hierarchical structure that permits the continuous simulation of a comprehensive range of hydrologic and water quality processes.

The HSPF currently contains three application moduli – PERLND, IMPLND, and RCHRES – and five utility moduli – COPY, PLTGEN, DISPLY, DURANL, and GENER (Figure 6). Basically the HSPF performs the simulation on a lumped parameter concept, whereby magnitudes of parameters must be determined by calibration.

Module PERLND simulates a pervious land segment with homogeneous hydrologic and climatic characteristics. Water movement is modeled along three flow paths - overland flow, interflow and groundwater flow - in the manner of the SWM. Erosion is modeled by the Negev (17) model. Water quality constituents can be simulated in a simple way as attached or adsorbed on the sediment or by a more complex model based on the adsorption-desorption equation. The former system was developed for the Nonpoint Pollution Source (NPS) model (31) while the latter approach was incorporated in the Agricultural Runoff Management Model (ARM) (32).

APPLICATION MODULES

PERLND	IMPLND	RCHRES
SNOW	SNOW	HYDRAULICS
WATER	WATER	CONSERVATIVE
SEDIMENT	SOLIDS	TEMPERATURE
QUALITY	QUALITY	SEDIMENT
PESTICIDE		NONCONSERVATIVE
NITROGEN		BOD/DO
PHOSPORUS		NITROGEN
TRACER		PHOSPHORUS
		CARBON
		PLANKTON

UTILITY MODULUES

COPY	PLTGEN	DISPLY
DATA TRANSFER	PLOT DATA	TABULATE AND SUMMARIZE

DURANI	GENER
DURATION ANALYSIS	TRANSFORM OR COMBINE

Fig. 6 Modules of the HSPF Hydrological Model (29)

Module IMPLND is designed to simulate impervious land segments with no infiltration. Water yield and movement is similar to PERLND except that no water movement occurs by interflow or groundwater flow. Solids are simulated using accumulation and removal relatonships in a manner similar to most urban runoff quality models (8). Water quality constituents are simulated using empirical relationships with solids and water yields.

Module RCHRES simulates the processes that occur in a single reach of an open channel or a completely mixed lake. Hydraulic behavior is modeled using the kinematic wave assumption. Water quality algorithms are similar to many other stream and lake models that have evolved in the past fifteen years.

HSPF's utility moduli are designed to give the users flexibility in managing simulation input and output. COPY is used to manipulate time series. The PLTGEN is a model used for plotting while DISPLY takes a time series and summaries it in a table. DURANL segment performs some elementary statistics on the time series such as the probability of exceedence or recurrence intervals.

The HSPF system has been in public use since 1980. It was used in a planning study of the Occoquon River Basin by the Northern Virginia Planning District Commission (33). In modeling land runoff quantity and quality, the simplified NPS option of the model was used in their study. Donigian et al. (30) describe an extensive modeling study with the HSPF on the Iowa River Basin. The HSPF model system is maintained and available from the U.S. EPA Environmental Research Laboratory, Athens, Georgia.

The Agricultural Runoff Management Model – ARM is an independent version of the NPS model that can be used specifically for simulation of nonpoint pollution from agricultural areas. The model simulates runoff (including snow accumulation and melt), sediment, pesticide, and nutrient loadings from a surface and subsurface sources.

The ARM model was tested on an experimental watershed located near Watkinsville, Georgia and other watersheds. The results showed fair agreement of simulated and measured data for paraquat and diaphenamid (34).

The ARM model (as the more complex HSPF model which incorporates ARM) requires extensive calibration. However, the results of a modeling study by Cornell University (35) indicates that it is possible to calibrate the model in such a way that it simulates dissolved nutrient losses with considerable accuracy.

The ARM model was also selected by the Battelle - Northwest Laboratories (36) for their methodology of assessment of pesticides. The methodology with the ARM model was tested extensively on data from experimnental watersheds in Iowa.

Field Scale Hydrologic Models

Cornell Nutrient Simulation - CNS. The CNS model consists of three basic components: daily water balance, a daily soil loss (erosion) calculation, and monthly N and P inventories (35).

The hydrologic model is based on the Soil Conservation Service runoff curve number runoff equation, with suitable extensions to handle snowmelt. A daily soil moisture inventory is maintained for the top 30 cm of soil, which is assumed to be homogeneous. Soil erosion is estimated by Onstad and Foster's modification of the Universal Soil Loss Equation (37). Daily runoff, percolation and soil losses are summed for each month and the monthly balances are then input to the nutrient and pesticide model.

Monthly inventories of mass balances are computed for the nutrients in the top 30 cm surface soil layer. Both absorbed and dissolved components are modeled. However, denitrification, ammonia volatilization and ammonium fixations are not modeled explicitly. Instead, 25% of the fertilizer N applications are subtracted to partially account for these losses. Soil P, is divided into available and fixed inorganic forms. The partitioning of available P into dissolved and adsorbed constituent is based on a linear equilibrium isotherm.

A number of small scale field watersheds in New York and Georgia were modeled to evaluate the accuracy of the model. Some problems were encountered in comparing the simulated and measured data.

Cornell Pesticide Model - CPM. This model consist of 5 components which detail the soil temperature and snowmelt, hydrology, soil loss, pesticide movement and statistics. The model uses both the SCS curve number equation and Green and Ampt (1) infiltration equation to determine the volume of surface runoff. The sediment submodel is similar to that of the CNS model (35).

The pesticide component consist of three parts: degradation and volatilization, downward displacement of pesticide, and pesticide in overland flow. Both CNS and CPM models are available from the Department of Agricultural Engineering, Cornell University, Ithaca, NY.

Chemical Runoff, and Erosion from Agricultural Management Systems -
CREAMS (37). The CREAMS model, similar to some of the previous
models, consists of three major components: hydrology, erosion/-
sedimentation, and chemistry.

The hydrology component has two options, depending upon availability
of rainfall data. Option one estimates storm runoff when only daily
rainfall data is available. This is accomplished by the SCS curve
number model. When hourly rainfall data is available, option two
estimates runoff by the Green-Ampt Equation.

The erosion component considers the basic processes of soil detachment transport, and deposition. The concept of the model presumes
that sediment load is controlled by the losses of transport capacity
or by the amount of sediment available for transport (38).

Detachment is described by a modification of the USLE for a single
storm event. The transport capacity of the overland and channel
flow is derived from Yalin's (39) sediment movement model.

The basic concepts of the nutrient component are that nitrogen and
phosphorus attached to soil particles are lost with the sediment
yield, soluble nitrogen and phosphorus are lost with surface runoff
and soil nitrate is lost by leaching with percolation, by denitrification or by plant uptake.

The pesticide component estimates concentrations of pesticides in
runoff (water and sediment) and total mass carried from the field
for each storm during the period of interest. Pesticide in runoff
is partitioned between the solution and the sediment phase using a
simplified linear isotherm model.

The CREAMS model has been tested with data from many locations. The
aerial size of the model units should be less than 100 ha and the
unit should be fairly uniform. In conclusion, CREAMS is a state-of-
the-art field size model for evaluation and consideration of water
quality of various farming practices. The CREAMS model is available from the U.S. Dept. of Agriculture, Science and Education
Administration, Tifton, Georgia.

Some Process Oriented Models

A number of research process oriented models have been developed in
the United States in the last 10 years. Some of the models were
incorporated in the larger watershed models discussed in the pre-
ceding section. A few promising concepts will be mentioned herein.

Nonpoint Source Model for Land Areas Receiving Animal Wastes. In a
series of articles the authors from the North Carolina Agricultural
Experimental Station (40-42) described a simple model capable of

simulating sediment and nitrogen accumulation and losses in the runoff from fields receiving manure.   The mineralization and immobilization of the nitrogen applied to soils was found to be governed by the carbon to nitrogen ratio of the applied waste. The losses of ammonia due to volatilization were related to CEC (Cation - Exchange Capacity) of the soils and to temperature. The soil erosion-sediment transport model was developed from a modified version of the USLE capable of estimating solids originating from soil and manure.

Models for runoff of pesticides. Baily et al.   (43) outlined the desirable characteristics of a pesticide model.   Pesticide models must be capable of simulating both phases, that is   adsorbed and dissolved, based on the pesticide mobility.   Thus the models must contain water (hydrological) and sediment components.   One of the first pesticide models was   published by Bruce et al. (44).   The hydrologic and sediment components were relatively simple, utilizing empirical or semi-empirical concepts. Testing of the model used the data from the Watkinsville, Georgia experimental watershed (used for verification of several other models). The simulated and measured water, sediment and pesticide data were very close. The model was a first generation model developed by the USDA Agricultural Research Service and was then followed by the   ARM, ACTMO and   CPM models presented in the previous section.

Modeling runoff contamination by phosphorus. Novotny et al. (28) proposed a model   for runoff contamination   by phosphorus.   The relationship between the adsorbed and dissolved phosphate in soils was described by a Langmuir adsorption isotherm that was found by several researchers to be the best representation of the soil phosphorus adsorption process.   The partition   coefficients   for the isotherm were determined from the soil, pH, clay and organic carbon content. The model consists of two components: I.   Free phase model (dissolved pollutant) and II. Sorbed phase model. A large hydrologic-sediment transport model supplies   the   necessary hydrological and sediment movement data. A simplified version of the concept was incorporated into the NONPT model discussed previously.

Several articles   from the   USDA-ARS Southern Plains Watershed Laboratory in Oklahoma presented   concepts and submodels for phosphorus behavior in soils (46, 47). The authors developed a nonlinear water diffusion model for the unsaturated zone with a soil adsorption-desorption component. They also   documented that a similar concept can be used for modeling pesticides.

PROBLEM AREAS AND RESEARCH NEEDS

Any model is only a crude approximation of the processes and phenomena taking place in the real world.   To expect a deterministic model to exactly reproduce measured tests would mean to deny
the inherent random variability component in the data and stochastic
nature of the modeled processes. For this reason, more complex and
detailed models may not necessarily provide better results than
simple models. The following discussion focuses on some problems
with models and their applications, and attempts to delineate some
areas of future research.

Lumped or distributed parameter models?

In one study a distributed parameter hydrological model — (the Distributed Parameter Model, DPM) was compared with a lumped parameter
model — USDAHL (48). It should be noted that the DPM model was
basically a research model, whereas USDAHL is an application model.
The distributed parameter model allows for a more detailed description of the watershed, including channel portions, while the lumped
parameter model typically allows the watershed to be described in a
limited (1 to 10) number of segments, usually without a channel
connection.

In this particular comparison the Distribution Parameter Model performed much better than the lumped parameter model; however, the
authors stated that the performance of the USDAHL model could have
been improved if enough data had been available for calibration.

Sediment and pollutant routing and delivery.

Not all sediment and pollution that is generated on a field during
a storm will reach the receiving water bodies. Many factors and
processes contribute to the fact that the upland erosion and
pollution potential differs, often significantly, from measured
sediment and pollutants yields in the receiving water bodies. Such
factors include redeposition of the particulates in the surface
water storage, trapping of the sediment by vegetation and its
residues, local scour and redeposition in rills and channels, and
possibly others yet unidentified factors (49).

The distributed parameter models with channel segments should be
capable of modeling the delivery adequately. The problem lies with
some lumped parameter models, whereby an arbitrarily defined factor
— the delivery ratio — has been introduced to account for the losses
of the sediment and pollutants between the source area and a drainage channel. This factor is usually ascertained by calibration.
Considering that several other important factors in lumped parameter

models must also be readjusted during the calibration process, proper attention should be given to study and estimation of the delivery factor.

Enrichment of pollutants in the runoff

Another arbitrary factor -- the enrichment ratio -- accounts for the difference between the pollutant concentrations in soils and on the sediment in the runoff. The magnitude of the enrichment factors is not adequately established.

APPLICATION OF HYDROLOGICAL MODELS

Most of the hydrologic models discussed in the preceeding section are limited in their areal extent. A typical size of the watershed would be in tens to hundreds of hectares. Thus, most of the applications have been limited to small experimental watersheds; techniques on how to expand the modeling to larger watersheds are still in evolution. Bearing in mind that only a small fraction of a typical large watershed is responsible for most of the nonpoint pollution, overview or screening model techniques should be made compatible with the detailed hydrologic models. The modeling thus should proceed in two stages: In the first step an overview -- screening modeling should identify the problem areas, to which a detailed hydrologic model should be applied in the second step.

As an example, the author found by modeling and analyzing field data that nonpoint pollutant loadings of particulate pollutants from fields located on medium texture soils is very small if the slope of the field is less than 3%. Furthermore it is known that the delivery ratio parameter in agricultural areas decreases rapidly with the distance of the source from the drainage system. Thus the objective of typical screening modeling process is to locate the "island" of active sources of nonpoint pollution such as high slope areas with a very high source strength as shown on Figure 7. Such modeling activities must consider both the source strength and the delivery.

Screening modeling is not generally designed for estimating accurate loading figures nor for studying the impact of various management practicies on the reduction of loadings under typical long-term or extreme meteorological conditions. This task is accomplished by a detailed hydrologic model. Calibration of such models is necessary if accurate loadings or pollutants are desired but less important if the effect of various management practices on the loadings is studied.

This approach leads to an hierarchical modeling process whereby the knowledge and extensive testing associated with some large and more complex agricultural (and urban) runoff-pollutant models can be

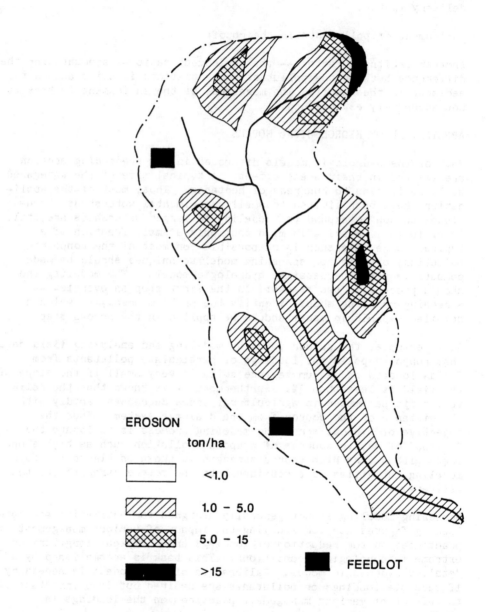

EROSION
ton/ha

	<1.0
	1.0 – 5.0
	5.0 – 15
	>15

FEEDLOT

Fig. 7 Example of Source Areas in Watershed Modeling

extrapolated to the practical local management models and approaches used for selection of hazardous critical segments and for selection of appropriate management practices. A detailed hydrologic model calibrated and verified by small plot field studies can be used to estimate unit loads for a screening overview model as shown by Novotny and Bannerman (50).

This concept is used by the Wisconsin Department of Natural Resources to identify the problem areas within so called "priority watersheds" -- that is, watersheds with serious nonpoint pollution problems.

The Model Enhanced Unit Load concept (MEUL) enables selecting the priority watersheds and the problem areas within them. The unit loadings were generated by a calibrated and verified hydrologic-nonpoint pollution model. In this study loadings were generated for 1 km (100 ha) areal unit located on four hydrologically uniform soil types with a uniform slope of 6%. A time series summarizing several years of meteorological data was input in the model and the results were statistically analyzed to provide long-term average loadings under different land use and crop conditions. The result was a simple matrix of unit loadings that could be extrapolated to other area sizes and other slopes using correction charts.

In an ongoing study by the Wisconsin Department of Natural Resources, ANSWERS is being applied to the problem areas within the priority watersheds to evaluate the effectiveness of various management practices on the pollution loadings.

Another example of an application of the hierarchical modeling approach is the Milwaukehe River Watershed. Therein, the unabated rural and urban inputs may negate the water pollution improvement of a $2.6 billion water pollution control program which is aimed primarily at reduction of point source inputs from the Milwaukee, Wisconsin metropolitan area. A similar situation exists for several other U.S. water bodies receiving both municipal and rural (agricultural) nonpoint pollution inputs. In the particular case of the Milwaukee River, the stream reaches in question and subjected to controversy are located in the lower portion of the rivers while the sources of pollution are scattered throughout the watersheds. The modeling process must address the following problems.

1. Location of sources and their strength.

2. Transmission coefficient (delivery ratio) of pollutants in the overland and channel flow between the sources and the stream reaches in question.

FUTURE DEVELOPMENTS

Forecasting of future developments is a risky business. However, certain trends are now emerging, which enables us to anticipate the near future developments.

Introduction of a variety of commercial minicomputers and high powered microcomputers means that almost every agency or consulting company in advanced countries will possess a computer. The computers that are now available are comparable to medium size computers of ten to fifteen years ago on which some of the hydrologic models were developed. There is now a potential for an explosion of the use of the models for estimating nonpoint pollution from agricultural sources and for design of control practices aimed at its reduction or elimination. However, adapation of hydrological models for use by practitioners is not an easy task.

The graphic capabilities of the small desktop computers are adding new possibilities and greatly improve comprehension of the results. By on-line interactions with the computer and the computing process the user can become directly involved in the modeling process, control its execution and selection of modeling alternatives. Thus, the bulk of future modeling advances will be in development of peripheral components of the models that will make them user interactive (user friendly) with advanced computer graphics capabilities.

ACKNOWLEDGMENT

The author wishes to express his gratitude to Margarita Morin de Gomez, School of Civil Engineering, Purdue University, for drawing the figures of this contribution from sketches.

REFERENCES

1. Green, W.H., and G.A. Ampt, "Studies in Soil Physics. I. The Flow of Air and Water Through Soils", J. Agr. Sci, 4:1-24, 1911.

2. Horton, S.E., "Estimating Soil Permeability Rates", Proc. Soil Sci. Soc. Amer., 5:399-417, 1939.

3. Philip, J.R., "An Infiltration Equation with Physical Significance", Soil Sci., 77:153-157, 1954.

4. Anon., "SCS National Engineering Handbook. Sec. 4 Hydrology", U.S. Soil Conservation Service, 1968, 548 pp.

5. Wischmeier, W.H. and D.D. Smith, "Predicting Rainfall-Erosion Losses from Cropland East of the Rocky Mountains - Guide for Selection for Practices for Soil and Water Conservation", Agr. Handbook No. 282, U.S. Department of Agriculture. 1965.

6. Crawford, N.H., and R.K. Linsley, "Digital Simulation in Hydrology: Stanford Watershed Model IV", Department of Civil Eng., Stanford University, Palo Alto, CA, 1966.

7. Barnwell, T.O. and P.A. Krenkel, "The Use of Water Quality Models in Management-Decision Making", Wat. Sci. Tech., Pergamon Press, 14:1095-1107, 1982.

8. Novotny, V., and G. Chesters, "Handbook of Nonpoint Pollution: Sources and Management", Van Nostrand-Reinhold Publ., New York, NY, 1981.

9. McElroy, A.D. et al., "Loading Functions for Assessment of Water Pollution fron Nonpoint Sources", Rep. No. 600/3-76-151, U.S. EPA, Washington, DC, 1976.

10. Johnson, M.G. et al., "Management Information Base and Overview Modeling", PLUARG Report. International Joint Commission U.S.-Canada, Windsor, Ontario, 1978.

11. Sonzogni, W.C. et al., "WATERSHED: A Management Technique for Choosing Among Point and Nonpoint Control Strategies", Proc. of the Seminar on Water Quality Management Trade-offs, U.S. EPA, Chicago, IL, 1980.

12. Haith, D.A. and L.J. Tubbs, "Watershed Loading Functions for Nonpoint Sources", Journal Env. Eng. Division, ASCE, 107:121-137, 1981.

13. Chow, V.T., "Hydrologic Modeling", Journal of the Boston Society of Civil Engineers, 60:1-27, 1972.

14. Bogardi, I., and L. Duckstein, "Input for Stochastic Control Model of P Loading", Ecological Modeling, 4:173-195, 1978.

15. Holtan, H.N., "A Concept for Nitrification Estimates in Watershed Engineering", Rep. No. ARS-41-51, Agr. Res. Service, U.S. Dept. of Agriculture, 1961.

16. Nash, J.E., "The Form of the Instantaneous Unit Hydrograph", Bull. Internat'l. Assoc. Sci. Hydrol., 111:114-121, 1957.

17. Negev, M., "A Sediment Model on a Digital Computer", Tech. Rep. No. 62, Dept. of Civil Eng., Stanford University, Palo Alto, CA, 1967.

18. Haan, C.T., H.P. Johnson, and D.K. Brakensiek, "Hydrologic Modeling of Small Watersheds", Monograph No. 5, Amer. Soc. Agr. Eng., St. Joseph, MT, 1982.

19. Beasley, D.B., and L.F. Huggins, "ANSWERS (Areal Nonpoint Source Watershed Environment Simulation) - User's Manual", Dept. of Agr., Eng., Purdue University, West Lafayette, IN, 1980.

20. Beasley, D.B., L.F. Huggins, and E.J. Monke, "ANSWERS: A Model for Watershed Planning", Transactions ASAE, 23:938-944, 1969.

21. Meyer, L.D., and W.H. Wischmeier, "Mathematical Simulation of the Processes of Soil Erosion by Water", Trans., ASAE, 12:754-758, 1969.

22. Foster, G.R., "Sedimentation, General", Proc. of the Symposium on Hydraulics and Sediment Control, University of Kentucky, Lexington, KY, 1976.

23. Beasley, D.B., L.F. Huggins, and E.J. Monke, "Modeling Sediment Yields from Agricultural Watersheds", Journal of Soil and Water Conservation, 37:113-117, 1982.

24. Free, M.H., C.A. Onstad, and H.N. Holtan, "ACTMO - An Agricultural Chemical Transport Model", Rep. No. ARS-H-3, Agricultural Research Service, U.S. Dept. of Agriculture, Washington, DC, 1975.

25. Holtan, H.N., and N.C. Lopez, "USDAHL - 70 Model of Watershed Hydrology", USDA-ARS Tech. Bull. No. 1453, U.S. Dept. of Agriculture, Washington, DC, 1971.

26. Foster, G.R., L.D. Meyer, and C.A. Onstad, "A Runoff Erosivity Factor and Variable Slope Length Exponents for Soil Loss Estimates", Transactions, ASAE, 20:683-687, 1977.

27. Holtan, H.N., "Procedures Manual for Sediment, Phosphorus, and Nitrogen Transport Computations with USDAHL", Rep. No. MP 943, Maryland Agr. Exp. Station, University of Maryland, College Park, Maryland, 1979.

28. Novotny, V., H. Tran, G.V. Simsiman, and G. Chesters, "Mathematical Modeling of Land Runoff Contamination by Phosphorus", Journal WPCF, 50:101-112, 1978.

29. Barnwell, T.O., Jr., and R. Johanson, "HSPF: A Comprehensive Package for Simulation of Watershed Hydrology and Water Quality", Proc. of the Seminar on Nonpoint Pollution Control - Tools and Techniques for the Future, Interstate Comm. on the Potomac River Basin, Rockville, Maryland, 1981.

30. Donigian, A.S., J.C. Imhoff, B.R. Bicknell, and J.L. Kittle, "Application Guide for Hydrological Simulation Program FORTRAN (HSPF)", Environmental Research Lab., U.S. EPA, Athens, GA, 1983.

31. Donigian, A.S., and N.H. Crawford, "Modeling Nonpoint Pollution from the Land Surface", Rep. No. EPA-600/3-76-083, U.S. EPA, Athens, GA, 1976.

32. Donigian, A.S., and H.H. Davis, "User's Manual for Agricultural Runoff Management (ARM) Model", Rep. No. EPA-600/3-78-080, U.S. EPA, Athens, GA, 1978.

33. Southerland, E., J.P. Hartigan, and C.W. Randall, "A Continuous Simulation Modeling Approach to Nonpoint Pollution Management", Proc. Industrial Wastes and Research Symp., 54th WPCF Conference, Detroit, MI, October, 1981.

34. Donigian, A.S., and N.H. Crawford, "Modeling Pesticides and Nutrients on Agricultural Lands", Rep. No. EPA-600/2-76-043, U.S. EPA, Athens, GA, 1976.

35. Haith, D.A., and R.C. Loehr, "Effectiveness of Soil and Water Conservation Practices for Pollution Control", Rep. No. EPA-600/3-79-106, U.S. EPA, Athens, GA, 1979.

36. Onishi, Y., et al., "Methodology for Overland and Instream Migration and Risk Assessment of Pesticides", Rep. No. EPA 600/3-82-024, U.S. EPA, Athens, GA, 1979.

37. Onstad, C.A., and G.R. Foster, "Erosion Modeling on a Watershed", Transactions ASAE, 18:288-292, 1975.

38. Knisel, W.G., "CREAMS: A Field-Scale Model for Chemicals, Runoff and Erosion from Agricultural Management Systems", Rep. No. 26, U.S. Department of Agriculture, 1980.

39. Yalin, Y.S., "An Expression for Bedload Transportation", J. Hydraulic Div., ASCE, 89:221-250, 1963.

40. Reedy, K.R. et al., "A Nonpoint Source Model for Land Areas Receiving Animal Wastes: II. Ammonia Volatilization", Transactions ASAE, 22:1339-1465, 1979.

41. Khaleel, R. et al., "A Nonpoint Source Model for Land Areas Receiving Animal Wastes: II. A Conceptual Model for Sediment and Manure Transport", Transactions ASAE, 22:1353-1360, 1979.

42. Khaleel, R., et al., "A Nonpoint Source Model for Land Areas Receiving Animal Wastes: IV. Model Inputs and Verification for Sediment and Manure Transport", Transactions ASAE, 22:1362-1368, 1979.

43. Bailey, G.W., R.R. Swane,Jr., and H.P. Nicholson, "Predicting Pesticide Runoff from Agricultural Land: A Conceptual Model", Journal of Environmental Quality, 3:95-102, 1974.

44. Bruce, R.R., et al., "A Model for Runoff of Pesticides from Small Upland Watersheds", Journal of Environmental Quality, 4:541-548, 1975.

45. Haith, D.A., "A Mathematical Model for Estimating Pesticide Losses in Runoff", Journal of Environmental Quality", 9:428-433, 1980.

46. Ahuja, L.R. et al., "Modeling the Release of Phosphorus and Related Adsorbed Chemicals from Soils to Overland Flow", Proc. Internat'l. Symposium on Rainfall-Runoff Modeling, Mississippi State University, 1981.

47. Ahuja, L.R., A.N. Sharpley, and O.R. Lehman, "Effect of Soil Slope and Rainfall Characterisitics on Phosphorus in Runoff", Journal of Environmental Quality, 11:9-13, 1982.

48. Yaramanoglu, M., J.E. Ayars, and G.K. Felton, "Evaluation of a Distributed Parameter and A Lumped Parameter Model for Use in Water Quality Modeling", Proc. Seminar on Nonpoint Pollution Control Tools and Techniques for the Future, Interstate Commission on the Potomac River Basin, Rockville, Maryland, 1981.

49. Novotny, V., "Delivery of Suspended Sediment and Pollutants from Nonpoint Sources During Overland Flow", Wat. Res. Bull., 16:1057-1065, 1980.

50. Novotny, V., and R. Bannerman, "Model Enhanced Unit Loadings of Pollutants from Nonpoint Sources", Proc. Hydraulic Transport Modeling Symp., Publ. No. 4-80, ASAE, St. Joseph, MI, 1980.

DEVELOPMENTS IN ECOLOGICAL MODELLING

Sven Erik Jørgensen

The Royal Danish School of Pharmacy, Dept. of Pharmaceutical Chemistry AD
2 Universitetsparken, DK−2100 Copenhagen, Denmark

Abstract

The wide use of ecological models for environmental management during the last 12-14 years has demonstrated that workable models can be developed, at least for a great number of environmental problems. The experience gained has provided us with guidelines for the setting up of ecological models. The guidelines are presented in 7 points in this paper.

The present trends in ecological modelling are also mentioned. Further development in ecological modelling will require the use of new methods to include description of what is called higher orders of dynamics, that is adaptation, changes in structure and changes in the genetic pool.

The following methods will be mentioned: the use of cybernetics, the thermodynamic function exergy and the catastrophy theory. Present day experience with these new approaches is quite limited but the results obtained up to now seem very promising.

Key words: ecological models, holism, parameter, estimation.

1. Introduction

The development of ecological modelling can be divided into four phases. The first phase was the appearance of Streeter-Phelps BOD/DO models and Lotka Volterras prey-predator models in the twenties. The second phase was the development of population dynamics in the fifties and the sixties, while the third phase started in the late sixties with the application of environmental management models initiated by Orlob and Chen, Thomann, Di Toro and O'Connors and others. This presentation will cope with a presentation of the development of ecological modelling during this third phase, but will also touch the fourth phase which is in its very beginning during these years.

2. The Development 1969 - 1984

The first models developed from the late sixties to the early seventies were characterized by a focus on environmental management problems, for instance, the oxygen balance of rivers and streams and on eutrophication of lakes.

Two trends in the sixties determined the further development of ecological models: the immense growth in computer technology and the strong growth in awareness of environmental problems. There was a strong need for environmental management, which was made possible by use of models due to modern computer technology.

However, most models developed in this early part of phase three were more complex than the ecological knowledge and the data could bear. It was rather easy to write equations for a great number of processes in the ecosystem and develop a computer program which represented the models but knowledge of the coefficients in the equations was not sufficient. A calibration was needed, but almost impossible due to quality and quantity of the data available.

The very nature of the modelling exercise, which tends to aggregate, average and smooth over the biological subtleties, leaves the more rigorous aquatic biologist somewhat disconcerted. He would prefer to concentrate efforts on a more correct representation of biological interactions, the kinetics of varying life stages, shifts in grazing preferences (Canale et al., 1976), ecological instabilities, etc. Some trade-offs are necessary simply because the model is an approximation of the real system. These seem to have occurred either by simplification of the aquatic ecosystem or by simplification of the circulatory and exchange processes of the impoundment. However, the scope is not to include more and more details and to build as complex a model as possible, but rather to build a model that gives a

quantitative description of what is in focus - to meet the aims
of the model. There is no such thing as a general ecological
lake model, but in every case study the goals and the resources
available must be balanced so that the right model can be selected.

During the seventies we have learned to select a better balance between problem and data on the one side and models on
the other.

In this context it must be stressed that a model cannot be
better than the data on which it is based. A very complex model
will contain more parameters to be calibrated, requiring more
observations. Furthermore, validation will require another independent set of observations. It is therefore not surprising
that the very comprehensive models are not validated or sufficiently well calibrated.

Model structure depends also on the accuracy required, so it
is important to consider the accuracy with which it is possible
to simulate a specific ecosystem. This brings up the question of
how much we can rely on the observed data. It is assumed that
ecological observations normally will have a standard deviation
of 10 - 25%, which must be taken into consideration when the accuracy of the model is estimated.

A determination of sufficient model complexity enters the
modelling process at two stages (Beck, 1978):

1) during the initial stage, when the analyst must choose a
   certain level of complexity before attempting to verify
   the model against field data, and

2) during the final phases, when the analyst must decide
   whether the model has been verified and has sufficient
   complexity for its intended application.

Jørgensen and Mejer (1977) suggested the use of a quantitative
index for the selection of model complexity. The idea is to use
a concept of sensitivity for identifying the model structure.
Basically it is an inverse "submodel sensitivity", called the
ecological buffer capacity, which measures the influence (sensitivity) that additional suggested submodels have on a particular state variable (e.g. phytoplankton concentration for eu-

trophication models) to see whether anything is changed by increasing the complexity.

Tapp (1978) examined and compared the use of simple and complex eutrophication models. He concluded that simple models can be used for first approximation analysis, but where data exist to establish a basis for a more complex model these should be used. This conclusion is in accordance with the state of the art (Jørgensen, 1979).

From the late sixties until today several types of models were developed and used as tools in environmental management: eutrophication models for lake management fishery models for assessment of fishery policy, models for prediction of distribution of air pollutants, models relating the CO_2-emission with the climate, models of the changes of the ozonlayer due to emission of freon and other air pollutants, models for forestry management, models used for the prediction of the distribution and effect of toxic substances in aquatic ecosystem, models of the cycling of nutrients in soil, models for the prediction of contamination of crops due to the presence of heavy metals in soil, models for management of national parks, models relating thermal pollution with its ecological consequences etc. to mention the most characteristic ecological models developed in the period from the late sixties to the early eighties.

For each type of model several alternative models were developed, and used in a number of case studies, which of course has given a wide experience in the development and application of ecological models.

3. What have we learned?

The experience gained during the last 12-14 years in the application of ecological models as an environmental management tool can be formulated in the following conclusions:

1) General models of ecosystem classes (lakes, rivers, forests etc.) are non-existent. All ecosystems have their distinctive character and a comprehensive knowledge of the system to be modelled is often needed to make a good start.

2) The predictive power of the model is strongly dependent on
   the quantity and quality of the data. If good data are not
   available and cannot be provided, it is necessary to set up
   a rather simple model corresponding to the quality of the
   data. A complex model will contain many parameters and as
   parameters are added to the model there will be an increase
   in uncertainty. The parameters must be estimated either by
   field observations, by laboratory experiments or by calibration, which again, as shown below, are based on field measurements. Parameter estimations are never error-free, but reflect
   to a certain extent the uncertainty of the data. Consequently,
   the errors of measurements will be carried through into the
   model and will contribute to the uncertainty of the prediction derived from the model. If high quality data are not
   available there seems to be a great advantage in reducing the
   complexity of the model - a consequence which is clearly reflected in the most recently published ecological models.

3) Calibration of the most crucial parameters, which can be found
   by a sensitivity analysis, is always required, because the
   many ecological processes not included in the model are taken
   into account by the fine tuning of the parameters. The model
   will of course contain the more important processes, but it
   must be remembered (see point 2 above) that if we add too many
   processes it will require more (and better) data for the parameter estimation. Furthermore, the parameter values obtained
   from the literature or laboratory experiments are only indicated as intervals. It is therefore recommended that 1) good
   literature values should be used for all parameters, 2) a sensitivity analysis of the parameters should be made before the
   calibration, and the most sensitive parameters should be selected, as an acceptable calibration of 4-6 parameters is
   possible with the present techniques.

4) Models have usually been calibrated on the basis of an annual
   measurement series with sampling frequency of for instance
   twice per month. This sampling frequency is, however, not

sufficient to describe the day to day dynamics of an ecosystem, See Fig. 1, which considers a lake.

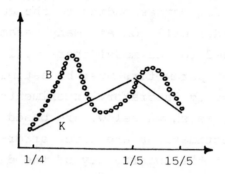

Figure 1 - Algae concentration plotted versus time
+ = sampling frequency twice a month
• = sampling frequency three times a week

The algae concentration is plotted versus time, April 1 to May 15, in a shallow hypereutrophic lake with a sampling frequency of K) twice per month and B) three times per week. As can be seen the two plots are significantly different and any attempt to get a realistic calibration on the day to day dynamics on the basis of K) will fail. K) should in principle only be used to simulate annual average concentrations of algae and at the most to indicate the approximate annual cycles of nutrients. This example illustrates that it is important not only to have data with low uncertainty but also data sampled with a frequency corresponding to the dynamics of the system to be modelled. This rule has often in the past been neglected in ecological modelling. Consequently, it is recommended to work along the following lines:

A: To use an intensive sampling programme in periods where the dynamics are most pronounced.

B: To use laboratory investigations to assess some of the crucial parameters.

C: To fine tune some of the most sensitive, already assessed parameters on the basis of the annual measuring programme.

5) It is recommendable to examine crucial submodels of the system either in situ or in the laboratory. For shallow lakes the sediment-water exchange processes are, for instance, of great importance, as a substantial part of the nutrient in the lake will be stored in the sediment. For further details see Kamp-Nielsen (1975, 1978) and Jørgensen, Kamp-Nielsen, Jacobsen (1975).

6) After calibration of the model it is important to validate it, preferably against a series of measurements from a period with changed conditions, e.g. with changed external loading or climatic conditions. Only through a validation of the model is it possible to indicate with what accuracy the model is able to predict changes in the ecosystem.

7) It is possible to set up a procedure for the development of ecological models, see Fig. 2. Other similar procedures are of course applicable, provided they contain the main elements shown in Fig. 2.

From the experience gained during the past years it can be concluded that if the points mentioned above are considered, it is possible to make workable models, at least for a great number of case studies.

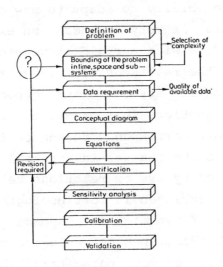

Fig. 2. Modelling procedure

4. Next Generation of Ecological Models

A modeller will always ask the question: Can I improve the model further? Experience has shown that in most cases it is hardly possible to improve the model by increasing the complexity, provided that the modeller already has considered the crucial processes in his model. On the other hand it is always recommendable to try to build some more details into the model and notify the changes in the modelling results. Through this exercise the modeller can record the sensitivity of the model to increased complexity (see also Jørgensen, Mejer, 1977). Many ecological processes have several alternative descriptions, see e.g. Jørgensen (1980), and for the most important processes it is recommendable to test these alternative descriptions, preferably in connection with the application of an intensive measuring period. The final results of these additional attempts to improve the model might often be better validation results than obtained at first. Generally, it can be recommended to examine the reactions of the model as much as possible to become acquainted with the very nature of the model. However, the model will still have some shortcomings.

We must acknowledge that the ecosystem has several characteristic features, which are not incorporated in the models of today. Species have the ability to adapt to new circumstances, and the species composition is very dependent on external factors. We use ecological models with a rigid structure and with fixed parameters. However, the real ecosystem is soft and flexible and is constantly changing its structure and composition.

Straskraba (1980) mentions four orders of dynamics in the ecosystem. The first order corresponds to the current changes in state variables due to the changes in external variables. These variations are modelled by use of relations between external and internal variables as included in most ecological models of today. The second order of dynamics corresponds to adaptation which in some cases might be described by replacing parameters with equations containing two or more parameters. The third order of dynamics is the current change in the structure (could also be denoted as the species composition) of the ecosystem. The fourth order of dynamics is the slow change of the genetic pool. In most environmental management models it seems unnecessary to include the last dynamics as it only becomes effective with a considerably longer time horizon than applied in most models.

Straskraba (1979) has suggested to overcome this gap between nature and model by introduction of varying parameters, controlled by use of a so-called goal function, which describes the consequences of one or more introduced ecological principle quantitatively.

Radtke and Straskraba (1980) use a maximization of the biomass as such governing principles, but admit that the application of other princples might be more appropriate as goal function.

Jørgensen and Mejer (1977, 1979, 1981a, 1981b) have suggested the use of thermodynamic function exergy as goal function. They show, by means of realistic and well examined models, that the ecosystem reacts to changes in the external factors by changing the structure and composition in such a way that the ecosystem gets higher exergy and becomes better buffered to meet such changes.

Exergy is defined as the maximum entropy-free energy a system is able to transfer to the environment on its way towards thermodynamic equilibrium. As seen from this definition exergy (Ex) is not a state variable as it is dependent on the environment. Exergy can be applied as a measure for thermodynamic order, since it measures the distance from thermodynamic equilibrium. This will be illustrated by a simple example.

Exergy is the thermodynamic information, I, multiplied by T:

$$Ex = I \cdot T$$

Many authors prefer the thermodynamic information as a measure for the ontogenetic order, which the ecosystems attempt to maintain, and even increase, by use of the solar radiation as energy input or source. However, exergy has a few pronounced advantages compared with thermodynamic information:

1) Exergy is conserved by transformation of entropy-free energy. If, for example, two reservoirs are exchanging entropy-free energy it is obvious that the resulting thermodynamic inform- ation is not generally zero, since:

$$I_1 + I_2 = \frac{\Delta Ex_1}{T_1} + \frac{\Delta Ex_2}{T_2}$$

However, the resulting change in exergy $[\Delta(Ex_1 + Ex_2)]$ is zero.

2) From the above mentioned example, we see that one bit of in- formation corresponds to $kT \ln 2$ in exergy. In other words, the temperature is important for the information. Information from a system having a high temperature can cause more con- structive changes in the environment than information from a low temperature system. Exergy is opposite to thermodynamic information in that it directly measures the amount of order that a system is able to induce in other systems.

It can be shown that exergy expressions for considered systems are independent of model details such as foodweb topology and rate formulas. For systems with inorganic net flows and passive organic outflows, it is easy to find a useful expression for the exergy. The expression is based on the flows of elements which take part in the biogeochemical reactions. If phosphorus is con- sidered the following expression is found: (see Mejer and Jørgen- sen (1979).

$$Ex = RT\Sigma_j a_j [P_j \ln(P_j/P_j^{eq})] kJ \ m^{-3}$$

a_j is a set of volume ratios (only different from unity for sedi- ment variables). P_j is the P-concentration in compartment j and P_j^{eq} the corresponding thermodynamic equilibrium concentration. R is the gas constant and T the absolute temperature.

A new concept called ecological buffer capacity, β, was also introduced (Jørgensen and Mejer (1977)), it is defined by the following equation:

$$\beta = \frac{\partial L}{\partial \psi}$$

where L is the loading (forcing function) and a state variable displaced by the influence of L (see Fig. 3). There are, in accordance with this definition, an infinite number of buffer capacities for all combinations of all possible forcing functions and all possible state variables, but the exergy, Ex, seems to be related to the buffer capacities by:

$$Ex = \sum_{i=1}^{i=n} \beta k$$

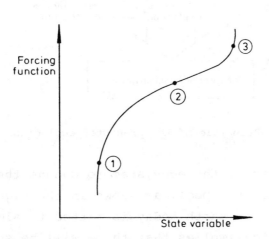

Figure 3. Ecological buffering capacity (β) showing the relationships between a forcing function and a state variable: $\beta = dn/dx$. At one and three, β is large; at two, β is small.

The principle introduced can therefore be formulated as follows: Changes in external factors will create new conditions for the ecosystems which the system will meet by changing the structure or composition in such a way that the exergy under these new circumstances is maximized. Exergy measures the ability of the ecosystem to be buffered against changes in the system caused by changes in external factors. Thermodynamically exergy also measures the organization or order of the system. The idea behind such models is illustrated in Fig. 4 (taken from Jørgensen, Mejer,

1981a).

The final results of these additional attempts to improve the model will often be a model which gives slightly better validation results than the model developed at first.

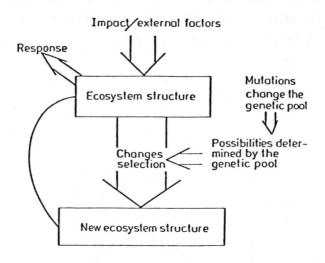

Figure 4. Principle of the presented ecological theory

The possibilities of the ecosystem to change the structure is dependent on the genetic pool, as shown in the figure. The wide spectrum of species on earth today is fitted to almost all natural conditions, which implies that there will be species in all ecological niches contributing to the total exergy ecosphere.
The thermodynamics of evolution states that an ecosystem evolves towards maximum biomass. It can easily be shown that this principle is included in the hypothesis on development (or evolution) towards maximum of exergy.

This implies that the response of ecosystems to new external conditions is linked to the evolution of ecosystems. It is a necessity for an ecosystem to meet perturbations by changing the structure in such a way that the ecological buffer capacity, that is, the ability to meet the perturbations, is increased, and that corresponds to a higher exergy level. This constant change in the external factors will, of course, also change the selection pressure on the species.

The selection is, however, not only serving the survival of the best fitted species but also the prevailing composition and structure of the ecosystem. The relatively quick changes in composition are caused by alterations in the external factors, which again are modifying the selection pressure. However, the pool of genetic material is simultaneously growing slowly and being modified. This opens for a new combination of selection pressure and possibilities to meet this pressure. Many mutations will not be better fitted to the steadily changing external factors, but there will always be a probability that some mutations, better fitted to a set of external factors valid at a given time, will occur.

As everything is linked to everything in an ecosystem, the evolution of species must work hand in hand with the selection of ecological composition and structure. As seen from this discussion, the principle of development of evolution towards higher exergy is able to explain how ecosystems react to perturbation as well as to evolution in the Darwin sense.

The use of a goal or control function for determination of the variations of essential parameters has up to now been quite limited, but an example should be given to illustrate the possibilities.

A current modification of the maximum growth of phytoplankton in our eutrophication model (Jørgensen, 1976; Mejer, Friis, 1978) was attempted. The exergy for a wide range of values for the maximum growth rate of phytoplankton was computed, and the value which gave the highest exergy was selected. The model was applied on a hypereutrophic lake and a 99% reduction of the phosphorus input was simulated. It was found that along with decreased phosphorus concentration and eutrophication the selected maximum growth rate increased, see Table 1.

This is in accordance with the observations that phytoplankton species in oligotrophic lakes are generally smaller (the specific surface is higher, giving a higher growth rate) than in eutrophic lakes.

Case	Max growth rate at highest exergy
Oligotrophic Lake $P_{total} < 0.05$ mg l^{-1}	3.3 day^{-1}
Eutrophic Lake $P_{total} \sim 0.5$ mg l^{-1}	2.2 day^{-1}
Hypereutrophic Lake $P_{total} \sim 1.5$ mg l^{-1}	1.6 day^{-1}

Table 1. Max growth rate for algae

These results have encouraged us to develop this parameter estimation method further. An automatic calibration method has been used (recently developed software called PSI). The method allows introduction of any type of optimization criteria. In addition to the generally applied procedure, which is based on best possible fit to observed data by use of

$$
S = \left(\frac{\sum_{i=1 j=1}^{i=n j=m} (x_{j,i} - x'_{j,i})^2}{m \cdot n - 1} \right)^{\frac{1}{2}}
$$

as objective function. $X_{j,i}$ are model results and $X'_{j,i}$ are observed data, n data points of m state variables, a certain number of crucial parameters will be selected in accordance with the maximum exergy principle. This will possibly give varying parameters, hopefully in accordance with the succession of species throughout the year. Further examination will be undertaken during the coming year.

Other suggestions to inclusion of higher order dynamics in ecological models are the application of the catastrophy theory (Kempf, 1980). A mathematical analysis of some models show that under certain circumstances various solutions are valid. The history of the system tells us which solution to use and also when the system will "jump" from one solution to another. Such drastic shifts have also been observed in nature, for instance when an aquatic ecosystem shifts from aerobic to anaerobic conditions, which implies that the system composition is changed entirely.

The catastrophy theory seems to be a powerful tool in describing sudden shifts in ecosystem, while the use of current modification of the important parameters seems a suitable tool for the description of relatively slow changes in species composition and ecosystem structure.

5. Conclusions

It seems feasible to conclude that it is today possible to build workable models for a number of environmental management problems. However, the experience in the use of such models for prediction of ecosystem changes caused by significant changes in the external variables is rather limited. It seems, furthermore, that under such circumstances it is required to expand the description of the model to include second and third order dynamics of the system, whereas the present models include only first order dynamics. Such models have been developed but only very few cases have been studied. It is too early to make a statement on the applicability of the approaches suggested so far although at this stage the methods seem promising.

REFERENCES

Canale, R.P., L.M. DePalma, and A.H. Vogel (1976). A plankton - based food web model for Lake Michigan. Modeling Biochemical Processes in Aquatic Ecosystems, ed. R.P. Canale (Ann Arbor, MI: Ann Arbor Science), pp. 33-74.

Jørgensen, S.E. (ed.)(1979). State of the art of eutrophication models. State of the Art in Ecological Modelling (Copenhagen: International Society for Ecological Modelling), pp. 293-298.

Jørgensen, S.E. (1980). Lake Management, Pergamon Press, Oxford.

Jørgensen, S.E., Mejer, H.F. (1977) Ecological Buffer Capacity, Ecological Modelling, 3, 39-61.

Jørgensen, S.E., Mejer, H.F. (1979). Holistic Approach to Ecological Modelling, Ecological Modelling, 7, 169-189.

Jørgensen, S.E., Mejer, H.F. (1981a). Exergy as Key Function in Ecological Models, in Mitsch, W. (ed.) Proceedings Exergy and Ecological Modelling, 587-590, Paper presented at the International Symposium Exergy and Ecological Modelling, Louisville, Kentucky, April 20-23.

Jørgensen, S.E., Mejer, H.F. (1981b). Application of Exergy in Ecological Models, in Dubois, D. (ed.) Progress in Ecological Modelling, Liege, 39-47.

Jørgensen, S.E., Kamp-Nielsen, L., Jacobsen, O.S. (1975). A Submodel for Anaerobic Mud-Water Exchange of Phosphate, Ecological Modelling, 1, 133-146.

Jørgensen, S.E., Mejer, H.F., Friis, M. (1978). Examination of a Lake Model, Ecological Modelling 4, 253-279.

Kamp-Nielsen, L. (1975). A Kinetic Approach to the Aerobic Sediment-Water Exchange of Phosphorus in Lake Esrom, Ecological Modelling, 1, 153-160.

Kamp-Nielsen, L. (1978). Modelling the Vertical Gradients in Sedimentary Phosphorus Fractions, Verh. Int. Verein. Limnol., 20, 720-727.

Radtke, E., Straskraba, M. (1980). Selfoptimization in a Phytoplankton Model, Ecological Modelling 9, 247-268.

Kempf, J. (1980). Multiple Steady States and Catastrophes in Ecological Models, ISEM Journal, 2, 55-81.

Mejer, H.F. and Jørgensen, S.E. (1979). Energy and ecological buffer capacity. State of the Art in Ecological Modelling, ed. S.E. Jørgensen (Copenhagen: International Society for Ecological Modelling), pp. 829-846.

Straskraba, M. (1979). Cybernetics in Ecological Modelling, Ecological Modelling, 6, 117-135.

Straskraba, M. (1979). Cybernetics in Ecological Modelling, Ecological Modelling, 6, 117-135.

Straskraba, M. (1980). Cybernetic - Categories of Ecosysten Dynamics, ISEM Journal, 2, 81-96.

Tapp, J.S. (1978). Eutrophication analysis with simple and complex models. Journal of Water Pollution Control Federation 50: 484-492.

RAINFALL ABSTRACTION AND INFILTRATION
IN NONPOINT SOURCE POLLUTION

J. W. Delleur
School of Civil Engineering, Purdue University
West Lafayette, Indiana 47907, U.S.A.

On Sabbatical Leave at
Institut de Mecanique de Grenoble
38402 Saint Martin d'Heres, France

INTRODUCTION

Hydrology plays an important role in the understanding of nonpoint source pollution. Hydrologic variables such as rainfall intensity, volume of rain, rainfall interception, infiltration, watershed storage and conveyance characteristics are strongly related to the pollutant loadings from agricultural areas. Thus the transport of pollutants and the occurrence and magnitude of nonpoint source pollutant loadings are, to a great extent, governed by the hydrologic characteristics of the watershed. Appropriate modifications of the hydrologic activity within a watershed may be an effective way of controlling pollution. On the other hand, uncontrolled modifications in a watershed may aggravate the pollution problem.

Simulation models of nonpoint source pollution are therefore based on watershed hydrology. This presentation is concerned with components of the rainfall-runoff process, mainly interception, depression storage and infiltration. As there are few models in the literature concerning interception and depression storage, only a few empirical values will be cited. In contrast, there is an extensive body of knowledge on infiltration which is the most important of these three mechanisms of rainfall abstraction in the runoff formation. For this reason, the process of infiltration will be reviewed in some detail. The review of infiltration starts with the physical properties of soils, continues with the soil moisture dynamics and several infiltration models, and concludes with a discussion of spatial variability of infiltration and of solute transport in unsaturated soil

More detailed information can be found in some general works such as Chow (1), and other treatments of the infiltration process are available in textbooks such as Hillel (2) and in recent reviews such as Vauclin (3), Smith (4), and Swartzendruber (5).

INTERCEPTION AND DEPRESSION STORAGE

A small fraction of the precipitation is intercepted by the vegetation and is stored on the vegetation surfaces. A film of water forms on the leaves and is eventually returned to the atmosphere by evaporation and is thus "lost" as far as the runoff formation is concerned. Precipitation striking the vegetation in excess of the interception capacity falls off the leaves or becomes stemflow. This throughfall and stemflow then reaches the ground. The amount of rainfall intercepted by the vegetation depends on the type of vegetation (species, age, density of canopy), season of the year and rainfall intensity. Lull (6) states that 0.5 to 2.5 mm of rain can be stored on the vegetation before appreciable drip occurs.

For interception by forest, Dunne and Leopold (7, p. 87) give a graphical linear relationship relating throughfall and stemflow to gross storm precipitation, from which approximately 82% of the rain on hardwood reaches the ground as throughfall and 7% as stemflow; for conifers the percentages are 75% as throughfall and 10% as stemflow. Lull (6) quotes from the Soil Conservation Service the following values of interception for various crops:

	during growing season	during low-veg. devel.
Alfalfa	36%	20%
Corn	16%	3%
Soybeans	15%	9%
Oats	7%	3%

For interception by grass Horton (8) gave the following relationship

$$I_L = 0.00042 \ H + 0.00026 \ HP,$$ (1)

where I_L is the interception, H is the grass height and P is the precipitation, all in depth (mm).

A model of total storm interception is (9)

$$I_L = S + KEt,$$ (2)

where I_L is the volume of interception loss (mm), S is the interception storage that is retained on the vegetation (typically 0.25 to 1.25 mm), K is the ratio of the surface area of the intercepting leaves to their horizontal projection area, E is the amount of water evaporated during the rain storm per unit of time and t is time

Another initial "loss" in the runoff formation is the depression storage. The rain reaching the ground must first wet the surface and fill the small depressions, forming puddles, before runoff can start. The water in depression storage

either infiltrates or evaporates. For urban areas Tholin and
Keifer (9) recommend 0.25 in (6.35 mm) for grassed areas and 1/16
in (1.5 mm) for impervious or paved areas. Depression storage
depends on the type of surface and on the agricultural practices
such as plowing, raking, and on vegetation. Often depression
storage and interception are lumped into a single value which can
be calibrated into the model by making use of field data. Viess-
man et al. (10) give a linear relationship between depression
storage loss and slope for impervious areas. The depression
storage decreases from approximately 28 mm for a 1% slope to 10
mm for a 3% slope. Novotny and Chesters (11, p. 83) give curves
of surface storage vs. percent slope for alfalfa, pasture, corn
and contour furrows. Maximum values of depression storage of 1
mm on steep, smooth hillsides to 50 mm on agricultural lands of
low gradient, that have been furrowed or terraced, are mentioned
by Dunne and Leopold (7, p. 259). Linsley et al. (12) give the
following model of depression storage

$$V = S_d (1-e^{-kPe}),\qquad\qquad(3)$$

where V is the depression storage at some time of interest, S_d is
the maximum storage capacity of the depressions, Pe is the rainfall excess and k is equal to $1/S_d$.

PHYSICAL PROPERTIES OF SOILS

Texture of Soils

Soils, sands and gravels are classified according to their
particle size. the classification developed by the U.S. Department of Agriculture (13) is commonly used in studies of soil physics and infiltration. This classification can be summarized as
follows:

Very coarse sand	1.2 mm
Coarse sand	0.5-1 mm
Medium sand	0.25-0.5 mm
Fine sand	0.1-0.25 mm
Very fine sand	50-100 μm
Silt	2-50 μm
Clay	0-2 μm.

The particle size distribution is determined by mechanical seive
analysis for material larger than 50 μm and for settlement velocity of particles in suspension (according to Stokes' Law) for
material smaller than 50 μm. The resulting particle size distribution is generally presented as a semi-logarithmic plot of the
percent of particles smaller than a given size vs. the logarithm
of the particle diameter. The particle size corresponding to 50%
is called the average particle size, D_{50}, and the size

corresponding to 10% smaller than or 90% larger than it (i.e. the sieve size retaining 90% of the material) is called the effective particle size, D_{90}. The uniformity ratio D_{50}/D_{90} is a measure of the uniformity of the material. The texture of a soil is classified in terms of the relative amounts of sand, silt and clay in accordance with the triangular chart of the Soil Conservation Service (13).

Structure of Clayey Soils

Gravels, sands and silts have a granular structure. They are essentially inert, that is, there are no electro-chemical reactions between soil grains and the cations in solution in the water surrounding the grains. In contradistinction, clays have a lamellar structure (14) with spaces between the plates or lamellae of the order of a water molecule. In the humidification of such soils, the water must widen the space between lamellae resulting in a swelling of the clays. These clays are very reactive from an electro-chemical point of view.

The structure of the clayey soils can be of two types, depending upon the arrangement of the clay particles: 1) the clay particles are dispersed or 2) the clay particles aggregate to form flocs of several millimeters in size. If the absorbed layer of cations surrounding the clay particles is small, the Van der Waals attractive forces are important and the clay is flocculated. If the absorbed layer is large, the repulsive electrostatic forces are important and the clay is dispersed.

When the absorbed cation layer consists primarily of sodium ions (Na^+), the clay particles cannot come close together because the Na^+ ions are surrounded by water molecules· and the positive charge of the monovalent Na^+ is not sufficient to compensate for the negative charge of the clay particles. As a result, the repulsive electrostatic forces between negatively charged clay particles exceed the Van der Waals attractive forces and the clay particles are separated or dispersed. Such clays have a low permeability and harden when drying.

On the other hand, when the cations surrounding the clay particles are bivalent calcium (Ca^{++}) or Magnesium (Mg^{++}), the clay particles can come closer together because they are not hydrated; and the positive charges of the bivalent cations effectively mask the negative charges of the clay particles. As a result, the Van der Waals attractive forces dominate, and the clay particles aggregate together to form flocs. These clays are more permeable and friable than the dispersed clays. It is possible to change the state of the clays from dispersed to flocculated by changing the absorbed ions from Na^+ to Ca^{++} or Mg^{++} and vice versa.

Characteristic Dimensions

The porosity of soils, n, is defined as

$$n = (V_t - V_s)/V_t = V_v/V_t,$$ (4)

where V_t is the total volume of an undisturbed sample, V_s is the volume of the solid phase of the sample, V_v is the volume of voids. It is sometimes expressed in percent. Typical values of the porosity are:

sand and gravel mixture	0.1 - 0.3
gravel	0.2 - 0.3
coarse sand	0.25 - 0.35
medium sand	0.35 - 0.40
fine sand	0.40 - 0.50
silts and clays	0.50 - 0.60

The Volumetric water content, θ, is defined as

$$\theta = V_w/V_t,$$ (5)

where V_w is the water volume in the undisturbed sample. The degree of saturation, S, is defined as

$$S = \theta/n.$$ (6)

The density of the dry soil, ρ_d, is defined as

$$\rho_d = M_d/V_t,$$ (7)

where M_d is the mass of the dry soil. The density of the grains, ρ_g, is defined as

$$\rho_g = M_g/V_g,$$ (8)

where M_g and V_g are the mass and volume of the grains, respectively. To define the pressure head, consider a capillary tube filled with water. The pressure balance between a point B, immediately below the meniscus and a point C at the free surface is written as

$$p_C = p_B + \rho_w gh,$$ (9)

Figure 1

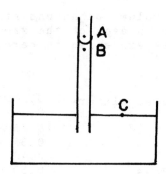

where h is the capillary rise. But the pressure at point C is atmospheric, as well as at point A immediately above the meniscus, thus $P_C = P_{atm} = P_A$, and

$$P_A = P_B + \rho_w gh \quad \text{or} \quad P_B = P_A - \rho_w gh. \tag{10}$$

Assuming that the pores of a granular medium behave similarly, and the air phase is at atmospheric pressure,

$$P_B - P_{atm} = -\rho_w gh = \psi, \tag{11}$$

where ψ is the "gauge" pressure in the porous media and the pressure head is

$$\psi/(\rho_w g) = -h. \tag{12}$$

The pressure head is seen to be negative. The absolute value of the pressure head, $|h|$, is called the suction or tension. As the water content in the soil decreases, the pressure head becomes more negative and the suction increases.

Soil-Water Characteristic Curves

The relationship $h(\theta)$, between the soil moisture θ and the pressure head, h, is called the soil-water characteristic. These relationships are different for differents soils, sandy or clayey for example, and they are not unique; specifically, they exhibit a memory effect called hysteresis. It is possible to obtain two primary curves, one for the drainage from a saturated state, and one for the wetting of a dry state. Any further modification (additional wetting or drying) will yield a curve between these extremes. Plots of h vs θ relations may be found, for example, in Watson (15) and in Vachaud and Thony (16).

MOISTURE DYNAMICS

Darcy's Law

The laminar flow discharge, Q, of a fluid of viscosity, μ, through a pipe of radius, R, under a pressure drop, Δp, in a length, L, is given by the Poiseuille formula as

$$Q = \frac{\pi R^4}{8 \mu} \frac{\Delta p}{L} \quad \text{or} \quad \frac{Q}{\pi R^2} = \frac{R^2}{8} \frac{\rho_w g}{\mu} \frac{\Delta p}{\rho_w g L}. \tag{13}$$

Similarly, for laminar flow in saturated porous media,

$$\overline{q} = -K_s \overline{\nabla} H, \tag{14}$$

where q is the flow rate per unit area, or apparent velocity, K_s is the hydraulic conductivity at saturation and H is the total head (pressure head plus elevation head). The hydraulic conductivity at saturation K_s can be written as the product of two quantities. The first depending only on the porous medium and called the intrinsic permeability, $k = R^2/8$, and a second quantity, $\rho_w g/\mu$, called the fluidity, designated by f. Then $K_s = kf$.

It is to be noted that, as K depends on μ, it will vary with temperature. Darcy's Law is extended to unsaturated media as

$$\overline{q} = -K(\theta) \overline{\nabla} H, \tag{15}$$

where $K(\theta)$ is the hydraulic conductivity at the soil moisture θ and $H = h(\theta) - z$, where $h(\theta)$ is the pressure head at the soil moisture θ and z is a distance measured positively downwards from the soil surface. The ratio $K(\theta)/K_s$ is called the relative permeability. The function $K(\theta)$ will vary for different types of soils. From the relationships $K(\theta)$ and $h(\theta)$ it is possible to obtain the function K(h) which will again exhibit hysteresis.

The Continuity Equation

The conservation of mass in an elementary cube is written as

$$\frac{\partial}{\partial t}(\rho_w \theta) = -\text{div}(\rho_w \overline{q}) - S' \tag{16}$$

$$= -(\frac{\partial}{\partial x}(\rho_w u) + \frac{\partial}{\partial y}(\rho_w v) + \frac{\partial}{\partial z}(\rho_w w)) - S',$$

where u, v, w are the x, y, z components of the flow velocity vector q, t is time and S is the source (or sink) of mass per unit volume and per unit time.

The Dynamic Equations

The motion of moisture in the unsaturated zone is, therefore, described in terms of two equations: an equation of motion or Darcy's Law:

$$\overline{q} = -K(\theta) \, \overline{\nabla} H, \tag{17}$$

where

$$H = h(\theta) - z, \tag{18}$$

and a mass conservation equation, which for constant water density is

$$\frac{\partial \theta}{\partial t} = -\text{div} \, \overline{q} - S, \tag{19}$$

where $S = S'/\rho_w$

Combining these two governing equations yields

$$\frac{\partial \theta}{\partial t} = \text{div} \, \{K(\theta) \, \overline{\nabla} \, (h(\theta) - z)\} - S, \tag{20}$$

or, assuming no source or sink,

$$\frac{\partial \theta}{\partial t} = \frac{\partial}{\partial x}\left[K(\theta) \, \frac{\partial h(\theta)}{\partial x}\right] + \frac{\partial}{\partial y}\left[K(\theta) \, \frac{\partial h(\theta)}{\partial y}\right] + \frac{\partial}{\partial z}\left[K(\theta) \, \frac{\partial h(\theta)}{\partial z} - 1)\right]. \tag{21}$$

For vertical infiltration the equation reduces to

$$\frac{\partial \theta}{\partial t} = \frac{\partial}{\partial z}\left[K(\theta)\left(\frac{\partial h}{\partial z} - 1\right)\right]. \tag{22}$$

It is customary to introduce the capillary diffusivity defined by

$$D(\theta) = K(\theta) \, \frac{dh}{d\theta} \tag{23}$$

with which the previous equation becomes

$$\frac{\partial \theta}{\partial t} = \frac{\partial}{\partial z}\left[D(\theta)\frac{\partial \theta}{\partial z} - K(\theta)\right], \tag{24}$$

which is known as the Fokker-Planck equation. In this equation

$$q(z,t) = K(\theta) - D(\theta) \, \frac{\partial \theta}{\partial z} \tag{25}$$

is the flow velocity.

By using the identity

$$\left(\frac{\partial \theta}{\partial t}\right)_z \left(\frac{\partial z}{\partial \theta}\right)_t = -\left(\frac{\partial z}{\partial t}\right)_\theta \tag{26}$$

the Fokker-Planck equation can also be written as

$$\frac{\partial z}{\partial t} = \frac{\partial}{\partial \theta}\left[-D(\theta)\frac{\partial \theta}{\partial z} + K(\theta)\right]. \tag{27}$$

An alternate form is obtained by integrating (27) with respect to θ as

$$\frac{\partial}{\partial t}\int_{\theta_0}^{\theta} z\,d\theta = -D\frac{\partial \theta}{\partial z} + (K - K_0), \tag{28}$$

where K_0 is the hydraulic conductivity at soil moisture θ_0. Introducing the capillary capacity

$$C(h) = \frac{\partial \theta(h)}{dh}, \tag{29}$$

equation (22) becomes

$$C(h)\frac{\partial h}{\partial t} = \frac{\partial}{\partial z}\left[K(h)\left(\frac{\partial h}{\partial z} - 1\right)\right] \tag{30}$$

which is a form of Richard's equation.

It is seen that there are two basic formulations of the vertical movement of the soil moisture; as a function of the moisture content Θ as in equation (24) or as a function of the pressure head h or suction (-h), as in equation (30). For a saturated homogeneous isotropic medium in steady-state equation (30) becomes Laplace's equation $\nabla^2 H = 0$.

It is important to emphasize the assumptions implied in the equations (24) and (30). There are: (a) the air contained in the soil is assumed to be at atmospheric pressure so that the flow is limited to a single water phase, (b) the medium is inert, isothermal and non-deformable, and (c) the effect of hysteresis is negligible.

ANALYTICAL SOLUTIONS OF THE INFILTRATION EQUATION

There exists no general analytical solution of equation (24) or its counterpart (30) because of their strong nonlinearity. However, some quasi-analytical solutions have been developed for specific boundary conditions. The quasi-analytical solution of Philip (17,18) is for ponding or concentration type (Dirichlet's problem) boundary conditions. Parlange (19,20) developed an iterative solution for the flux boundary condition (Newman's problem). Another approach, also requiring successive approximation for flux boundary conditions, has been given by Philip (21) and Philip and Knight (22). Babu (23) presents a method based on perturbation theory, and Tolikas et al. (24) present a solution for the case that the relations $D(\theta)$ and $K(\theta)$ are exponential. Only the Philip solution is summarized here.

The Philip solution of the absorption equation.

Horizontal infiltration is called absorption. Equation (21) reduces to

$$\frac{\partial \theta}{\partial t} = \frac{\partial}{\partial x} \left[D \left(\theta \left(\frac{\partial \theta}{\partial x} \right) \right) \right]$$ (31a)

The gravity effect is absent and the phenomenon is governed by the diffusion of moisture. This is solved for the boundary conditions

$$t < 0, \ x \geqslant 0, \ \theta = \theta_0, \ \text{and} \ t \geqslant 0, \ x = 0, \ \theta = \theta_1,$$ (31b)

or initially, an infinitely long column is at the uniform soil moisture θ_0, and at $t = 0$, and thereafter the entry plane at $x = 0$ is maintained at θ_1. With Boltzman's transformation, $\phi_1 = xt^{-1/2}$, the partial differential equation (31a), subject to (31b), reduces to an ordinary differential equation (32a) with new boundary conditions (32b):

$$- \frac{\phi_1}{2} \frac{d \theta}{d \phi_1} = \frac{d}{d \phi_1} \left(D \frac{d \theta}{d \phi_1} \right),$$ (32a)

$$\phi_1 = 0, \ \theta = \theta_1 \quad ; \quad \phi_1 \to 0, \ \theta = \theta_0.$$ (32b)

The solution of (31a,b) is then $x(\theta,t) = \phi_1(\theta) \, t^{1/2}$. The wetting front is thus seen to advance proportionally to the square root of time. Integration of equation (32a) with respect to θ yields

$$\int_{\theta_0}^{\theta} \phi_1 \, d\theta = -2D \frac{d \theta}{d \phi_1} = -\frac{2D}{\phi_1}.$$ (33)

The cumulative absorption of water going through the entry plane is

$$I = \int_{\theta_0}^{\theta_1} x(\theta,t) \, d\theta = \left[\int_{\theta_0}^{\theta_1} \phi_1 \, d\theta \right] t^{1/2} = St^{1/2},$$ (34)

where the integral inside the brackets, designated by S, is called the sorptivity.

The Philip solution of the Vertical Infiltration.

Philip gave a quasi-analytical solution to the Fokker-Planck equation (24) subject to the boundary conditions

$$t < 0, \ z \geqslant 0, \ \theta = \theta_0; \quad t \geqslant 0, \ z = 0, \ \theta = \theta_1,$$ (35)

where θ_0 is the initial soil moisture and θ_1 is the soil moisture imposed at the surface. That is, the soil is at a constant wetness θ_0, and at time $t = 0$, it is submerged under a shallow layer of water which maintains the wetness to a new value θ_1 near saturation. The solution is given, as a generalization of the previous case, as an infinite series of powers of $t^{1/2}$.

$$Z(\theta,t) = \phi_1 t^{1/2} + \phi_2 t + \phi_3 t^{3/2} + \ldots \qquad (36)$$

By introducing (36) into (28) and by equating coefficients of equal powers of $t^{1/2}$, one obtains a series of integro-differential equations for ϕ_1, ϕ_2, ϕ_3, etc. The first of these equations is equation (34) which represents the diffusion effect, and the gravitational effects are contained in the following terms. The solution (36) shows that the moisture θ initially advances as $t^{1/2}$ as adsorption, but at larger times the advance tends to a constant $(K_1 - K_0)/(\theta_1 - \theta_0)$.

The cumulated infiltration per unit surface is

$$I(t) = K_0 t + \int_{\theta_0}^{\theta_1} z(\theta,t) \, d\theta = K_0 t + \sum_i \left[\int_{\theta_0}^{\theta_1} \phi_i(\theta) \, d\theta \right] t^{i/2}$$

$$= K_0 t + S t^{1/2} + A_2 t + A_3 t^{3/2} + A_4 t^2 + \ldots, \qquad (37)$$

where S is the sorptivity and $A_i = \int_{\theta_0}^{\theta_1} \phi_i \, d\theta$, and the infiltration rate at the surface is given by

$$i(t) = \frac{dI(t)}{dt} = \frac{1}{2} S t^{-1/2} + (A_2 + K_0) + \frac{3}{2} A_3 t^{1/2} + \ldots \qquad (38)$$

The solution of equation (36) does not converge for very large time, and Philip (18) has given an empirical radius of convergence as

$$t_{grav} = \left(\frac{S}{K_1 - K_0} \right)^2. \qquad (39)$$

For $t > t_{grav}$, Philip gave an asymptotic solution of the form

$$z(\theta,t) = Z_\infty(\theta) + u(t) \left[t - t_{grav} \right], \qquad (40)$$

where $z_\infty(\theta)$ is the moisture profile assumed to occur at $t = t_{grav}$ and $u(t) = (K_1 - K_0)/(\theta - \theta_0)$ is the translation velocity of this profile. The asymptotic profile $z_\infty(\theta)$ is obtained by introducing equation (40) into equation (28) and making use of the relation for $u(t)$, as

$$Z_\infty = (\theta_1 - \theta_0) \int_{\theta}^{\theta_1 - \Delta\theta} \frac{D(\theta) \, d\theta}{(K_1 - K_0)(\theta - \theta_0) - (K - K_0)(\theta_1 - \theta_0)}, \qquad (41)$$

where $\Delta\theta$ is an infinitesimally small, positive quantity.

Numerical Solutions of the Infiltration Equation

As the analytical solutions are available only for specific boundary conditions, one has to resort to numerical solutions for many practical problems. The finite difference treatment of the infiltration equations has been given in detail by Vauclin et al. (25) and by Haverkamp (26).

There are three starting equations, two of which are the Fokker-Planck equation (24) and the Richard's equation (30). A third formulation known as the Kirchhoff equation uses the flow potential defined by

$$V(h) = \int_{-\infty}^{h} K(h) \, dh \qquad (42)$$

with which Richard's equation becomes

$$\frac{C(h)}{K(h)} \frac{\partial V}{\partial t} = \frac{\partial^2 V}{\partial z^2} - \frac{1}{K} \frac{\partial K}{\partial h} \frac{\partial V}{\partial z}. \qquad (43)$$

These equations are generaly treated in dimensionless form by introducing the following dimensionless variables defined by

$$z^* = \frac{z}{L}, \quad t^* = \frac{t}{T}, \quad h^* = \frac{h-h_0}{h_1-h_0}, \quad L = h_1-h_0, \quad T = \frac{(h_1-h_0)(\theta_1-\theta_0)}{K_s}. \qquad (44)$$

In using a numerical solution, one must consider the choice of the starting equation, the choice of the discretization time step and distance step, the choice of an explicit, implicit or Krank-Nicholson discretization scheme, the problem of time interpolation (linearization) of the value of C at some instant between t^* and $t^* + \Delta t^*$, the problem of space interpolation of the hydraulic conductivity $K^*(h)$ at the time $t^* + \Delta t^*$ for implicit schemes, and the choice of the values (weighting) of $K^*(h)$ at points $z^* - \Delta z^*/2$ and $z^* + \Delta z^*/2$.

The stability and convergence characteristics have been studied in detail by Vauclin et al. (25) and by Haverkamp (26) for about 40 different numerical schemes and compared to an improved Philip solution. Some of their conclusions are the following: 1.) an important parameter is $M^* = \Delta t^*/(\Delta z^*)^2$ which defines the relation between the time step and the distance step. To avoid the accumulation of computer truncation errors it is necessary to have $M^* > 5$, although this may vary somewhat with computer types. 2.) For an optimal precision, they recommend the use of the totally implicit Kirchhoff type model (which eliminates the weighing problem); however, it is not applicable to stratified media nor to cases with evaporation. 3.) For the models based on the Richard's equation, the estimation of the internodal hydraulic conductivity has an important influence on the accuracy of the results. The geometric mean gave the best results. 4.) The totally implicit schemes appear to be preferable to explicit schemes. For equal precision the former have fewer stability and computing time constraints.

SIMPLIFIED INFILTRATION MODELS

Approximations often used for small times, t, are the two parameter equations for the accumulated infiltration, I, and the infiltration rate, i, based on Philip's derivation. For this case equations (37) and (38) can be simplified to

$$I(t) = St^{1/2} + Kt \quad \text{and} \quad i(t) = 1/2\ St^{-1/2} + K, \quad (45)$$

where K is the soil's upper layer hydraulic conductivity and S is the sorptivity. It is seen that for large times $i(t) = K$. One criticism of this equation is that it yields an infinite infiltration rate at time zero.

Another simplified solution can be obtained by omitting the gravity term $\partial K(\theta)/\partial z$ in equation (24) and assuming that the capillary diffusivity is a constant. The dynamic equation reduces to

$$\frac{\partial \theta}{\partial t} = D\ \frac{\partial^2 \theta}{\partial z^2}. \quad (46)$$

For the boundary conditions

$$\theta = \theta_i, \quad z < 0, \quad t = 0 \quad \text{and} \quad \theta = \theta_0, \quad z = 0, \quad t > 0.$$

Eagleson (27) shows that one obtains the commonly used Horton equation

$$i = i_c + (i_0 - i_c)\ e^{-Dm^2 t} \quad (47)$$

where the quantity Dm^2 is generally represented as a single constant, k, which determines the rate of decrease of the infiltration rate from the initial value i_0 to the final constant value i_c. In practice the three constants i_0, i_c, and k must be evaluated experimentally. For this reason Horton's equation is usually regarded as empirical.

One of the earliest infiltration equations was that of Green and Ampt. In contradistinction with the Horton approach, the Green-Ampt approach assumes that the diffusive effects associated with the capillary forces are negligible compared to the gravity forces. There is thus a distinct wetting front that progresses downwards. This is similar to the motion of the equilibrium soil moisture profile z_∞ of Philip's asymptotic solution, but now the profile becomes a rectangle and there is a piston type flow. Above the wetting front the soil is uniformly wet and of constant hydraulic conductivity. The hydraulic conductivity-pressure head curve is thus assumed to be discontinuous, changing abruptly from the high value prevailing above the wetting front to a lower value at higher suctions (low heads) prevailing below the wetting front. The hydraulic conductivity vs. (negative) pressure head (K vs h) curve is thus represented by a step function. The value of the pressure head at which the step occurs is called the critical pressure head, h_{cr}. Bouwer (28, p. 239) selects h_{cr} such that the area under the step function is equal to the area under the K-h curve which it approximates. With a ponding head of H_0, the distance between the soil surface and the wetting front designated by L_f, Darcy's Law yields

$$i = K (H_0 + L_f - h_{cr})/ L_f .\tag{48}$$

Substituting the rate of advance of the wetting front $dL_f/dt = i/(\theta_1 - \theta_0)$ into the above equation and integrating yields

$$\frac{Kt}{\theta_1 - \theta_0} = L_f - (H_0 - h_{cr})\ln[1 + L_f/(H_0 - h_{cr})] .\tag{49}$$

The magnitude of the second term in the right-hand side of equation (49) increases much more slowly than L_f for large values of t. Thus for large t

$$L_f \simeq Kt/(\theta_1 - \theta) + C_1 \tag{50}$$

and

$$I \simeq (\theta_1 - \theta) L_f = Kt + C_2 \quad \text{and} \quad i = K \tag{51}$$

where C_1 and C_2 can be taken as constants. Bouwer suggests to take K as the resaturated hydraulic conductivity, roughly equal to half the saturated hydraulic conductivity, and alternatively h_{cr} and K can be measured in the field with an air-entry permeameter. Equation (49) can be extended to soil with increasing or decreasing permeability by applying it to a succession of small layers of constant hydraulic conductivity (Bouwer 29, 30). Philip (31) has shown that the Green-Ampt equation can be obtained from equation (24) by assuming that the diffusivity $D(\theta)$ be represented by $D = S^2 (\theta_1 - \theta_0) \delta(\theta_1 - \theta)/2$ where δ is the Dirac-delta function. Thus D is large near $\theta = \theta_1$ (the imposed soil moisture at the surface) and is negligible elsewhere.

Other Empirical Infiltration Equations

Holtan (32) proposed an equation of the type

$$i = i_c + a(S - I)^n \qquad 0 \leqslant I \leqslant W$$

$$i = i_c \qquad I \geqslant W, \tag{52}$$

where i_c, s, S and n are constants, S is the maximum water storage capacity of the soil above the first impending layer (in equivalent depth unit). The values of a and n are well documented for different soil type and vegetation.

The Kostiakov (33) equation is $i = Bt^{-n}$, where B and n are empirical constants. It is seen that i tends to infinity as t = 0. It should be remembered that the Green-Ampt and Philip equations apply under ponding conditions, whereas the Horton, Holtan and Kostiakov equations apply under flux conditions. The simplified Philip, Horton, Holtan and Kostiakov models do not explicitly take into account the initial soil moisture, but the Green-Ampt model does.

Infiltration of Rainfall-Ponding Time

If uniform rain falls with an intensity r less than the potential infiltration rate i, then all the rain will infiltrate. But if the rainfall rate r is larger than the infiltration rate i, then runoff will be generated at the rate r-i. As the infiltration rate decreases with time, all the rain may initially infiltrate until such a time that the two rates are equal. This time is called the ponding time. The accumulated infiltrated depth up to ponding time I_p and the depth of the wetting front are related by $L_f = I_p /(\theta_1 - \theta_0)$. The ponding time is then given by $t_p = I_p /r$. Substitution of these relationships into equation (48) with $H_0 = 0$ gives the ponding as

$$t_p = - \frac{h_{cr} \, (\theta_1 - \theta_0)}{r \, (r - K)} \cdot K \qquad (53)$$

Under variable rainfall, the time history is generally represented by a hyetogram or bar diagram which shows the rainfall intensity remaining constant for each time interval, but varying from one interval to the next. For such a case Morel-Seytoux (34) extends the above formula as

$$t_p = t_{j-1} - \frac{1}{r_j} \left[\frac{h_{cr} \, (\theta_1 - \theta_0) \, K}{(r_j - K)} - \sum_{\nu=1}^{j-1} r_\nu \, (t_\nu - t_{\nu-1}) \right] \qquad (54)$$

starting with j = 1, then j = 2, etc. until $t_p < t_j$.

To calculate the infiltration after ponding, Morel-Seytoux (34) proposes an extension of Philip's simplified equation as

$$i = \frac{1}{2} S \, (I_p, \theta_0) \, \frac{1}{(t - t_p + t_r)^{1/2}} + \frac{K}{\beta}, \qquad (55)$$

where $S \, (I_p, \theta_0)$ is called the rainfall sorptivity given by

$$S(I_p, \theta_0) = \left\{ 2K \, (S_f + I_p)^2 / \beta \left[S_f + I_p \, (1 - \frac{\beta}{\beta}) \right] \right\}^{1/2}, \qquad (56)$$

69

where

$$S_f = -h_{cr}(\theta_1 - \theta_0)$$

and β_p and β are values of viscous correction factor at ponding time and at equilibrium. Finally, t_r in equation (55) is given by

$$t_r = \frac{1}{2}(S_f + I_p)^2 / \left\{ \frac{K}{\beta}(\beta\frac{r}{K} - 1)^2 \left[S_f + I_p(1 - \frac{\beta_p}{\beta}) \right] \right\}. \qquad (57)$$

A numerical example of infiltration calculation and an example of parameter estimation are given in Morel-Seytoux (34).

An alternate method to estimate the rainfall excess is the SCS (Soil Conservation Service) method which is of difficult application outside the United States because it requires a soil classification which may not be available elsewhere. The foundations of the method have recently been criticized. For these reasons it is not reviewed here. However, the hydraulics characteristics of soils have been related to the soil classification and a computer program for the calculation of the rainfall excess has been given by Verdin and Morel-Seytoux (35).

Spatial Variability of Infiltration

So far the discussion has been limited to infiltration in a homogeneous medium. At the parcel or watershed level soils are heterogeneous and their properties vary spatially. The spatial variability of soil properties has been reviewed, among others, by Warrick and Nielsen (36) and by Vauclin (37). At the parcel level, 1) The state variables (water content, soil water storage) and the static parameters (soil density, textural composition) exhibit a low to average variability with a coefficient of variation less than 50% and are usually normally distributed; 2) The dynamic parameters (infiltration rate, hydraulic conductivity) exhibit a large variability with a coefficient of variation larger than 50% and are usually log-normally distributed; and 3) The chemical properties are more variable than the physical properties, and the coefficient of variation may reach 300%.

Soil properties are not purely random over a field, but statistically they exhibit a spatial structure that must be considered. Two methods of estimating the degree of spatial dependence among the varying parameters are the autocorrelation function and the semi-variogram. The spatial dependence measured by these parameters does not provide, however, any information on the frequency distribution of the variables. The autocovariance function $C(h)$ of a random variable $z(x)$ with mean μ and variance $\sigma^2 = \text{var}[z(x)]$ is defined as

$$C(h) = E\{[z(x) - \mu][z(x + h) - \mu]\}, \qquad (58)$$

70

where h is the lag or distance separating the points where the variable z is measured, and E is the mathematical expectation. The autocorrelation function ρ (h) then is defined as

$$\rho(h) = \frac{C(h)}{C(0)} = \frac{C(h)}{\sigma^2}. \tag{59}$$

The autocorrelation function decreases as h increases, and eventually becomes smaller than a confidence interval. The corresponding lag shows the distance of spatial dependence between neighboring observations. Examples of the autocorrelation function for steady-state infiltration rates have been given by Sisson and Wierenga (38). The semi-variogram function is defined by (Delhomme, 39)

$$\gamma(h) = 1/2 \text{ var} \left[z(x) - z(x + h) \right]. \tag{60}$$

If the mean of z(x) and the covariance of z(x) and z(h + h) do not depend on the location (i.e. for second order stationarity), then

$$\gamma(h) = \frac{1}{2} E \left\{ \left[z(x) - z(x + h) \right]^2 \right\} \tag{61}$$

and there is a simple relationship between the autocorrelation function and the semi-variogram.

$$\rho(h) = 1 - \gamma(h)/\sigma^2. \tag{62}$$

Semi-variogram functions for the steady-state infiltration rate have been given by Vieira et al. (40) and for soil temperature by Vauclin et al. (41). The semi-variogram is an increasing function of the lag h and eventually reaches a plateau indicating the range of lags over which there is a measurable spatial dependence. The semi-variogram is used in kriging (39) for finding the unbiased, least variance estimate of a variable, such as infiltration, $z(x_0)$ at point x_0 as a weighted average of n measured value $z(x_i)$, as

$$z(x_0) = \sum_{i=1}^{n} \lambda_i \, z(x_i). \tag{63}$$

An application of kriging to the construction contour maps of infiltration rate has been done by Vieira et al. (40). The calculation of the λ's involves the inversion of a matrix of the semi-variogram values for the several pairs of observation points (42).

An alternate method to treat spatial variability of infiltration and to coalesce large volumes of data is through scaling, which has been summarized by Miller (43). Warrick and Amoozegar-Fard (44) used spatially scaled hydraulic properties to calculate infiltration and drainage. Introducing a scaling parameter α_i, the scaled depth z and scaled pressure head h at

71

site i are given by $z = \alpha_i z$, $h = \alpha_i h$. Since the intrinsic permeability is proportional to the square of the pore radius, and assuming invariance of the fluid properties, the hydraulic conductivities at site i, K_i, and at the reference site, K, are related by $K_i = \alpha_i^2 K$. Introducing a scaled time T defined by $T = (\alpha_i^3/\theta_s)t$, the relative degree of saturation $S = \theta/\theta_s$ and the scaled soil water diffusivity, $D = K(dh/dS)$, the Fokker-Planck equation (24) is rewritten as

$$\frac{\partial S}{\partial T} = \frac{\partial}{\partial z}\left[D\frac{\partial S}{\partial z} - K\right].$$ (64)

With the initial and boundary conditions $S(z, 0) = S_i$ and $S(0, T) = S_f$, Philip's solution, summarized above, applies.[1] The numerical solution is performed only once, and the variability is introduced for the specific sites through the factor α_i.

SOLUTE TRANSPORT

The transport of solutes through unsaturated soil is of importance in nonpoint source pollution problems. The equation of the mass balance for a conservative solute dissolved in the liquid phase of the unsaturated soil has been given, among others, by Bear (45). For vertical infiltration it reduces to

$$\frac{\partial}{\partial\theta}(\theta C) + \frac{\partial}{\partial z}(qC) = \frac{\partial}{\partial z}(\theta D_s(\theta)\frac{\partial C}{\partial z}),$$ (65)

where C is the solute concentration and D_s is the solute diffusion coefficient. By replacing in equation (65) the flow velocity q by its expression given in (25), by using the water content distribution given in equation (24) and by assuming that D_s is a function of θ only, the equation for solute is obtained as

$$\theta\frac{\partial C}{\partial t} = \frac{\partial}{\partial z}\left[\theta D_s(\theta)\frac{\partial C}{\partial z}\right] + D(\theta)\frac{\partial\theta}{\partial z}\frac{\partial C}{\partial z} - K(\theta)\frac{\partial C}{\partial z}.$$ (66)

The equations describing the water and solute transport are thus the Fokker-Planck equation (24) and equation (66) respectively. These equations have to be solved for a set of boundary conditions, for example, equation (35) with respect to water and equation (67) with respect to the solute.

$$t < 0, \ z \geqslant 0, \ C = C_0; \quad t \geqslant 0, \ z = 0, \ C = C_1.$$ (67)

Elrick et al. (46) extended Philip's power series solution for water to solute. The solution is thus of a form similar to equation (36), specifically.

$$z(C,t) = \phi_{1s}(C)t^{1/2} + \phi_{2s}(C)t + \phi_{3s}(C)t^{3/2} + \phi_{4s}(C)t^2 + \ldots$$ (68)

Elrick et al. (46) give a table which lists the differential equations that must be solved to obtain ϕ_{1s}, ϕ_{2s}, ϕ_{3s}, ϕ_{4s}. A numerical procedure was used to obtain these solutions.

It is apparent that the water and solute movements can be predicted, at least for the assumptions of homogeneous soil, isothermal conditions and the given initial and boundary conditions. Van Genuchten (47) compared several numerical solutions of solute transport equations and reviewed their efficiency, accuracy and stability. Making use of a simpler representation, that of piston flow, Jury (48) obtained a transfer function model for the prediction of solute movement through variable field systems. Bressler and Dagan (49) show that with this piston flow assumption it is possible to correctly estimate the mean and variance of the concentration distribution in heterogeneous fields.

CONCLUSIONS

It is apparent that there is a large variety of models describing the infiltration phenomenon. Simple models such as the Green-Ampt model, or empirical models such as Holtan's model are often successfully used to describe this facet of the hydrologic cycle. More complete solutions, such as that of Philip and other numerical solutions, are usually needed to gain a better understanding of the infiltration and of the accompanying water redistribution in the soil. The stochastic variation of the flow parameter, principally the hydraulic conductivity, is generally recognized, but simple solutions of the spatial variation of the infiltration at the field scale or at the watershed scale are difficult to obtain and are the subject of current research. However, it appears that variability in space may outweigh the variability in time, that is, the points in a field or a watershed that are the wettest or the driest, consistently keep this characteristic in time. A more detailed understanding of the fluid mechanics of infiltration would require the consideration of the pressure of air. Consideration of viscous resistance to flow due to air, air counterflow and air compression all add complexity to the problem and was not included in this review. However, Morel-Seytoux (50) and Morel-Seytoux and Vauclin (51) claim that the added insight is worth the increase in complexity. The transport of solute can be modeled, but only with added mathematical effort and the amount of validation of the models is less extensive than for the case of water motion only.

The initial abstractions such as interception and depression storage are less important and have not been the subject of fundamental research as infiltration has. As a result the modeler is limited in his choice to a few empirical methods for estimating these initial losses.

REFERENCES

1. Chow, V.T., (Ed.), Handbook of Applied Hydrology, McGraw-Hill, New York, 1964.

2. Hillel, D., Fundamentals of Soil Physics, pp. 413, Academic Press, New York, 1980.

3. Vauclin, M., "Infiltration in Unsaturated Soil", Lecture at NATO Advanced Study Institute on "Mechanics of Fluids in Porous Media New Approach in Research", University of Delaware, July 1982, pp. 50.

4. Smith, R.E., Rational Models of Infiltration Hydrodynamics, pp. 107-180, "Modeling Components of the Hydrologic Cycle", Proceedings of the International Symposium on Rainfall Run-off Modeling, Mississippi State University, May 1981, ed. by V. P. Sing, Water Resources Publications, 1982.

5. Swartzendruber, D., The Flow of Water in Unsaturated Soils, pp. 215-292 in Flow Through Porous Media, ed. R. J. M. DeWiest, Academic Press, New York, 1969.

6. Lull, H.W., Ecological and Silvicultural Aspects, Chapter 6 in Handbook of Applied Hydrology, ed. V. T. Chow, McGraw-Hill, New York, 1964.

7. Dunne, T., and Leopold, L.B., Water in Environmental Planning, Freeman and Co., San Francisco, 1978.

8. Horton, R.E., Rainfall Interception, Monthly Weather Review, Vol. 47, 1919, pp. 193-206.

9. Tholin, A.L., and Keifer, C.J., Hydrology of Urban Runoff, Transactions, American Society of Civil Engineers, Paper No. 3061, 1960.

10. Viessam, W., Knapp, J.W., Lewis, G.L., and Harbaugh, T.E., Introduction to Hydrology, Dun-Donnelly, New York, 1977.

11. Novotny, V., and Chesters, G., Handbook of Nonpoint Pollution, Van Nostrand, New York, 1981.

12. Linsley, R.K., Kohler, M.A., and Paulhus, J.L.H., Applied Hydrology, McGraw-Hill, New York, 1949.

13. Soil Survey Staff, Soil Survey Manual, U.S. Department of Agriculture Handbook, No. 18, 1951.

14. Baver, L.D., Gardner, D.H., and Gardner, W.R., Soil Physics, J. Wiley and Sons, New York, 1972.

15. Watson, K.K., Some Applications of Unsaturated Flow Theory, in Drainage of Agriculture, J. van Schilfgaarda, ed., Agronomy Monograph No. 17, American Society of Agronomy, 1974, pp. 359-405,

16. Vachaud, G., and Thony, J.L., "Hysteresis During Infiltration and Redistribution in a Soil Column at Different Initial Water Contents", Water Resources Research, 7, 1971, pp. 111-127.

17. Philip, J.R., "The Theory of Infiltration", Soil Science 83, 1957, pp. 345-357.

18. Philip, J.R., "Theory of Infiltration", in Advance in Hydrosciences, ed. V.T. Chow, Vol. 5, 1969, pp. 215-296.

19. Parlange, J.Y., "Theory of Water Movement in Soils: 2. One-dimensional Infiltration", Soil Science 111, 1971, pp. 170-174.

20. Parlange, J.Y., "Theory of Water Movement in Soils: 8. One-dimensional Infiltration with Constant Flux at the Surface", Soil Science 14, 114, 1972, pp. 1-4.

21. Philip, J.R., "On Solving the Unsaturated Flow Equation: 1. The Flux-Concentration Relation", Soil Science 116, 1973, pp. 328-335.

22. Philip, J.R., and Knight, J.H., "On Solving the Unsaturated Flow Equation: 3. New Quasi-analytical Technique", Soil Science, 119, 1, 1974, pp. 1-13.

23. Babu, D.K., "Infiltration Analysis and Perturbation Methods, 3: Vertical Infiltration", Water Resources Research 12, 5, 1976, pp. 1019-1024.

24. Tolikas, P.K., Tolikas, D.K., and Tzimopoulos, C.D., Vertical Infiltration of Water into Unsaturated Soil with Variable Diffusivity and Conductivity, Journal of Hydrology, 62, 4, 1983, pp. 321-332.

25. Vauclin, M., Haverkamp, R., and Vachaud, G., Application a l'infiltration de l'eau dans les sols non satures." Presses Universitaires de Grenoble, 1979, pp. 183.

26. Haverkamp, R., "Resolution de l'equation de l'infiltration de l'eau dans le sol, Approches analytiques et numeriques", Thesis for the title of Docteur es-Sciences Physiques, Universite Scientifique et Medicale de Grenoble and Institut National Polytechnique de Grenoble, 1983, pp. 250.

27. Eagleson, P.S., "Dynamic Hydrology", McGraw-Hill, New York, 1970.

28. Bouwer, H., "Ground Water Hydrology", McGraw-Hill, New York, 1978, pp. 480.

29. Bouwer, H., "Infiltration of Water into Nonuniform Soil", Journal of Irrig. Drain., Div. American Society of Civil Engineers, 95, (IR 4), 1969, pp. 451-462.

30. Bouwer, H., "Infiltration into Increasingly Permeable Soils", Journal of Irrig. Drain., Div. American Society of Civil Engineers, 102 (IR 1), 1976, pp. 127-136.

31. Philip, J.R., "Absorption and Infiltration in Two-and Three-dimensional Systems", in "Water in the Unsaturated Zone", R.E. Rijtema and Wassink, eds., Vol. 2m, 1966, pp. 503-525, IASH/UNESCO Symposium, Wageningen.

32. Holtan, H.N., Concept for Infiltration Estimates in Watershed Engineering, U.S. Department of Agriculture, Ag. Res. Serv. Public, 1961, pp. 41-51.

33. Kostiakov, A.N., On the dynamics of the coefficient of water-percolation in soils and on the necessity of studying it from a dynamic point of view for purposes of ameliora-tion, Trans. Com. Int. Soc. Soil Sci., 6th, Moscow, Part A, 1932, pp. 17-21.

34. Morel-Seytoux, H., "Application of Infiltration Theory for the Determination of Excess Rainfall Hyetograph", Water Resources Bulletin 17, 1981, pp. 1012-1022.

35. Verdin, J.P., and Morel-Seytoux, H., User's Manual for XSRAIN, a FORTAN IV Program for the Calculation of Flood Hydrographs for Ungaged Watersheds, Colorado State Univer-sity Report CER80-81-JPV-HJM17, Nov. 1980, pp. 176.

36. Warrick, A.W., and Nielsen, D.R., "Spatial Variability of Soil Physical Properties in the Field", Chapter 13 in "Application of Soil Physics", by D. Hillel, Academic Press, 1980, pp. 319-344.

37. Vauclin, M., "Methodes d'etudes de la variabilite Spatiale des proprietes du sol", in "Variabilite Spatiale des pro-cessus de transfert dans les sols", Avignon, France, June 1982, Les Colloques de l'INRA, No. 15, 1983, pp. 9-43.

38. Sisson, J.R., and Wierenga, P.J., "Spatial Variability of Steady-State Infiltration Rates on a Stochastic Process", Journal of Soil Science, Soc. Am. 45, 1981, pp. 699-704.

39. Delhomme, J.P., "Kriging in Hydroscience", Centre d'Informatique Geologique, Fontainebleau, France, 1976.

40. Vieira, S.R., Nielsen, D.R., and Biggar, J.W., "Spatial Variability of Field-Measured Infiltration Rates", Soil Science, Am. J. Vol. 45, 1981, pp. 1040-1048.

41. Vauclin, M., Vieira, S.R., Bernard, R., and Hatfield, J.L., "Spatial Variability of Surface Temperature along Two Transects of Bare Soil", Water Resources Research 18.6, 1982, pp. 1677-1686.

42. Burgess, T.M., and Webster, R., "Optimal Interpolation and Isorithmic mapping of Soil Properties: 1. The Semi-variogram and Punctual Kriging", Soil Science 31, 1980, pp. 315-331.

43. Miller, E.E., "Similitude and Scaling of Soil-Water Phenomena", Chapter 12 in "Applications of Soil Physics, by D. Hillel, Academic Press, 1980, pp. 300-318.

44. Warrick, A.W., and Amoozegar-Fard, "Infiltration and Drainage Calculation Using Spatially Scaled Hydraulic Properties", Water Resources Research 15, 1979, pp. 1116-1120.

45. Bear, J., "Dynamics of Fluids in Porous Media", Elsevier, New York, 1972.

46. Elrick, D.E., Laryea, K.B., and Groenevelt, P.H., "Hydrodynamic Dispersion During Infiltration of Water into Soils", Soil Science Am. J., Vol. 43, 1979, pp. 856-865.

47. Van Genuchten, M. Th., "A Comparison of Numerical Solutions of the One-dimensional Unsaturated Flow and Mass Transport Equations", Advances in Water Resources, 5, 1982, pp. 47-55.

48. Jury, W.A., "Simulation of Solute Transport Using a Transfer Function Model", Water Resources Research 18, 1982, pp. 363-368.

49. Bressler, E., and Dagan, G., "Unsaturated Flow in Spatially Variable Fields, Three Solute Transport Models and Their Application to Two Fields", Water Resources Research, 19, 1983, pp. 429-435.

50. Morel-Seytoux, H., "Infiltration Affected by Air, Seal, Crusts, Ice and Various Sources of Heterogenity", Proceedings of the National Conference on Advances in Infiltration, Am. Soc. of Agric. Engineering, Dec. 1983, pp. 132-146.

51. Morel-Seytoux, H., and Vauclin, M., "Superiority of Two-Phase Formulation for Infiltration", Proceedings of the National Conference on Advances in Infiltration, Am. Soc. of Agric. Engineering, Dec. 1983, pp. 34-37.

»LATERAL MOVEMENTS OF VADOSE WATER IN LAYERED SOILS«

Aldo Giorgini and Martinus Bergman
School of Civil Engineering, Purdue University
West Lafayette, Indiana 47907, U.S.A.

Abstract

Since groundwater contamination is becoming recognized as a serious problem both in the United States and abroad, and since the contamination processes involve the vadose zone below the soil surface, the question on whether infiltration may occur in directions other than vertical acquires a place of utmost importance. In this paper it is shown that, under particular conditions, which are chosen so as to emphasize the importance of the problem, lateral infiltration occurs and it can be more predominant than vertical infiltration.

Introduction

In 1957, E.C. Childs prefaced his paper on anisotropic hydraulic conductivity of soil with the following statement. "It has long been recognized that soils may have the property of anisotropic hydraulic conductivity, that is to say they may conduct water more readily in certain directions than in others. The absence until recently of a ready means of measuring anisotropic conductivity in the field has resulted in a situation where on the one hand speculation tended to lay great emphasis on the frequency and magnitude of anisotropy and yet, on the other hand, quantitative practical matters such as the design of drainage systems have inevitably proceeded without any reference to such a property."

This statement could be repeated now, mutatis mutandis, for the situation of the hydraulic conductivity of unsaturated anisotropic soils, with a curious twist; on the one hand there exists a strong practical urgency for the use of anisotropic hydraulic conductivity in unsaturated flows to better explain lateral subsurface flows on hillslopes (Atkinson, 1978; Zaslavsky and Sinai, 1981), and yet, on the other hand, no recent literature can be found that presents actual solution, however elementary, of the differential equations provided by theory.

As for E.C. Childs' statement, while true at the time of its writing, it was rendered moot by the appearance of a lucid study by Maasland (1957) on the effects of soil anisotropy on land drainage systems design. This study has been more or less directly the model for all subsequent treatises of textbooks which deal with flow through saturated anisotropic soils, and it contains the most extensive bibliography of the contemporary state-of-the art on the flow of water in saturated anisotropic soils. In the following presentation, we shall draw substantially from Maasland's (1957) study as far as references introduction is concerned. Maasland's (1957) approach to the solution of applied problems relies on the fact that the equation of motion of a liquid in a homogeneous anisotropic porous medium, when written in Cartesian coordinates coincident with the principal directions of anisotropy, can be reduced to the Laplace equation by suitable deformation of each coordinate. Maasland presents his theory of fluid flow through anisotropic media in the form of five general theorems which are used throughout his study and in the applications. We shall present the theorems here in their original formulation, since doing so will simplify their development's attributions in the annotated references.

Theorem I. A porous medium, consisting of any number of arbitrarily directed sets of parallel, elementary flow tubes, can always be replaced by an equivalent, fictitious, porous medium of equal size with three, mutually perpendicular, uniquely directed systems of pore tubes. In this fictitious medium, the net flow per unit area is the same in every direction as in the actual medium, provided that the hydraulic head is the same everywhere in the fictitious medium as in the actual medium.

Theorem II. The effect of an anisotropy in the hydraulic conductivity is equivalent to the effect of shrinkage or expansion of the coordinates of a point in the flow system. That is, one can, by suitably shrinking or expanding the coordinates of each point in an anisotropic medium, obtain an equivalent, homogeneous, isotropic system.

Theorem III. The hydraulic conductivity, k, for the equivalent homogeneous isotropic medium into which the anisotropic medium may be expanded or shrunk is related to the hydraulic conductivities of the actual anisotropic system by the relation

$$k = (k_x k_y k_z / k_0)^{1/2},$$

where k_0 is an arbitrary constant, and k_x, k_y, and k_z are the hydraulic conductivities for the principal directions of the actual anisotropic medium.

Theorem IV. If the square root of the directional hydraulic conductivity (that is, the hydraulic conductivity in the flow direction) is plotted in all the corresponding directions at a point of an anisotropic medium, then one obtains an ellipsoid; that ellipsoid is called the ellipsoid of direction.

Theorem V. The equipotentials in an anisotropic medium are conjugate to the flow lines with regard to the ellipsoid of direction.

80

Pertinent Literature Review

As it is clear from the above statements, the theory of fluid flow through anisotropic media leading to them does not utilize a general tensor formulation, and it is therefore of little use for generalization to moisture flows in unsaturated anisotropic soils. Toward this end the theory developed by Ferrandon (1948, 1954) is far more useful and it has historically fathered that.

The most significant contribution to the theory of fluid flow through anisotropic saturated media are those of Versluys (1915), Samsioe (1931), Schaffernak (1933), Dachler (1933, 1936), Vreedenburgh (1935, 1936, 1937), Vreedenburgh and Stevens (1936), Muskat (1937), Aravin (1937), Yang (1948, 1949, 1953), Ferrandon (1948, 1954), Ghizetti (1949), Litwiniszyn (1950), Irmay (1951), Schneebeli (1953), Scheidegger (1954, 1955, 1960), Maasland and Kirkham (1955), Hall (1956), Edwards (1956), Childs (1957a, 1957b), Maasland (1957), Liakopoulos (1962, 1965a, 1965b), Bear (1972), Bouman (1979), and Falade (1981).

The relationship between micro-stratification and anisotropy has been explored experimentally and theoretically by Dachler (1933), Schaffernak (1933), Vreedenburgh (1937), and Maasland (1957) who has furthermore presented the law of refraction in two anisotropic media. Stevens (1936, 1938) has given a general example of the transformation of an anisotropic two-layer system and he has developed an electrical analog for seepage problems.

The factors affecting soil micro-stratification, be it natural or artificial, have been analyzed by Graton and Frazer (1935), Frazer (1935), Russell and Taylor (1937), Dapples and Rominger (1945), Johnson and Hughes (1948).

Other early research on anisotropic porous media has been performed by researchers of the textile industry like Fowler and Hertel (1940), Sullivan and Hertel (1940), and Sullivan (1941).

As for measurement techniques and actual measurements in the field or in the laboratory, evidence of directional preference (usually horizontal) have been found by Thiem (1907), Fraser (1935), Muskat (1937), Russell and Taylor (1937), Reeve and Kirkham (1951), Johnson and Breston (1951), Childs (1952), Childs, Cole, and Edwards (1953), Yang (1953), DeBoodt and Kirkham (1953), Maasland and Kirkham (1955, 1959), Childs, Collis-George, and Holmes (1957), Marcus (1962), Wilkinson and Shipley (1972), and Irmay (1980).

For the case of fluid flow in unsaturated anisotropic porous media, the literature is very scarce and is limited to the contribution of Liakopoulos (1964), Childs (1969), Burejev and Burejeva (1969), Whistler and Klute (1969), Cisler (1972), Shul'gin (1973), Sawhney, Parlange, and Turner (1976), Dirksen (1978), Akan and Yen (1981), Yeh and Gelhar (1982), and Yen and Akan (1983). On the other hand, in the closely related field of dispersion in porous media, work done with the tensor formulation is rather substantial. It suffice to mention here the contributions of Bear (1961), Scheidegger (1961), De Josselin De Jong (1961), and Bachmat and Bear (1964).

We shall not present here a discussion of the literature on infiltration in isotropic porous media because that subject matter is extensively covered by the paper by Delleur (1984) to which this contribution could be considered a complement.

The only references which address the occurrence of lateral flow in unsaturated soils, that is, the occurrence of nonvertical infiltration are those of Atkinson (1978) and Zaslavsky and Sinai (1981). We purposely omit mentioning the few contributions to the study of the lateral movement of vadose water in the capillary fringe just above the phreatic surface.

The fact that traditionally infiltration is assumed to take place vertically stems from the fact that it is usually assumed that the porous medium is isotropic. Even when the soil is recognized to be anisotropic, if one of the principal axes of anisotropy coincides with the direction of gravity, then infiltration takes place vertically. It is when the layers of stratification have been disturbed from their original horizontal direction that lateral flows are possible.

Since groundwater contamination, particularly from hazardous wastes, has recently been recognized as a very serious national problem [RCRA (1976), USEPA (1978, 1980), Winograd (1981), Wood, Ferrara, Gray and Pinder (1984)], it is of utmost importance to know whether special configurations of soil layers through which contaminants percolate allow lateral flows in the vadose zone and to estimate the amount of such lateral flows.

This contribution, based on Giorgini, Bergman, Hamidi, and Pravia (1984) wants to address that problem by presenting the general tensor formulation of the flow within an unsaturated anisotropic soil under a sloping surface, and by focusing on a special case with the purpose of attracting attention to the occurrence of lateral flows.

Problem Formulation

Our interest will be confined to two-dimensional problems definable in a plane containing the gravitational field vector. The (vertical) ordinate, directed against the gravitational field, is called z and the (horizontal) abscissa is called s.

We define further an axis x which forms an angle α with the axis s and an axis y orthogonal to x. The axis x (y = 0) defines the boundary between porous medium (free surface) and the ambient. The coordinate systems z,s and y,x are congruent.

Another coordinate system is furthermore defined as y', x', coherent with the previous systems, such that the angle between x and x' is ϕ. The axis x' (y' = 0) defines the directions of the soil layers.

The three coordinate systems are illustrated in Figure 1.

The coordinate system x,y defines the symmetry of the problems we wish to consider. In fact we will agree that the only type of problems we will examine in this report are problems where all physical quantities are independent of x. This means that boundaries which are not at infinity must be parallel to the x axis.

Notice that the x axis is not horizontal and that there is therefore a "driving force" due to gravity in both x and y directions.

The coordinate system x', y' coincides with the principal axes of the hydraulic conductivity tensor K'_{ij}. With this notation the hydraulic conductivity along the x' axis is, K'_{11}, the hydraulic conductivity along the y' axis is K'_{22}.

We will assume, without loss of generality, that the K'_{11} is larger than the K'_{22} component. In fact we will assume that $K'_{11} = \lambda K'_{22}$ where λ is at most a function of the moisture content θ.

The tensor K'_{ij} is therefore expressible as

$$K'_{11} = \lambda K$$

$$K'_{22} = K \tag{1}$$

$$K'_{12} = K'_{21} = 0.$$

We will recall, for reference ease, the definitions and the equations which we shall use in the remainder of this article. They are:

1) The generalization of Darcy's law

$$q_i = - K_{ij} \frac{\partial \Psi}{\partial x_j}, \tag{2}$$

where q_i is the specific discharge vector, x_i is any orthogonal Cartesian coordinate system, and Ψ is the piezometric head $z + p/\gamma$ for saturated soils and $z + \psi$ for unsaturated soils. The physical quantity ψ is the negative pressure head due to the capillary effects of the soil porosity, and it is called interchangeably as capillary pressure head, moisture potential, moisture suction, or negative pressure

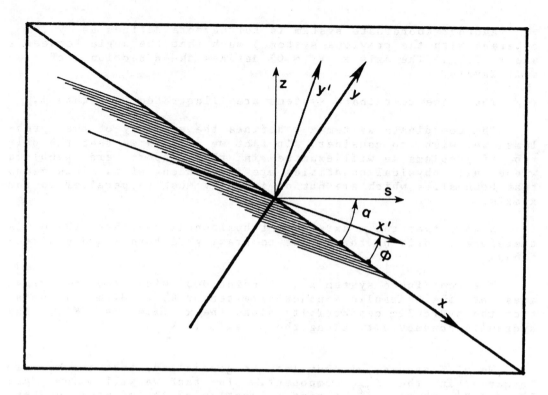

Figure 1. The three coordinate systems defined in the problem.

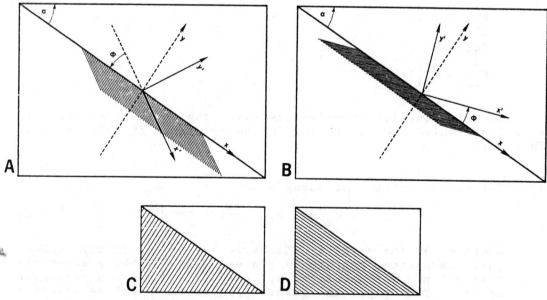

Figure 2. A.) Case of positive lateral flow due to absorption;
B.) Case of negative lateral flow;
C.) and D.) Cases of null lateral flow.

head.

2) The continuity equation

$$\frac{\partial \theta}{\partial t} + \frac{\partial q_i}{\partial x_i} = 0.$$
(3)

Upon insertion of (2) into (3) we obtain

$$\frac{\partial \theta}{\partial t} = \frac{\partial}{\partial x_i}\left[K_{ij}\frac{\partial \psi}{\partial x_j}\right],$$
(4)

or

$$\frac{\partial \theta}{\partial t} = \frac{\partial}{\partial x_i}\left[K_{ij}\frac{\partial \psi}{\partial x_j}\right] + \frac{\partial}{\partial x_i}\left[K_{ij}\frac{\partial z}{\partial x_j}\right].$$
(5)

The first term on the right-hand side of (5) represents the capillary effects on the fluid motion, the second term represents the effects of gravity.

If we now assume that the x_1, x_2 coordinate system coincides with x,y, we obtain

$$\frac{\partial \theta}{\partial t} = \frac{\partial}{\partial x}\left[K_{11}\frac{\partial \psi}{\partial x} + K_{12}\frac{\partial \psi}{\partial y}\right] + \frac{\partial}{\partial y}\left[K_{21}\frac{\partial \psi}{\partial x} + K_{22}\frac{\partial \psi}{\partial y}\right]$$

$$= \frac{\partial}{\partial x}\left[K_{11}\frac{\partial z}{\partial x} + K_{12}\frac{\partial z}{\partial y}\right] + \frac{\partial}{\partial y}\left[K_{21}\frac{\partial z}{\partial x} + K_{22}\frac{\partial z}{\partial y}\right],$$
(6)

and the expressions for the specific discharge along x and y are

$$q_x = -K_{11}\frac{\partial \psi}{\partial x} - K_{12}\frac{\partial \psi}{\partial y} - K_{11}\frac{\partial z}{\partial x} - K_{12}\frac{\partial z}{\partial y}$$
(7)

$$q_y = -K_{21}\frac{\partial \psi}{\partial x} - K_{22}\frac{\partial \psi}{\partial y} - K_{21}\frac{\partial z}{\partial x} - K_{22}\frac{\partial z}{\partial y},$$
(8)

where q_x is the lateral specific flowrate and q_y is the normal specific flowrate.

Since the elevation z can be expressed in terms of x and y as

$$z = y \cos \alpha - x \sin \alpha,$$
(9)

we can rewrite (6), (7) and (8) as follows

$$\frac{\partial \theta}{\partial t} = \frac{\partial}{\partial x}\left[K_{11}\frac{\partial \psi}{\partial x} + K_{12}\frac{\partial \psi}{\partial y}\right] + \frac{\partial}{\partial y}\left[K_{21}\frac{\partial \psi}{\partial x} + K_{22}\frac{\partial \psi}{\partial y}\right]$$

$$+ \cos \alpha \left[\frac{\partial K_{12}}{\partial x} + \frac{\partial K_{22}}{\partial y}\right] - \sin \alpha \left[\frac{\partial K_{11}}{\partial x} + \frac{\partial K_{21}}{\partial y}\right]$$
(10)

and

$$q_x = - K_{11} \frac{\partial \psi}{\partial x} - K_{12} \frac{\partial \psi}{\partial y} + K_{11} \sin \alpha - K_{12} \cos \alpha \qquad (11)$$

$$q_y = - K_{21} \frac{\partial \psi}{\partial x} - K_{22} \frac{\partial \psi}{\partial y} + K_{21} \sin \alpha - K_{22} \cos \alpha. \qquad (12)$$

The above equations are drastically simplified by our condition that all the fields be x independent: equations (10), (11), and (12) become, respectively

$$\frac{\partial \theta}{\partial t} = \frac{\partial}{\partial y} \left[K_{22} \frac{\partial \psi}{\partial y} \right] + \cos \alpha \frac{\partial K_{22}}{\partial y} - \sin \alpha \frac{\partial K_{21}}{\partial y} \qquad (13)$$

$$q_x = - K_{12} \frac{\partial \psi}{\partial y} + K_{11} \sin \alpha - K_{12} \cos \alpha \qquad (14)$$

$$q_y = - K_{22} \frac{\partial \psi}{\partial y} + K_{21} \sin \alpha - K_{22} \cos \alpha. \qquad (15)$$

The components K_{ij} of the hydraulic conductivity tensor, can be written in terms of the principal components $K_{11}' = \lambda K$, $K_{22}' = K$ as follows

$$K_{11} = \frac{K_{11}' + K_{22}'}{2} + \frac{K_{11}' - K_{22}'}{2} \cos 2\phi \qquad (16)$$

$$K_{22} = \frac{K_{11}' + K_{22}'}{2} - \frac{K_{11}' - K_{22}'}{2} \cos 2\phi \qquad (17)$$

$$K_{21} = K_{12} = - \frac{K_{11}' - K_{22}'}{2} \sin 2\phi \qquad (18)$$

or

$$K_{11} = K \left[\frac{\lambda+1}{2} + \frac{\lambda-1}{2} \cos 2\phi \right] = \xi K \qquad (19)$$

$$K_{22} = K \left[\frac{\lambda+1}{2} - \frac{\lambda-1}{2} \cos 2\phi \right] = \eta K \qquad (20)$$

$$K_{21} = K_{12} = K \left[\frac{\lambda-1}{2} \sin 2\phi \right] = \chi K \qquad (21)$$

The above equations give implicitly the definitions of the functions ξ, η, and χ in terms of λ and ϕ.

Substituting (19), (20) and (21) into (13, (14) and (15) we obtain, assuming λ independent of θ,

$$\frac{\partial \theta}{\partial t} = \eta \frac{\partial}{\partial y} \left[K \frac{\partial \psi}{\partial y} \right] + \eta \cos \alpha \frac{\partial K}{\partial y} - \chi \sin \alpha \frac{\partial K}{\partial y} \tag{22}$$

and

$$q_x = - \chi K \frac{\partial \psi}{\partial y} + (\xi \sin \alpha - \chi \cos \alpha) K$$

$$q_y = - \eta K \frac{\partial \psi}{\partial y} - (-\chi \sin \alpha + \eta \cos \alpha) K \tag{23}$$

or

$$q_x = \frac{\chi}{\eta} q_y + \frac{\lambda}{\eta} \sin \alpha K. \tag{24}$$

If we define

$$\mu = - \chi \sin \alpha + \eta \cos \alpha$$

$$\nu = \xi \sin \alpha - \chi \cos \alpha, \tag{25}$$

equations (22) and (23) can be rewritten as

$$\frac{\partial \theta}{\partial t} = \eta \frac{\partial}{\partial y} \left[K \frac{\partial \psi}{\partial y} \right] + \mu \frac{\partial K}{\partial y} \tag{26}$$

$$q_x = - \chi K \frac{\partial \psi}{\partial y} + \nu K = q_{ax} + q_{gx}$$

$$q_y = - \eta K \frac{\partial \psi}{\partial y} - \mu K = q_{ay} + g_{gy}, \tag{27}$$

where we have implicitly defined q_{ax}, q_{ay} as the absorption contribution to the specific flowrate and q_{gx}, q_{gy} as the gravitation contributions to the specific flowrate.

If we introduce the definitions

$$\overline{K} = \frac{\mu^2}{\eta} K$$

$$Y = \frac{\mu}{\eta} y, \tag{28}$$

(26) can be rewritten as

$$\frac{\partial \theta}{\partial t} = \frac{\partial}{\partial Y} \left[\overline{K} \frac{\partial \psi}{\partial Y} \right] + \frac{\partial \overline{K}}{\partial Y}, \tag{29}$$

which has the same form as the equation of vertical infiltration (through a horizontal surface) for isotropic media.

Once (28) is solved, then q_x and q_y can be found via (23) and (24).

The absorption contribution $q_{ax} = - \chi K\, \partial\psi/\partial y$ to the lateral flow is null when $\chi = 0$, that is either when the porous medium is isotropic or when $\sin 2\phi = 0$, that is when the principal axes of anisotropy coincide with the y,x axes. For other values of ϕ, it can be positive or negative. For the particular case where $\partial\psi/\partial y > 0$ (which corresponds to situations where the moisture is gradually decreasing from the surface of the soil downward) q_{ax} is positive for $-\pi/2 < \phi < 0$ and it is negative for $0 < \phi < \pi/2$. These two cases are illustrated in Figure 2, where the direction of the hatching coincides with the principal direction with maximum hydraulic conductivity.

The absorption contribution $q_{ay} = - \eta K(\partial\psi/\partial y)$ to the normal flow is always negative when $\partial\psi/\partial y > 0$ and has extreme values (corresponding to $\eta = 1$ and to $\eta = \lambda$) for $\phi = 0$ and for $\phi = \pi/2$ respectively.

The two components of the absorption contribution to the flow can be found by using the geometrical construction shown in Figure 3 (Mohr circle). From the construction it can be seen that there are instances in which $|q_{ax}|$ can be larger than $|q_{ay}|$. This occurs only for

$$\lambda > 3 + 2\sqrt{2} = 5.83. \tag{30}$$

For values of λ satisfying (30), the range of the angle ϕ for which $|q_{ax}| > |q_{ay}|$ is given by the two values

$$\phi_{1,2} = \frac{1}{2} \sin^{-1}\left[\frac{\lambda + 1 \mp \sqrt{\lambda^2 - 6\lambda + 1}}{2(\lambda - 1)} \right] \tag{31}$$

and the angle ϕ^* for which the ratio $|q_{ax}|/|q_{ay}|$ is a maximum is

$$\phi^* = \frac{1}{2} \cos^{-1}\left[\frac{\lambda - 1}{\lambda + 1} \right]. \tag{32}$$

As for the gravitation contribution to the specific discharge, its geometrical construction can be more conveniently carried out in terms of the horizontal component q_{gs} and of the vertical component q_{gz}. The general expressions in terms of lateral and normal components are given by

$$q_s = q_x \cos\alpha + q_y \sin\alpha$$

$$q_z = -q_x \sin\alpha + q_y \cos\alpha, \tag{33}$$

which lead to the particular expressions

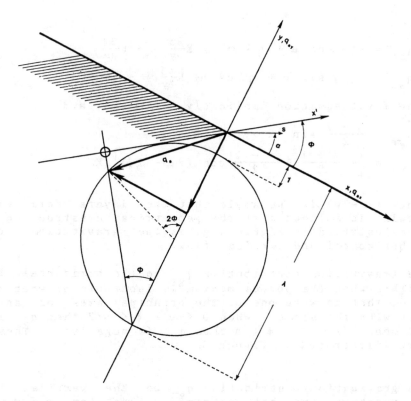

Figure 3. Geometrical construction for the absorption contribution to the specific flowrate.

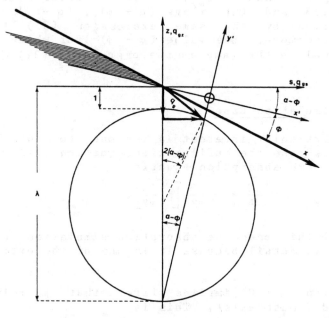

Figure 4. Geometrical construction for the gravity contribution to the specific flowrate.

$$q_{as} = -(\chi \cos \alpha + \eta \sin \alpha) \, K\frac{\partial \psi}{\partial y} = -\tau K\frac{\partial \psi}{\partial y}$$

$$q_{az} = -(-\chi \sin \alpha + \eta \cos \alpha) \, K\frac{\partial \psi}{\partial y} = -\mu K\frac{\partial \psi}{\partial y}, \tag{34}$$

where the first equation implicitly defines τ, and

$$q_{gs} = \frac{\lambda - 1}{2} \sin (2\alpha - 2\phi) \, K$$

$$q_{gz} = \left[-\frac{\lambda + 1}{2} + \frac{\lambda - 1}{2} \cos (2\alpha - 2\phi) \right] K. \tag{35}$$

Since $(\alpha - w)$ is the angle that the layers form with the horizontal, it follows that the geometrical construction (Mohr's circle) illustrated in Figure 4 gives the gravitation contribution to horizontal and vertical flow.

The gravitation contribution q_{gs} to the horizontal flow is null either when the porous medium is isotropic or when $\alpha = \phi$ or $\alpha = \psi + \frac{\pi}{2}$, that is when one of the principal axes of anisotropy coincides with the slope. When $0 < \alpha - \phi < \pi/2$ then q_{gs} is positive and when $\pi/2 < \alpha - \phi < \pi$ then q_{gs} is negative. These four cases are illustrated in Figure 5.

The gravitation contribution q_{gz} to the vertical flow is always negative and has extreme values for $\alpha - \phi = 0$ and $\alpha - \phi = \pi/2$ respectively. Anologously to what has been shown for the absorption contribution to the flow, there are instances when $|q_{gs}|$ is larger than $|q_{gz}|$. This occurs only for $\lambda > 3 + 2\backslash\overline{2} = 5.83$ and the range $(\alpha - \phi)_1$, $(\alpha - \phi)_2$ for which this occurs is given by the same expression (31) found for ϕ_1 and ϕ_2. Furthermore, the value $(\alpha - \phi)^*$ for which the ratio $|q_{gs}/q_{gy}|$ is maximal is the same as the expression (32) for ϕ^*.

The Absorption Simulation Case

A case of particular interest is the one for which $\mu = 0$. In fact in this case the infiltration equation (26) reduces to the much simpler pure absorption equation

$$\frac{\partial \theta}{\partial t} = \eta \frac{\partial}{\partial y} \left[K \frac{\partial \psi}{\partial y} \right]. \tag{36}$$

We will call this case the absorption simulation case, and we will study it in detail because it is one of the extreme cases of lateral flow.

The condition $\mu = 0$ implies first that a relationship between α, ϕ, and λ must exist. This is

$$\tan \alpha = \frac{\eta}{\chi} = \frac{(\lambda + 1) - (\lambda - 1) \cos 2\phi}{(\lambda - 1) \sin 2\phi}. \tag{37}$$

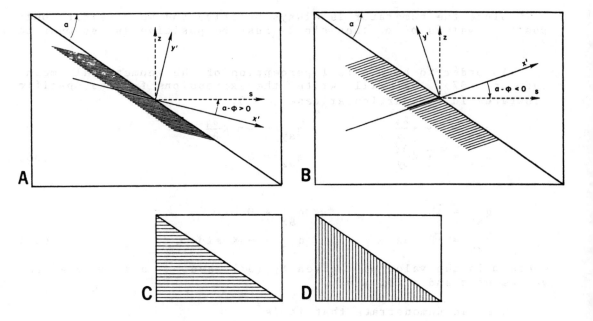

Figure 5. A.) Case of positive horizontal flow due to gravity;
B.) Case of negative horizontal flow;
C.) and D.) Cases of null horizontal flow.

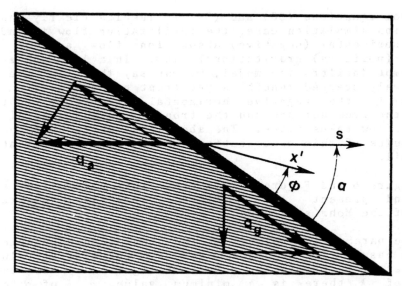

Figure 6. In the absorption simulation case, the absorption contribution to
the flow is purely horizontal and the gravitational contribution
to the flow is purely lateral.

Since the numerator is always positive and we consider only positive values of α, then $\sin 2\phi$ must be positive and so must be ϕ.

In order to have a full perception of the geometrical meaning of (37), we will write the expressions for the specific flowrates in the particular case $\mu = 0$.

$$q_{ax} = -\chi K \frac{\partial \psi}{\partial y} \qquad\qquad q_{ay} = -\eta K \frac{\partial \psi}{\partial y}$$

$$q_{as} = -\bar{\tau} K \frac{\partial \psi}{\partial y} \qquad\qquad q_{az} = 0 \qquad\qquad (38)$$

$$q_{gx} = \bar{\nu} K \qquad\qquad\qquad q_{gy} = 0$$

$$q_{gs} = \bar{\nu} K \cos \bar{\alpha} \qquad\qquad q_{gz} = -\bar{\nu} K \sin \bar{\alpha}, \qquad\qquad (39)$$

where $\bar{\alpha}$ is the value of α given by (37) and $\bar{\tau}$ and $\bar{\nu}$ are the values of τ and ν for $\alpha = \bar{\alpha}$.

One can demonstrate that it is

$$\bar{\tau} = \sqrt{\chi^2 + \eta^2} = \sqrt{\frac{\lambda^2 + 1}{2} - \frac{\lambda^2 - 1}{2} \cos 2\phi}, \qquad\qquad (40)$$

$$\bar{\nu} = \frac{\xi \eta - \chi^2}{\sqrt{\chi^2 + \eta^2}} = \lambda/\bar{\tau}. \qquad\qquad (41)$$

From equations (38) and (39) it results clearly that in the absorption simulation case, the infiltration flow is made up of a purely horizontal (negative) absorption flow and of a purely lateral (positive) gravitational flow. In a language echoing the Green Ampt infiltration model, we can say that in the case of impulsively started runoff[1] a wet front propagates from the free surface in the negative horizontal direction. In the region within the free surface and the front, a purely lateral gravitational flow takes place. The above statements contain in embryo the results of this contribution, which will be presented later in detail.

Figure 6 and Figure 7 illustrate the absorption simulation case and present the pertinent geometrical constructions by means of the Mohr's circles.

The parametric analysis of the absorption simulation case is rather easy. If we plot α versus ϕ for several values of the parameter λ, we obtain the curves of Figure 8. For each given value of λ there is a minimum value α_{min} of α below which

---

1. Impulsively started runoff problem is the slope analog of the instantaneously started ponding for horizontal surfaces.

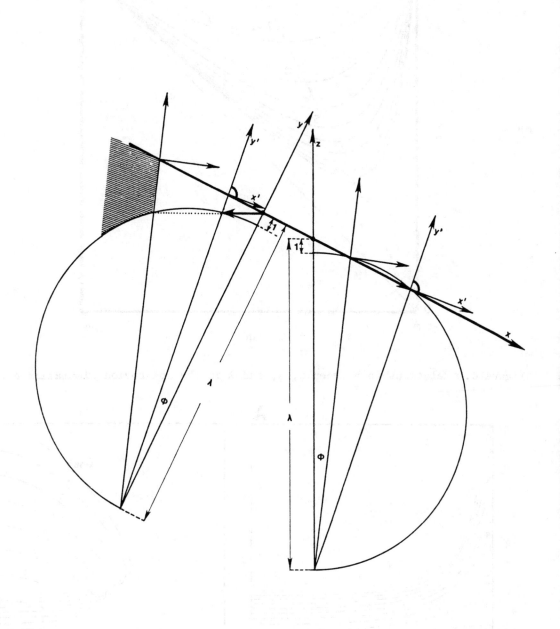

Figure 7. Geometrical constructions for the absorption simulation case.

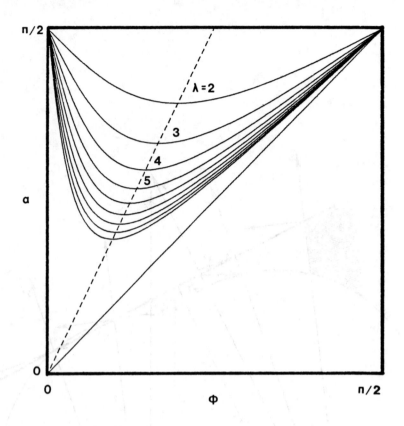

Figure 8. Relationship between α, φ, and λ in the absorption simulation case.

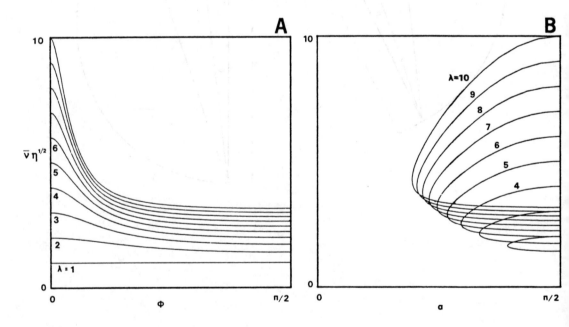

Figure 9. Magnification factor for lateral flow due to gravity.

absorption simulation cannot occur. This value is given by

$$\alpha_{min} = \tan^{-1} \frac{2\sqrt{\lambda}}{\lambda - 1}, \tag{42}$$

and the locus of the minima is the straight line $\alpha = 2\phi$. On the other hand, given any value of α, there is a value of λ^* of λ below which no simulation of absorption can occur. This value is

$$\lambda^* = \frac{1 + \cos\alpha}{1 - \cos\alpha}. \tag{43}$$

For $\lambda > \lambda^*$ there are two possible values ϕ_1^* and ϕ_2^* of ϕ for which absorption simulation occurs. These two values satisfy the following inequality

$$0 < \phi_1^* < \frac{\alpha}{2} < \phi_2^* < \alpha$$

and are given by

$$\phi_1^*, \ \phi_2^* = \tan^{-1}\left[\frac{\tan\alpha}{2} \frac{\lambda - 1}{\lambda} \mp \sqrt{\left[\frac{\tan\alpha}{2} \frac{\lambda - 1}{\lambda}\right]^2 - \frac{1}{\lambda}}\right]. \tag{44}$$

Of the two cases, the one corresponding to $\phi = \phi_1^*$ is the most interesting for the purpose of lateral flow. In fact, that case has a rather large lateral flow with respect to the absorption flow.

Subsurface Lateral Flow in the Case of Absorption Simulation

We formulate our problem in the following fashion. We choose an arbitrary value of $\lambda > 1$, and then a value of ϕ that satisfies the following inequality

$$0 < \phi < \frac{1}{2}\tan^{-1} \frac{2\sqrt{\lambda}}{\lambda - 1}, \tag{45}$$

so that we are sure that we are in the most propitious conditions for lateral flow, then we calculate the value of α by means of (37). The absorption simulation case is therefore assured.

We set as initial condition

$$t = 0, \ y > 0, \ \theta = \theta_0 \tag{46}$$

and as boundary condition

$$t \geqslant 0, \ y = 0, \ \theta = \theta_1 \tag{47}$$

The initial condition implies an initial uniform flow, due to gravitation, $q_{xo} = \overline{\nu} K(\theta_0)$.

Following the suggestion of Philip (1969), we define the cumulative absorption into the porous medium as

$$i = \int_{-\infty}^{0} (\theta - \theta_0)\, dy = - \int_{-\infty}^{0} y\, d\theta$$

$$i = \eta^{1/2}\, t^{1/2} \int_{\theta_0}^{\theta_1} \Phi(\theta)\, d\theta, \tag{48}$$

where $y(\theta) = -\eta^{1/2}\, t^{1/2} \Phi(\theta)$ and where the integral appearing in (48) is the sorptivity $S(\theta_0, \theta_1)$ of an isotropic porous medium of hydraulic conductivity K, from θ_0 to θ_1. The absorption rate, in the y direction, is then

$$v = -\frac{1}{2}\, \eta^{1/2}\, t^{-1/2}\, S(\theta_0, \theta_1). \tag{49}$$

The novel feature of the hillslope case on an anisotropic medium is constituted by the lateral flow. If we call Q_x the excess of lateral flow with respect to the initial condition, we obtain

$$Q_x = \int_{-\infty}^{0} (q_x - q_{xo})$$

or

$$Q_x = \int_{-\infty}^{0} \left[-\chi D \frac{\partial \theta}{\partial y} + \bar{v}(K - K(\theta_0)) \right] dy$$

or

$$Q_x = -\chi \int_{\theta_0}^{\theta_1} D(\theta)\, d\theta + \bar{v}\eta^{1/2}\, t^{1/2} \int_{\theta_0}^{\theta_1} \Phi(\theta)\, \frac{\partial K}{\partial \theta} d\theta. \tag{50}$$

It is to be noted that Q_x is made up of two contributions: a negative absorption contribution which is time independent and which is conceptually justified by the fact that a negative q_{ax} proportional to $t^{-1/2}$ takes place in a region of thickness proportional to $t^{1/2}$, and a positive gravity induced contribution proportional to $t^{1/2}$.

We will call the integral

$$C(\theta_0, \theta_1) = \int_{\theta_0}^{\theta_1} D(\theta)\, d\theta \tag{51}$$

the (non-directional) absorption conveyance of the medium and the integral

$$G(\theta_0, \theta_1) = \int_{\theta_0}^{\theta_1} \Phi(\theta)\, \frac{\partial K}{\partial \theta}\, d\theta \tag{52}$$

the (non-directional) gravitation conveyability of the medium.

With these definitions, (50) can be expressed in terms of λ and ϕ, as

$$Q_x = -\frac{\lambda - 1}{2} \sin 2\phi \, C(\theta_0, \theta_1)$$

$$+ \lambda \sqrt{\left|\frac{\lambda+1 - (\lambda-1)\cos 2\phi}{\lambda^2+1 - (\lambda^2-1)\cos 2\phi}\right|} \, G(\theta_0, \theta_1) \, t^{1/2}$$

or

$$Q_x = -\chi C(\theta_0, \theta_1) + \overline{\nu} \, \eta^{1/2} \, G(\theta_0, \theta_1) \, t^{1/2}. \tag{53}$$

Figure 9.A presents the coefficient $\overline{\nu}\eta^{1/2}$ in terms of λ and Figure 9.B presents $\overline{\nu}\eta^{1/2}$ in terms of λ and α.

If we consider i as the thickness of the region where flow occurs, we can define the average velocity v_x at any time t as

$$v_x = \frac{Q_x}{i} = -\chi \, \eta^{-1/2} \, t^{-1/2} \, \frac{C}{S} + \overline{\nu} \, \frac{G}{S}, \tag{54}$$

and if we respect the directions of the q_a (horizontal) and q_g (lateral) found before, we obtain

$$v_s = -\frac{\chi}{\cos \alpha} \, \eta^{-1/2} \, t^{-1/2} \, \frac{C}{S} + \overline{\nu} \, \frac{G}{S} \cos \alpha \tag{55}$$

$$v_z = -\overline{\nu} \, \frac{G}{S} \sin \alpha$$

which, once integrated in t, yield

$$s = s_0 - 2 \frac{\chi}{\cos \alpha} \, \eta^{-1/2} \, t^{-1/2} \, \frac{C}{S} + \overline{\nu} \, \frac{G}{S} \cos \alpha \, t \tag{56}$$

$$z = z_0 - \overline{\nu} \, \frac{G}{S} \sin \alpha \, t$$

as the approximate trajectory of a particle that at time $t = 0$ leaves the point s_0, z_0 on the surface of the incline.

A measure of the closeness of the trajectories to the incline is given by the elongation E, defined as the distance between the point of departure s_0, z_0 and the next point in the trajectory where $s = s_0$. The elongation is given by

$$E = \frac{C^2}{SG} \frac{1}{\lambda} \left[\frac{\lambda^2 + 1 - (\lambda^2 - 1)\cos 2\phi}{(\lambda - 1) \sin 2\phi}\right]^2. \tag{57}$$

Figure 10 illustrates the trajectories described above for three values of the parameter E.

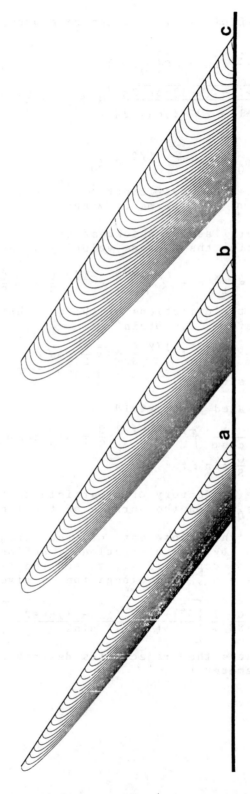

Figure 10. Approximate trajectories of fluid particles released at time t=0 from the incline's surface. The three cases illustrated correspond to elongations in the ratio 1:2:4.

Conclusions

It has been shown that in the case of nonisotropic soils, infiltration through an incline may lead to lateral flow. For those geometrical combinations of parameters which lead to absorption simulation the lateral flow can be calculated in terms of those geometrical parameters, of Philip's sorptivity, and of two new properties of porous media which have been called absorption conveyance and gravitation conveyability.

Annotated References

1. Akan, A.O., and B.C. Yen, 1981, Mathematical Model of Shallow Water Flow over Porous Media, ASCE Journal Hyd. Division, 107, 4, 479-494.

2. Aravin, V.N., 1937,, Trudy Leningrad industr. in-ta, Rasdel. gidrotekh., 19, 2.
    Early theoretical investigations. The essence of the first four theorems is present in this paper.

3. Aronovici, V.S., 1947, The Mechanical Analysis as an Index of Subsoil Permeability, Proc. Soil Sci. Soc. America, 11, 137-141.

4. Atkinson, T.C., 1978, Techniques for Measuring Subsurface Flow on Hillslopes, Hillslope Hydrology, M.J. Kipkby Ed., Wiley, New York.

5. Bachmat, Y., and J. Bear, 1964, The General Equations of Hydrodynamic Dispersion in Homogeneous, Isotropic, Porous Mediums, Journal Geophys. Research, 69, 12, 2561-2567.

6. Bear, J., 1972, Dynamics of Fluids in Porous Media, American Elsevier Pub. Co., New York, New York.

7. Bear, J., 1961, On the Tensor Form of Dispersion in Porous Media, Journal Geophys. Research, 66, 4, 1185-1197.

8. Bhattacharya, R.N., V.K. Gupta, and G. Sposito, 1976, On the Stochastic Foundations of the Theory of Water Flow Through Unsaturated Soil, Water Resources Research (USA), 12, 3, 503-512.

9. Boumans, J.H., 1979, Drainage Calculations in Stratified Soils Using the Anisotropic Soil model to Simulate Hydraulic Conductivity Conditions, Proc. Int. Drainage Workshop, 16-22, Wageningen, Holland.

10. Burejev, L.N., and Z.M. Burejeva, 1969, Some Numerical Methods for Solving Problems of Nonsteady Seepage in Non-Homogeneous Anisotropic Soils, Water in the Unsaturated Zone, (P.E. Rijtema and H. Wassink, Editors), Proceedings of the Wagenigen Symposium, IASH/AIHS - Unesco.

11. Carman, P.C., 1956, Flow of Gases Through Porous Media, Butterworth Scientific Publications, London, 5-6.
    Brief discussion about non-isotropic flow problems.

12. Childs, E.C., 1952, The Measurement of the Hydraulic Permeability of Saturated Soil in situ, I. Proc. Roy. Soc., A215, 525-535.
    It develops practicable field techniques for

measurement of hydraulic conductivity without assumption of presence or absence of anisotropy.

13. Childs, E.C., 1957, The Anisotropic Hydraulic Conductivity of Soil, J. Soil Science, 8, 1, 42-47.
    It contends that the frequency of occurrence of appreciably anisotropic conductivity may have been overestimated in past speculations.

14. Childs, E.C., 1957a, The Physics of Land Drainage, Drainage of Agricultural Lands, Ed. by J.N. Luthin, Amer. Soc. of Agronomy, Madison, Wisconsin.
    (48-55) An outline of the diadic form of nonisotropic conductivity is given. A demonstration of Theorem I is given for two-dimensional flow.

15. Childs, E.C., 1957b, The Anisotropic Hydraulic Conductivity of Soil, Jour. Soil. Sci., 8, 1, 42-47.
    He gives a proof that the principal axes of anisotropic soils are mutually orthogonal, regardless of the orientation of the elementary "fissures" of the soil. He proves that the hydraulic conductivity tensor is a symmetric tensor.

16. Childs, E.C., 1969, An Introduction to the Physical Basis of Soil Water Phenomena, Wiley, New York, New York.

17. Childs, E.C., A.H. Cole and D.H. Edwards, 1953, The Measurements of the Hydraulic Permeability of Saturated Soil in situ. II. I. Proc. Roy. Soc., A216, 72-89.
    It develops practicable field techniques for measurement of hydraulic conductivity without assumption of presence or absence of anisotropy.

18. Childs, E.C., N. Collis-George, and J.W. Holmes, 1957, Permeability Measurements in the Field as an Assessment of Anisotropy and Structure Development, J. Soil Sci., 7, 27-41.

19. Cisler, J., 1972, On the Tensor Concept of Unsaturated Anisotropic Hydraulic Conductivity, Water Resources Research, 8, 2, 525-528.

20. Dachler, R., 1933, Uber Sickerwasserstroemungen in Geschichten Material, Wasserwirtschaft, 2, 13-16.
    He gives a proof of Theorem II for two-dimensional flow.

21. Dachler, R., 1936,, Grundwasserstromung, Julius Springer, Vienna.
    He represents a new derivation of Theorem II for two-dimensional flows which he had derived in his earlier paper (1933).

22. Dapples, E.C. and J.F. Rominger, 1945, Orientation Analysis of Fine-grained Clastic Sediments: a Report on Progress, Journal Geol., 53, 246-261.
   They show that soil grains from laboratory fluvial and eolian environments exhibit a pronounced preferred elongation parallel to the direction of flow of the depositing agent and a marked tendency to lie with their larger ends up current.

23. De Boodt, M.F., and D. Kirkham, 1953, Anisotropy and Measurement of Air Permeability of Soil Clods, Soil Sci., 76, 127-133.

24. De Josselin De Jong, G., 1961, Discussion of Paper by Jacob Bear, "On the Tensor Form of Dispersion in Porous Media", Journal Geophys. Research, 66, 10, 3623-3624.

25. Delleur, J.W. 1984, "Rainfall Abstractions and Infiltration in Nonpoint Source Pollution", Proceedings of the Conference on Prediction of Agricultural Nonpoint Source Pollution: Model Selection and Application, Venice, Italy.

26. Dirksen, D., 1978, Transient and Steady Flow From Subsurface Line Sources at Constant Hydraulic Head in Anisotropic Soil, Trans. ASAE, 21, 5, 913-919.

27. Edwards, D.H., 1956, Water Tables, Equipotentials, and Streamlines in Drained Soil with Anisotropic Permeability, Soil Sci. 81, 3-18.
   He solves some practical problems of drainage for anisotropic soils.

28. Falade, G.K., 1981, Mathematical Analysis of Fluid Flow in Porous Media with General Anisotropy, Water Resources Research, 17, 4, 1071-1074.

29. Ferrandon, J., 1948, Les Lois de L'ecoulement de Filtration, Genie Civil, 125, 24-28.
   Ferrandon is the first developer of a comprehensive tensor theory on fluid flow through anisotropic media. His presentation is substantially the one that Irmay (1951) and Scheidegger (1954) re-present later, and the one that researcher following Ferrandon have used, after suitable polishings of few aspects that were left unexplored (like the one of the symmetric nature of the conductivity tensor.)

30. Ferrandon, J., 1954, Mecanique des Terrains Permeables, Houille Blanche, 9, 466-480.

31. Fowler, J.L., and K.L. Hertel, Flow of a Gas Through Porous Media, Journal Appl. Phys., 11, 496-502.
   It presents evidence of directional flow preference in

textiles.

32. Fraser, H.J., 1935, Experimental study of the Porosity and Permeability of Clastic Sediments, Jour. Geol., 43, 8, 910-1010.
     This paper is the geological follow up of the paper by Graton and Fraser (1935). Fraser studies the effects of mineral grain orientation with respect to currents existing at the time of deposition. He shows that normal wave action along a beach tends to orient sand grains with their long axes at right angles to the direction of wave movement.

33. Ghizetti, A., 1949,, Ann. Soc. Polon. Math, 22, 195.
     One of the first developers of a comprehensive tensor theory of fluid flow in anisotropic porous media.

34. Giorgini, A., M. Bergman, A. Hamidi, and J. Pravia, 1984, Lateral Flow in Unsaturated Anisotropic Porous Media, Purdue University, Water Resources Research Center Report.

35. Gould, J.P., 1949, Analysis of Pore Pressure and Settlement Observations at Logan International Airport, Harvard Soil Mech. Ser. No. 34, Dept. Eng., Harvard Univ., Cambridge, Massachusetts.

36. Graton, L.C. and H.J. Fraser, 1935, Systematic Packing of Spheres with Particular Relation to Porosity and Permeability. Journal Geol., 43, 8, 785-909.
     A major work on the geometry and fluid mechanics of porous media made up of uniform spheres. Several arrangements are analyzed. The unit void of each case is thoroughly explored. Geometry of the intersphere voids receives particular attention as affecting fluid flow through them. Effect on assemblage orientation on flow is emphasized. "Since permeability is of vectorial quality, every systematic assemblage of spheres is anisotropic with respect to permeability; therefore, if a single value is to be used for permeability, it must be the mean value." The paper is an introduction to a paper by Fraser (1935).

37. Hall, W.A., 1956, An Analytic Derivation of the Darcy Equation, Trans. Amer. Geophys. Un., 37, 185-188.
     He casts the theory of tensor conductivity in diadic form.

38. Hvorslev, M.J., 1951, Time Lab and Soil Permeability in Groundwater Observations, Bul. 36, Waterways Exp. Sta., Corps Eng., U.S. Army, Vicksburg, Missouri.

39. Irmay, S., 1951, Darcy Law for Non-isotropic Soils, Assoc. Intern. Hydrol. Sci. (U.G.G.I.), Assemblee Gen. Bruxelles, 2, 178.

> He reviews what has been done up to 1951 and presents the tensor theory of fluid flow in anisotropic porous media along the path set by Ferrandon (1948).

40. Irmay, S., 1980, Piezometric Determination of Inhomogeneous Hydraulic Conductivity, Water Resources Research, 16, 4, 691-694.

41. Johnson, W.E. and J.N. Breston, 1951, Directional Permeability Measurements on Oil Sandstones from Various States, Producers' Monthly, 14, 10-19.

42. Johnson, W.E. and R.V. Hughes, 1948, Directional Permeability Measurements and Their Significance, Producers' Monthly, 13, 17-25.

> Basing their finds upon observation of a Pennsylvanian oil field, they state that the direction of maximum hydraulic conductivity is in the direction of the larger axis of the sand grains, but that environmental factors, subsequent the deposition, like solution, cementation and compaction may alter that state.

43. Liakopoulos, A., 1962, On the Tensor Concept of the Hydraulic Conductivity, Review of Engineering, Am. Univ. of Beirut, No. 4, 35-42.

44. Liakopoulos, A., 1965a, Variation of the Permeability Tensor Ellipsoid in Homogeneous Anisotropic Soils, Water Resources Research, 1, No. 1, 135-141.

45. Liakopoulos, A.C., 1965b, Darcy's Coefficient of Permeability as Symmetric Tensor of Second Rank, Int. Ass. Sci. Hydrol. Bull., 10, 3, 41-48.

> He proves that the permeability tensor for anisotropic soils is a symmetric tensor.

46. Liakopoulos, C., 1964, Theoretical Aspects of the Flow of Water Through Anisotropic Unsaturated Soils, Int. Ass. Sci. Hydrol. Bull., 9, 1, 62-70.

47. Litwiniszyn, J., 1950, Stationary Flows in Heterogeneously Anisotropic Media, Ann. Soc. Polon. Math, 22, 185-199.

> One of the first developers of a comprehensive tensor theory of flow in anisotropic porous media.

48. Maasland, M. and D. Kirkham, 1955, Theory and Measurement of Anisotropic Air Permeability in Soil, Soil Sci. Soc. Amer. Proc., 19, 395-400.

> They present the first three Theorems and apply the theory based on them to soil clods. They modify

slightly Muskat's (1937) treatment of Theorem II and give a shorter proof of Theorem III based on Vreedenburgh's (1936).

49. Maasland, M. and D. Kirkham, 1959, Measurement of the Permeability of Tri-axially Anisotropic Soils, Jour. Soil Mech. and Foundation Div., Proc. ASCE, 85, 3, 25-34.

50. Maasland, M., 1957, Soil Anisotropy and Land Drainage, Drainage of Agricultural Lands, Ed. by J.N. Luthin, American Soc. of Agron., Madison, Wisconsin.
He presents a very comprehensive state of the science of the theory of nonisotropic porous media (the non-tensorial form) in terms of the "five theorems". A large number of applications and thorough reference commentary is included, which has been used in the annotated bibliography of this contribution.

51. Marcus, H. and D.E. Evenson, 1961, Directional Permeability in Anisotropic Porous Media, Contr. No. 31, Water Res. Center, Univ. of California, Berkeley, California.

52. Marcus, H., 1962, The Permeability of a Sample of an Anisotropic Porous Medium, Journal of Geoph. Res., 67, No. 13, 5215.

53. Muskat, M., 1937, The Flow of Homogeneous Fluids Through Porous Media, McGraw-Hill, New York; or reprinted 1946, J.W. Edwards, Ann Arbor, Michigan.
(225-227) Gives a simple proof at Theorem II and presents the result of Theorem III without proof. He derives formulae related to Theorem IV. Page 111 mentions that out of 65 samples of sand, more than two thirds had larger hydraulic conductivity in the direction parallel to the bedding flow than normal to it. In the former case the ratio k_h/k_v reached 42, and in the latter the ratio k_v/k_h was as high as 7.3.

54. Philip, J.R., 1969, Theory of Infiltration, Advances in Hydroscience, Ven Te Chow Editor, Academic Press, New York, New York.

55. Reeve, R.C. and D. Kirkham, 1951, Soil Anisotropy and Some Field Methods for Measuring Permeability, Trans. Amer. Geophys. Un., 32, 582-590.

56. RESOURCE CONSERVATION AND RECOVERY ACT, Amendment to Solid Waste Disposal Act, Public Law 94-580, 90 Stat. 2795, 42 U.S.C., Paragraph 6901 et seq., 1976.

57. Russell, R.D. and R.E. Taylor, 1937, Roundness and Shape of Mississippi River Sands, Journal Geol., 45, 225-267.
The authors bring evidence that in fluvial transport

larger sand grains tend to become slightly more angular
as they travel downstream. Also, there is a decrease
in the roughness and sphericity with decreasing size of
grain. The direction of maximum hydraulic conductivity
due to depositional environment can be determined only
after careful geological study.

58. Samsioe, A.F., 1931, Einfluss von Rohrbrunnen auf die
    Bewegung des Grundwassers, Zeitsch. angew. Math. und Mech.,
    11, 124-135.
        He derives Theorem II for two-dimensional flow.

59. Sawhney, B.L., J.Y. Parlange, and N.C. Turner, 1976, Deter-
    mination of Soil-Water Diffusivity for Anisotropic Strati-
    fied Soils, Soil Science Soc. of America Journal, 40, 1, 7-
    9.

60. Schaffernak, F., 1933, Erforschung der physikalischen
    Gesetze, nach welchen die Durchsickerung des Wassers durch
    die Talsperre oder durch den Untergrund stattfindet,
    Wasserwirtschaft, 30, 399-405.
        He writes the tensor form of Darcy law in terms of the
        principal axes of anisotropy and recognizes that the
        coordinate deformation yields the Laplace equation.
        The essence of Theorem II is implicit in this paper.

61. Scheidegger, A.E., 1953, Statistical Hydrodynamics in Porous
    Media, Journal Appl. Phys., 25, 994-1001.

62. Scheidegger, A.E., 1954, Directional Permeability of Porous
    Media to Homogeneous Fluids, Geofisica Pura e Applicata,
    Milano, 28, 75-90.
        Re-presents the Ferrandon theory, recalculating the
        polar representation of k. He derives formulas related
        to Theorem IV.

63. Scheidegger, A.E., 1955, General Statistical Hydrodynamics
    in Porous Media, Geofisica Pura e Applicata, Milano, 30,
    17-26.

64. Scheidegger, A.E., 1960, The Physics of Flow Through Porous
    Media, The Macmillan Company, New York, New York.

65. Scheidegger, A.E., 1961, General Theory Dispersion in Porous
    Media, Journal Geophys. Research, 66, 10, 3273-3278.

66. Schneebeli, G., 1953, Sur la Theorie des Ecoulements de Fil-
    tration, La Houille Blanche, 1, 80-86.
        He presents some formulas related to the tensor theory
        of fluid flow in anisotropic porous media as developed
        by Ferrandon (1948).

67. Shul'gin, D.F., 1973, Water Movement in a Stratified Soil Under a Hydraulic Head During Systematic Irrigation, Sbornik Trudov po Agronomicheskoi Fizike, 31, 88-94.

68. Smythe, W.R., 1939, Static and Dynamic Electricity, Ed. I., McGraw-Hill Book Company, Inc., New York.
   He derives the substance of Theorem I for the analogous problem of anisotropic dielectrics.

69. Sposito, G., 1978, Statistical Mechanical Theory of Water Transport Through Unsaturated Soil, 1. The Conservation Laws, 2. Derivation of the Buckingham-Darcy Flux Law, Water Resources Research, 14, 3, 474-478, 479-484.

70. Sposito, G., V.K. Gupta, and R.N. Bhattacharya, 1979, Foundation Theories of Solute Transport in Porous Media: a Critical Review, Advances in Water Resources, 2, 2, 59-68.

71. Stevens, O.B., 1936, Discussion of a Paper by Vreedenburgh, Proc. Intern. Conf. on Soil Mech. and Foundation Eng., 3, 165-166.

72. Stevens, O.B., 1938, Electrical Determination of the Line of Seepage and Flow Net of a Groundwater Flow Through Joint Regions with Different Anisotropy, De Ingenieur in Ned. Indie, 9, 205-212.

73. Sullivan, R.R. and K.L. Hertel, 1940, The Flow of Air Through Porous Media, J. Appl. Phys., 11, 761-765.
   It presents evidence of directional flow preference in textiles.

74. Sullivan, R.R., 1941, Further Study of the Flow of Air Through Porous Media, J. Appl. Phys., 12, 503-508.
   The laminar flow of air through highly porous wads of textile fibers is studied. The rate of flow is found to be twice as great for fiber parallel to flow as for fibers perpendicular to flow.

75. Thiem, G., 1907, Lagerungszustande und Durchlaessigkeit der Geschiebe, Jour. Gasbeleucht. und Wasserversorg., 50, 377-382.
   The earliest investigation on the dependence of hydraulic conductivity on the orientation of solid particles. Measurements are done both along the larger axis and across it.

76. UNITED STATES ENVIRONMENTAL PROTECTION AGENCY, Damages and Threats Caused by Hazardous Material Sites, Oil and Materials Control Division Report, 1980.

77. UNITED STATES ENVIRONMENTAL PROTECTION AGENCY, Land Disposal and Hazardous Wastes: Proceedings of the Fourth Annual

Research Symposium, EPA-A-600/9-78-016, Cincinnati, Ohio, 1978.

78. Versluys, J., 1915, De Onbepaalde Vergelijking der Permanente Beweging van het Grondwater, Verh. Geol.-Mijnbouw. Genoot. Ned. en Kolonien., Geol. Serie 1, 349-360.
He derives Theorem I for any combination of arbitrarily directed sets of parallel, non-intersecting capillaries. He assumes that the results can be applied to more general porous media such as soils. A somehow more general deviation for an analogous problem in electricity, the one of anisotropic dielectrics is presented very clearly in Smythe (1939).

79. Vreedenburgh, C.G.F., 1935, Over de Stationnaire Waterbeweging door grond met Homogeen Anisotrope Doorlaatbaarheid, Ingen. in Ned. Indie, 11, 140-143.
He derives Theorem II for the three-dimensional case.

80. Vreedenburgh, C.G.F., 1936, On the Steady Flow of Water Percolating through Soils with Homogeneous-Anisotropic Permeability, Proc. Intern. Conf. Soil Mech. and Foundation Eng., 1, 222-225.
He re-presents a derivation of Theorem II for three-dimensional flows, which was derived in his earlier paper (1933), and derives Theorem III, Theorem IV, and Theorem V.

81. Vreedenburgh, C.G.F., 1937, De Parallelstroming door Grond Bestaande uit Evenwijdige Regelmatig Afwisselende Lagen van Verschillende Dikte en Doorlaatbaarheid, Ingen. in Ned. Indie, 8, 111-113.

82. Whistler, F.D., and A. Klute, 1969, Analysis of Infiltration into Stratified Soil Columns, Water in the Unsaturated Zone, (P.E. Rijtema and H. Wessink, Editors), Proceedings of the Wageningen Symposium, IASH/AIHS, Unesco, Vol. I.

83. Wilkinson, W.B., and E.L. Shipley, 1972, Vertical and Horizontal Laboratory Permeability Measurements in Clay Soils, Fundamentals of Transport Phenomena in Porous Media, Developments in Soil Science 2, IAHR, Elsevier Pub. Co.

84. Winograd, I.Y., 1981, Radioactive Waste Disposal in Thick Unsaturated Zones, Science, Vol. 212, 4502, p. 1462.

85. Wood, E.F., A.F. Ferrara, W.G. Gray, and G.F. Pinder, 1984, Groundwater Contamination from Hazardous Wastes, Prentice-Hall, Englewood Cliffs, New Jersey.

86. Yang, S.T., 1948, On the Permeability of Homogeneous Anisotropic Soils, Proc. 2nd Intern. Conf. Soil Mech. and Foundation Eng., Rotterdam, 2, 317-320.

Yang derives relationships between the hydraulic conductivity components in different directions at a given point. He elaborates on the applicability of Mohr circle to the problem.

87. Yang, S.T., 1949, Seepage Toward a Well by the Relaxation Method, Thesis, Harvard University Library, Cambridge, Massachusetts.

88. Yang, S.T., 1953, On the Permeability of Homogeneous Anisotropic Soils, Proc. 2nd Intern. Conf. on Soil Mechanics, Rotterdam, 2, 317-320.

89. Yeh, T.C.J., and L.W. Gelhar, 1982, Unsaturated Flow in Heterogeneous Soils, Role of the Unsaturated Zone in Radioactive and Hazardous Waste Disposal, J.W. Merces, P.S. Rao, and I.J. Marine, Editors, Ann Arbor Sci. Publ., Ann Arbor, Michigan.

90. Yen, B.C., and A.O. Akan, 1983, Effects of Soil Properties on Overload Flow and Infiltration, Journal Hyd. Res., 21, 2, 153-173.

91. Zaslavsky, D., and G. Sinai, 1981a, Surface Hydrology: I - Explanation of Phenomena, J. Hyd. Div., ASCE, 107, 1, 1-16.

92. Zaslavsky, D., and G. Sinai, 1981b, Surface Hydrology: II - Distribution of Raindrops, J. Hyd. Div., ASCE, 107, 1, 17-35.

93. Zaslavsky, D., and G. Sinai, 1981c, Surface Hydrology: III - Causes of Lateral Flow, J. Hyd. Div., ASCE, 107, 1, 37-52.

94. Zaslavsky, D., and G. Sinai, 1981d, Surface Hydrology: IV - Flow in Sloping, Layered Soil, J. Hyd. Div., ASCE, 107, 1, 53-64.

95. Zaslavsky, D., and G. Sinai, 1981e, Surface Hydrology: V - In-Surface Transient Flow, J. Hyd. Div., ASCE, 107, 1, 65-93.

EROSION AND SEDIMENT TRANSPORT PROCESSES
FOR AGRICULTURAL WATERSHEDS

G. R. Foster
Hydraulic Engineer, USDA-Agricultural Research Service
Associate Professor, Department of Agricultural Engineering
National Soil Erosion Research Laboratory, Purdue University
W. Lafayette, Indiana 47907
USA

INTRODUCTION

Sediment eroded from agricultural land can be a major pollutant and a carrier of polluting chemicals such as pesticides and plant nutrients. Furthermore, excessive sedimentation in water conveyance structures reduces their capacity and utility, and excessive erosion on the landscape reduces the productive potential of cropland. Processes involved in erosion, sediment transport, and sedimentation by water on agricultural watersheds are discussed in this paper.

BASIC CONCEPTS

Erosion by water is a process of detachment and transport of soil particles by raindrop impact and surface runoff. Detachment is the removal of soil particles from the soil mass, while transport is the movement of sediment, detached soil particles, to a location away from the point of detachment. Principal detaching agents are impacting raindrops and surface runoff while the principal transport agent is surface runoff.

Raindrops vary in size from about 0.25 to 5 mm and impact the earth at velocities up to about 10 m/s. An impacting raindrop creates intense shear and pressure forces along the soil surface, which can detach large quantities of sediment. The very high radial velocity created as raindrops impact soil splashes sediment horizontally and vertically as far as one meter (1). Splash, a local sediment transport process, occurs in all directions and therefore moves little sediment downslope. Most downslope transport of sediment is by surface runoff.

Surface runoff occurs after rainfall fills the interception storage on plants and retention storage on the soil, and rainfall rate exceeds the soil's infiltration rate. This flow applies shear stress to the soil and detaches sediment when runoff rate and slope steepness combine to increase shear stress to where it exceeds the critical shear stress of the soil. Therefore, detachment from runoff is greatest on steep and long slopes.

Sediment transport by surface runoff is described by the continuity equation (2):

$$\partial q_s/\partial x + \rho_s \partial(cy)/\partial t = D_1 + D_f \qquad [1]$$

where: q_s = sediment load (mass/width * time)
        x = distance

111

ρ_s = mass density of sediment particles (mass/volume)
c = concentration of sediment in flow (volume of sediment/volume of flow)
y = flow depth
t = time
D_1 = lateral inflow of sediment (mass/area * time)
D_f = detachment or deposition by flow (mass/area * time)

The term $\partial q_s/\partial s$ is the buildup (or reduction) of sediment load with distance, the term $\rho_s \partial(cy)/\partial t$ is the rate of change of storage of sediment within the flow, D_1 is the contribution of sediment by an outside source, and D_f is the contribution of sediment detached by flow or a loss of sediment deposited by flow.

Since rainfall is unsteady, both runoff and erosion processes are unsteady. Except for a few special cases (3,4), equation 1 is usually solved numerically at many points in time and space. However, steady conditions are assumed for this discussion to reduce equation 1 to:

$$dq_s/dx = D_1 + D_f \qquad [2]$$

Equation 2 is integrated:

$$q_s = \int(D_1 + D_f)dx \qquad [3]$$

to give sediment load q_s at any location x along a slope. Sediment load at the end of a slope or at the outlet of a watershed is called sediment yield.

A fundamental principle frequently used to describe erosion is that sediment load is limited by either detachment or transport capacity. This concept is illustrated by erosion on a concave slope having a steep slope at its upper end and a flat slope at its lower end. At the steep upper end, transport capacity of the flow is usually so great that it exceeds the amount of sediment added to the flow by detachment. As runoff moves down a concave slope, distance adds flow that tends to increase transport capacity. Concurrently, flow is moving down a slope of continually decreasing steepness that tends to decrease transport capacity. The net effect is that transport capacity increases with downslope distance until it reaches a maximum, and then it decreases further downslope.

Sediment load continues to increase after transport capacity begins to decrease. At some point, the two become equal, which is where deposition begins and continues through the end of the slope. Transport capacity limits sediment load over the lower reach of slope where deposition occurs. In simple erosion models, sediment load is set equal to transport capacity at locations where transport capacity limits sediment load.

A more complex consideration of the processes assumes a simple, linear relation between detachment or deposition and sediment load. For detachment, the relationship is (5):

$$D_f/D_c + q_s/T_c = 1 \qquad [4]$$

where:

D_c = detachment capacity for flow (mass/area * time)
T_c = transport capacity of flow (mass/width * time)

The concept of equation 4 is that energy of the flow is distributed between detachment and transport. The idea is that as sediment load fills transport capacity, less energy is available for flow to detach sediment.

Detachment capacity D_c and transport capacity T_c are defined by flow hydraulics, sediment properties, and soil conditions at a location. When the sediment load is small, most of the flow's energy is available to detach sediment and the detachment rate D_f nearly equals the detachment capacity rate D_c. Rearrangement of equation 4 more clearly shows the relation of D_f to D_c:

$$D_f = D_c(1 - q_s/T_c) \qquad [5]$$

Another arrangement of equation 4 is:

$$D_f = (D_c/T_c)(T_c - q_s) \qquad [6]$$

which shows that detachment by flow is linearly proportional to the difference between transport capacity and sediment load. If both D_c and T_c are written as simple power functions involving shear stress, the ratio D_c/T_c is a function of soil conditions (6).

When sediment load exceeds transport capacity, the term $(T_c - q_s)$ becomes negative, indicating deposition. Equation 6 is rewritten for deposition as:

$$D_f = \alpha(T_c - q_s) \qquad [7]$$

where α is defined by (7):

$$\alpha = \beta V_f/q \qquad [8]$$

where:

β = a factor reflecting flow turbulence and other flow disturbances
     that hinder deposition
V_f = fall velocity of sediment particles
q = discharge rate for surface runoff (volume/width * time)

Deposition rate is small when α is small, which occurs when β or V_f is small or q is large. For example, β is small for shallow flow that is highly disturbed by raindrop impact, fall velocity is small for fine particles such as clay-sized sediment, and q is large when either flow velocity V or flow depth y is large since $q = V y$.

The significance of equations 7 and 8 is illustrated for the simple conditions of $T_c = (dT_c/dx)x, q = \sigma x$, and $\psi = \beta V_f/\sigma$ where dT_c/dx is constant with x and σ = runoff rate (volume/area * time). The solution to equations 4, 7, and 8 is (8):

$$q_s=(\psi T_c + D_1 x)/(1 + \psi) \qquad [9]$$

One extreme is the case for very small particles where V_f approaches zero and therefore ψ approaches zero, which gives $q_s = D_1 x$. That is, sediment load of the very fine particles equals the amount of them added to the flow by lateral inflow. The other extreme is the case of large particles where V_f is large, which gives a large ψ and $q_s \sim T_c$. That is, sediment load equals transport capacity for the coarse particles. This transport of very fine particles is similar to the concept of washload in classical concepts for sediment transport by rivers, but as discussed in a following section, more deposition of fine particles occurs for shallow flow than is expected in river flow. The sediment load of coarse particles is assumed to closely follow transport capacity, typical in classical sediment transport concepts for river flow (9). Equations 7 and 8 provide a method for dealing with intermediate-sized particles.

SEDIMENT SOURCES

Almost no soil surface is so smooth or straight that shear stress of the flow or the susceptibility of the soil to detachment is uniform across slope. Therefore, detachment by flow is almost always nonuniform across slope, even on a practically smooth soil. Small eroded channels evolve at points where local detachment by flow is maximum and soil resistance to detachment is minimum. As erosion progresses, distinct areas of flow concentration and incised channels evolve. These eroded channels, called rills, are numerous across slope, frequently about one per meter (10), and they are small, about 200 mm wide by 100 mm deep. Areas between rills are called interrill areas. In a mature rill-interrill system, runoff on the interrill area is very shallow and moves predominantly laterally toward the rills, while flow in rills is predominantly downslope.

In principle, equations 1 and 6 (or 7) apply to both rill and interrill areas. However, in practice, interrill areas are usually defined such that detachment by flow on them is negligible and rills are defined such that detachment by raindrop impact in them is negligible. However, flow on interrill areas transports most of the sediment eroded on them to the rills.

A well defined rill-interrill system is ridges for crop rows separated by furrows. The ridge sideslopes are interrill areas while the furrows are rill areas. If the furrows are on a steep grade, rill erosion (detachment by flow), occurs in them, and if they are on a flat grade, deposition occurs in them. A rough plowed field has many small depressional areas where deposition occurs while the protruding soil clods are interrill source areas. Rill-interrill areas are not distinct on many soil surfaces, but the rill-interrill concept is useful in erosion modeling.

The topography of many fields causes overland flow to concentrate in a few major, natural waterways before leaving the fields. Erosion in these areas is called concentrated flow erosion. These channels are tilled each year, and unlike rills, which tillage obliterates, concentrated flow occurs in the same location after tillage. Eroded channels within fields too deep for crossing with farm implements are called gullies.

The profiles of many landscapes and concentrated flow channels are concave causing deposition. Also, flow is often ponded at the outlet of many fields, causing deposition, and some conservation practices such as tile-outlet-terraces pond runoff to reduce sediment load by causing deposition. Consequently, sediment yield from many fields is much less than the amount of sediment eroded upslope (11).

Although this discussion mainly considers field-sized areas, a large watershed is a system of channels that extends upstream through rills. Unfortunately, practical limitations prevent consideration of each individual channel and rill in most erosion analyses. Areas upstream from the smallest channel that can be considered are treated as overland flow areas where broad, sheet flow is assumed. Accurate modeling of erosion on a watershed requires a watershed representation that represents key elements and minimizes distortion of parameter values in the hydrologic and erosion equations.

DETACHMENT PROCESSES

Detachment rate is a function of the magnitude of fluid forces applied to the soil relative to the resistivity of the soil to erosion. Detachment can be easily studied in ideal, simple systems, but field situations are much more complex, and are considered in the following discussion.

Interrill Detachment

Detachment on interrill areas is by raindrop impact. The fluid forces in a raindrop impact vary rapidly in time and space (12). Detachment from a single drop impact could be modeled by integrating a detachment equation involving these temporally and spatially varying forces (13). Furthermore, the distribution of raindrop sizes, impact velocities, and nonperpendicular impact angles could be considered, and perhaps will in the future with improved computational capabilities. Most current erosion models use empirical terms such as kinetic energy, momentum, and rainfall intensity, as measures of raindrop erosivity (14). In general, interrill detachment varies with the square of rainfall intensity (15), and detachment by a single raindrop varies with kinetic energy (16).

Measures of soil erodibility are also highly empirical. Fundamentally, erodibility should be measured by the forces that bind soil particles, but these forces are not easily measured. Soil strength as measured by a fall-cone penetrometer describes detachment by raindrop impact on a given soil over a range of soil strength, but it has not adequately explained differences between soils (16). Other indirect measures of soil erodibility include sodium, iron oxide, aluminum oxide, clay, and organic matter contents; soil density, and aggregate stability (17).

Soil erodibility factor values from empirical erosion equations such as the Universal Soil Loss Equation (USLE) (19) are not fundamental measures of soil erodibility. They are often regression coefficients in lumped equations that consider more than detachment by either raindrop impact or surface runoff. For example, the soil erodibility factor in the USLE is low for both sand and clay soils. It is low for sand soils because they produce little runoff, while it is low for clay soils because particles are not easily detached from a clay soil. However, soil erodibility is high for a sand soil when the relation of detachment to runoff is considered. Therefore, erodibility factor values are usually functions of the structure of empirical erosion equations, and their rank varies among equations (20).

Erodibility is frequently defined as erosion rate for a base condition, but field conditions require consideration of how plant growth and soil management affect erodibility (17). For example, soil following a soybean crop is about 40 percent more erodible than it is following a corn crop. Undisturbed soil is less erodible than freshly tilled soil. Plant roots in soil near the surface mechanically bind soil particles, and incorporation of crop residue reduces soil erodibility. Unfortunately, the effect of many field factors that modify soil erodibility have not been extensively quantified in equations that can be used in erosion modeling (14, 18).

Field conditions also modify the erosive forces of raindrops. Plant cover intercepts rainfall, and if the plant canopy is close to the soil, impact velocity of waterdrops falling from the canopy will be low, causing little erosion even though these drops are usually larger than raindrops (14). Cover on the soil surface provides greater protection than does canopy. Drops striking ground cover have no fall height to regain energy, and surface cover slows runoff increasing flow depth. The erosive forces are slight from water drops striking a water layer deeper than three to six drop diameters (14).

Rill and Concentrated Flow Detachment

Detachment processes in channels are similar across a wide range of channels, which allows application of the same fundamental analysis to both single rills and concentrated flow channels (21). Three distinct types of detachment occur in rills (22). One is the highly localized detachment that occurs at headcuts. Numerous headcuts can form at regular intervals along a rill. Their geometry and advance rate depend on discharge rate, grade of the rill, soil strength, surface cover, and buried material like crop residue, roots, and gravel. Another type of rill erosion is the undercutting and sloughing of short reaches of about 1/4 meter of rill sidewall. When the weight of the overhanging soil exceeds soil strength, it sloughs, and flow in the rill quickly cleans out the sloughed soil. The third type is a somewhat uniform removal of soil from a rill's wetted perimeter by shear stress of the flow.

Very little analytical information exists for the first two types of detachment in rills, but recent theory has been developed for the third type (21). This theory assumes that detachment around the wetted perimeter of a rill can be described by:

$$D_p = K(\tau - \tau_c) \qquad [10]$$

where:

D_p = detachment rate at a point on the wetted perimeter (mass/area * time)
K = a soil erodibility factor (mass/force * time)
τ = shear stress of flow at a point on the wetted perimeter (force/area)
τ_c = critical shear stress of soil (force/area)

With the assumption of a distribution of shear stress around a wetted perimeter, evolution of an eroding channel can be calculated. Furthermore, the channel evolves to an equilibrium shape that moves downward at a steady rate for a steady discharge. Geometry of the equilibrium channel can be analytically derived as a function of discharge rate, grade of the channel, hydraulic roughness, and critical shear stress of the soil.

When an eroding channel reaches a nonerodible layer, it begins to widen, and as it widens, its erosion rate decreases. It continues to widen at a decreasing rate until the shear stress at the intersection of the channel sidewall and the channel bottom equals the critical shear stress of the soil. This theory has been successfully used to describe rill erosion, concentrated flow erosion, and several morphological characteristics of rivers (21, 23).

hile this theory seems generally applicable, little is known about the parameters K and τ_c as functions of soil conditions. Clearly, tillage reduces τ_c for several soils and leaves them much more erodible than undisturbed soils (14). Frequently, untilled soil immediately below the tilled layer acts as a nonerodible layer, and channels in these soils are wide and shallow. However, tillage does not have this distinct effect on other soils, where channels are deep and narrow. Some soils are especially susceptible to detachment by flow when they are thawing. Buried residue and roots can significantly reduce detachment by flow by acting as miniature grade control structures. Many of the same soil conditions that affect detachment by raindrop impact also affect detachment by flow.

Cover on the soil surface reduces the erosive forces that flow exerts on the soil (14). This effect can be analyzed by treating flow shear stress as two components, one acting on the soil and the other acting on the cover. The theory for this division is the same one used in open channel hydraulics to separate shear stress into components for grain and form roughness (9).

Rill erosion can be analyzed by considering a typical eroded channel for ridge-furrow systems, a concept which is beginning to be used in erosion models (6). Also, rill patterns could be generated by combination deterministic-stochastic models, and typical rills from the pattern could be analyzed as individual channels. However, rill erosion is usually analyzed assuming hydraulic relationships for broad, sheet flow. Since flow is very nonuniform across a slope, this assumption leads to considerable distortion of parameter values. A distortion also occurs when excessively long overland flow slope lengths are assumed; overland flow is usually collected within 100 m in concentrated flow areas.

SEDIMENT TRANSPORT CAPACITY

Sediment transport capacity is a function of both flow hydraulics and sediment characteristics. Flow depths range from less than a millimeter on interrill areas to less than a meter in concentrated flow areas. Sediment is nonuniform in size, density, and shape. Its diameter ranges from clay-size of less than 0.001 mm to gravel-size of greater than 10 mm.

Sediment Characteristics

Since most agricultural soils are cohesive, they erode as a nonuniform mixture of primary particles (sand, silt, and clay) and aggregates (conglomerates of primary particles and organic matter). The aggregates can be much larger than their primary particles, and they can have specific gravities that range from 1.5 to 2.65 depending on their composition and degree of saturation (14). Aggregation and distributions of size, density, and shape at the point of detachment depend on texture, organic matter, type of clay, and recent history of the matrix soil and whether the particles are detached by raindrop impact or runoff (14). Equations for sediment composition as a function of soil texture have been derived for nonpoint source pollution analyses (24).

Hydraulics

Many of the hydraulic variables such as velocity, discharge rate, shear stress, and turbulence that are important in sediment transport in rivers are also important in sediment transport on field sized areas. However, important differences exist (25); for example, the ratio of flow depth to particle diameter is much less on fields than in rivers. Even though flow tends to be broad and shallow over depositional areas in fields, it is still quite nonuniform across slope. Flow can be highly disturbed from raindrop impact and large scale form roughness, and flow can go through numerous regime changes as it moves over irregular bottom profiles. Raindrops impacting the surface of shallow flow significantly increase the flow's transport capacity (26). Particle density and shape are important factors in sediment transport by shallow flow. Often particle diameter is on the same order as flow depth, and flow transports some particles by rolling them (26).

Cohesive soil may contain large-sized sand and gravel that are separate from the cohesive soil matrix. Initiation of movement of particles from cohesive soil is controlled by detachment processes related to the critical shear stress for detachment. Initiation of movement of the large sand and gravel is related to the critical shear stress for transport, and therefore, the armoring process of cohesive soil in fields differs from that for noncohesive sediment typical of rivers.

Sediment from cohesive field soils seems to move very much like noncohesive sediment. However, once this sediment is deposited, no significant interchange seems to occur between sediment in the flow and sediment in the bed as occurs with noncohesive beds (26). Clearly, this interchange does not occur in areas where detachment is occurring.

No sediment transport equation has been derived specifically for the shallow flow typical of agricultural fields. Therefore, equations from stream flow literature have been adopted, but with limited success (27). For shallow flow, the Yalin equation works best of several common equations, especially for aggregates of low density and clay and silt-sized particles. It has been modified for nonuniform sediment (14).

DEPOSITION

Deposition is important in off-site sedimentation and nonpoint source pollution analyses because deposition reduces sediment yield and changes sediment properties. Although deposition within fields is both site and storm specific, as much as 80 percent of the sediment produced by sheet and rill erosion can be deposited without leaving the field (28).

Deposition is a selective process and enriches the sediment load in fines, especially important in nonpoint source pollution analyses because most soil associated pollutants are adsorbed on fine sediment particles. Concentration of a pollutant on sediment is as much as seven times that in the soil producing the sediment (29). This enrichment increases with increased deposition and with poorly aggregated sediment (17).

Deposition is usually spatially selective. Coarse particles are deposited upslope while fine particles are deposited downslope. However, particles as small as 0.01 mm are more readily deposited by shallow flow than expected (26). Also, small particles of low density in a nonuniform sediment mixture having arge particles can be deposited under hydraulic conditions that would not ause deposition for the same small sized particles in uniform sediment.

rosion reduces soil depth, and if the reduction is more rapid than soil ormation, long term loss of productivity occurs. Also, erosion selectivity emoves soil fines leaving a less productive, coarser texture soil. An nresolved issue is whether this selectivity occurs during detachment or eposition. An examination of erosion processes on the point scale suggests hat this selectivity might be related to local transport and deposition rocesses (17).

UMMARY

rosion by water is a process of detachment and transport of soil particles by aindrop impact and surface runoff. Climate and hydrology drive erosion and ediment transport by providing rainfall and surface runoff. Erosion is a unction of the erosivity of the eroding agents relative to the erodibility of he soil. Many factors including cover and management of the soil affect both rosivity and erodibility.

fundamental concept that is frequently used in erosion analyses is that ither detachment or transport capacity may limit sediment load. Detachment imits sediment load by the amount of sediment that it makes available for ransport, while transport capacity limits sediment load by the amount of ediment that flow can transport. Deposition usually occurs when transport apacity limits sediment load. Deposition is a selective process that ncreases the concentration of fine particles and soil adsorbed pollutants in he sediment load.

low on field-sized areas is nonuniform across slope. Some of the onuniformity results from tillage and some from natural irregularities. When low erodes soil, the result is numerous, small eroded channels called rills. reas between rills are called interrill areas, where soil is predominantly etached by raindrop impact. Flow on these areas is lateral toward rill areas, nd its transport capacity is greatly enhanced by raindrop impact. Most ownslope sediment transport is by flow in rill areas. Surface runoff within ost fields collects in a few natural waterways before leaving the fields. rosion processes occurring in these concentrated flow areas are very much like hose occurring in rills.

any hydraulic concepts can be transferred from sediment transport theory for oncohesive beds in rivers to erosion processes on fields. However, onuniformity of flow and sediment characteristics, soil cohesiveness, and the ffects of cover and management cause major differences and emphasizes the need o give special consideration to erosion and sediment transport processes on ields.

REFERENCES

1. Ellison, W.D., "Studies of Raindrop Erosion", Agricultural Engineering. Vol. 25, No. 4, 1944, pp. 131-136.

2. Bennett, J.P., "Concepts of Mathematical Modeling of Sediment Yield", Water Resources Research. Vol. 10, No. 3, 1974, pp. 485-492.

3. Croley, E.T., "Unsteady Overland Sedimentation", Journal of Hydrology. Vol. 56, No. 3/4, 1982, pp. 325-346.

4. Singh, V.J., "Analytical Solutions of Kinematic Equations for Erosion on a Plane", Advances in Water Resources. Vol. 6, June, 1983, pp. 88-95.

5. Foster, G.R. and Meyer, L.D., "Mathematical Simulation of Upland Erosion by Fundamental Erosion Mechanics", Present and Prospective Technology for Predicting Sediment Yield and Sources. ARS-S-40, USDA-Agricultural Research Service, Washington, D.C., 1975, pp. 190-197.

6. Foster, G.R., Smith, R.E., Knisel, W.G., and Hakonson, T.E., "Modeling the Effectiveness of On-Site Sediment Controls", Paper No. 83-2092. American Society of Agricultural Engineering, St. Joseph, Michigan, 1983.

7. Foster, G.R., Lane, L.F., Nowlin, J.D., Laflen, J.M., and Young, R.A., "Estimating erosion and Sediment Yield on Field Sized Areas", Transactions of the American Society of Agricultural Engineers. Vol. 24, No. 5, 1981, pp. 1253-1262.

8. Renard, K.G. and Foster, G.R., "Soil Conservation: Principles of Erosion by Water", Dryland Agriculture. Agronomy Monograph No. 23, American Society of Agronomy, Madison, Wisconsin, 1983, pp. 155-176.

9. Graf, W.H., Hydraulics of Sediment Transport. McGraw-Hill Book Col, New York, New York, 1971, 544 pp.

10. Meyer, L.D., Foster, G.R., and Romkens, M.J.M., "Source of Soil Eroded by Water from Upland Slopes", Present and Prospective Technology for Predicting Sediment Yields and Sources. ARS-S-40, USDA-Agricultural Research Service, Washington, D.C., 1975, pp. 177-189.

11. American Society of Civil Engineers, Sedimentation Engineering. American Society of Civil Engineering, New York, New York, 745 pp.

12. Huang, C., Bradford, J.M., and Cushman, J.H., "A Numerical Study of Raindrop Impact Phenomena: The Rigid Case", Soil Science Society of America Journal. Vol. 46, No. 1, 1982, pp. 14-19.

13. Mutchler, C.K., and Young, R.A., "Soil Detachment by Raindrops", Present and Prospective Technology for Predicting Sediment Yields and Sources. ARS-S-40, USDA-Agricultural Research Service, Washington, D.C., 1975, pp. 113-117.

14. Foster, G.R., "Modeling the Erosions Process", Hydrologic Modeling of Small Watersheds. ASAE Monograph No. 5, American Society of Agricultural Engineers, St. Joseph, Michigan, 1982, pp. 296-380.

15. Meyer, L.D., "How Intensity Affects Interrill Erosion", Transactions of the American Society of Agricultural Engineers. Vol. 25, No. 6, 1981, pp. 1472-1475.

16. Al-Durrah, M.M., and Bradford, J.M., "Parameters for Describing Soil Detachment Due to Single Waterdrop Impact", Soil Science Society of America Journal. Vol. 46, No. 4, 1982, pp. 836-840.

17. Foster, G.R., Young, R.A., Romkens, M.J.M., and Onstad, C.A., "Processes of Soil Erosion by Water", Soil Erosion and Crop Productivity. R. F. Follett and B. A. Stewart, Editors. American Society of Agronomy, Inc., Crop Science Society of America, Inc. Soil Science Society of America, Madison, Wisconsin, USA, 1985. pp. 137-162.

18. Laflen, J.M., Foster, G.R., and Onstad, C.A., "Simulation of Individual-storm Soil Loss for Modeling Impact of Soil Erosion on Crop Productivity", Soil Erosion and Conservation. Soil Conservation Society of America, Ankeny, Iowa, pp. 285-295.

19. Wischmeier, W.H., and Smith, D.D., Predicting Rainfall Erosion Losses. Agricultural Handbook No. 537, USDA-Science and Education Administration, Washington, D.C., 1978, 58 pp.

20. Foster, G.R., Lombardi, F., and Moldenhauer, W.C., "Evaluation of Rainfall-Runoff Erosivity Factors for Individual Storms", Transactions of the American Society of Agricultural Engineers. Vol. 25, No. 1, 1982, pp. 124-129.

21. Foster, G.R., and Lane, L.J., "Erosion by Concentrated Flow in Farm Fields", Proceedings of the D.B. Simons Symposium on Erosion and Sedimentation. Colorado State University, Ft. Collins, Colorado, 1983, pp. 9.65-9.82.

22. Meyer, L.D., Foster, G.R., and Nikolov, S., "Effect of Flow Rate and Canopy on Rill Erosion", Transactions of the American Society of Agricultural Engineers. Vol. 18, No.5, 1975, pp. 905-911.

23. Osterkamp, W.R., Lane, L.J., and Foster, G.R., An Analytical Treatment of Channel-Morphology Relations. Professional Paper 1288, U.S. Geological Survey, Washington, D.C., 1983, 21 pp.

24. Foster, G. R., Young, R.A., and Neibling, W.H., "Composition of Sediment for Nonpoint Source Pollution Analyses", Transactions of the American Society of Agricultural Engineers. Vol. 28, No. 1, 1985, pp. 133-139, 146.

25. Foster, G.R., and Meyer, L.D., "Transport of Soil Particles by Shallow Flow", Transactions of the American Society of Agricultural Engineers. Vol. 15, No. 1, 1972, pp. 99-102.

26. Neibling, W.H., and Foster, G.R., "Transport and Deposition of Soil Particles by Shallow Flow", Proceedings of the D.B. Simons Symposium on Erosion and Sedimentation. Colorado State University, Ft. Collins, Colorado, 1983, pp. 9.43-9.64.

27. Alonso, C.V., Neibling, W.H., and Foster, G.R., "Estimating Sediment Transport Capacity in Watershed Modeling", Transactions of the American Society of Agricultural Engineers. Vol. 24, No. 5, 1981, pp. 1211-1220, 1226.

28. Piest, R.F., Kramer, L.A., and H.G. Heinemann, "Sediment Movement from Loessial Watersheds", Present and Prospective Technology for Predicting Sediment Yield and Sources. ARS-S-40, USDA-Agricultural Research Service, Washington, D.C., 1975, pp.130-136.

SIMULATION OF SEDIMENT YIELD FROM ALPINE WATERSHEDS

George Fleming^ and Sergio Fattorelli^^

Department of Civil Engineering, University of Strathclyde
John Anderson Building, 107 - Rottenrow, Glasgow G40NG, Scotland
Cattedra di Idrologia Forestale, Ist. di Meccanica Agraria
Università di Padova, via Loredan, 20 - 35131 Padova, Italy

INTRODUCTION: This paper concerns itself with the practical application of simulation techniques for the assessment of the range of sediment yields from two river basins in the Italian Alps.

Three concepts are involved. The first is the concept of treating the total water and sediment response of a river basin by simulating the dynamic interaction between the hydrology and sediment processes for the complete basin. The development of this approach is already documented in the literature (Fleming 1969, 1972, Fleming & Walker 1980). The second concept is the use of the simulation technique in the development of sediment response curves for a particular basin. This concept is also documented in the existing literature (Fleming 1981, Fleming & Fattorelli 1981, Fleming & Al Kadhimi 1982). The third concept is to apply the technique to the problem of Alpine Watersheds. This problem is made more difficult than general hydrology and sediment problems due to the extreme variation in hydrological and sediment response of steep sloping torrent watersheds; the inclusion of snow accumulation and melt processes; the high level of water resource development and control and the increasing pressure to'further utilise the water and sediment resource in these areas. General aspects of the problem are documented by Glen (1982.).

THE RIVER BASIN MODEL

The river basin model used consists of three elements as shown in figure 1.   The elements include the Watershed model (Fleming & McKenzie 1984) based on the Stanford model IV (Crawford & Linsley 1966), the Erosion model (Fleming 1983) and the Routing model (Fleming 1983).   The present model represents the current state of a continuously evolving system.   Earlier versions of the model developed over the last 18 years are described in previous references.   The Watershed model includes the ability to simulate the snow accumulation and melt process amd prepares output data files for either the erosion model or directly to the flow and sediment routing model.

The structure of the erosion model is shown in figure 2, where the erosion process is subdivided into two components - detachment and transport.   These components act on the erosion by rainfall and overland flow.   The balance between the rate of detachment given a variable supply of soil and the capacity to transport within the given hydro-logical conditions is continuously assessed.   The influence of physical catchment conditions such as vegetation cover, topography, soil type and disturbance levels is related to the physics of the detatchment and transportation by a series of empirical equations.   These equations have been developed based on component process research such as that of Bagnold (1966) for sediment transport;   Farmer (1978) for inter-particle bond in organic soils;   Mutchler & Hansen (1970) on the rain splash process;   and Chen (1976) on the flow resistance of grassed surfaces.

An important part of the model is the computation of the mass balance equation by particle size.   The detailed particle size distribution is input as part of the physical data base and as each rainfall or over-land flow event takes place it is related to the available size of particle and hence their mass.   Selective detachment is allowed in the top soil layer hence the armouring process can be simulated. Further, when a surface disturbance takes place, such as ploughing, the size distribution of the top removed soil layer is replaced by the size distribution of the parent soil.   Hence more detailed simulation is possible of the well observed phenomena of high variation in sediment loads for the same successive rainfall or flow.

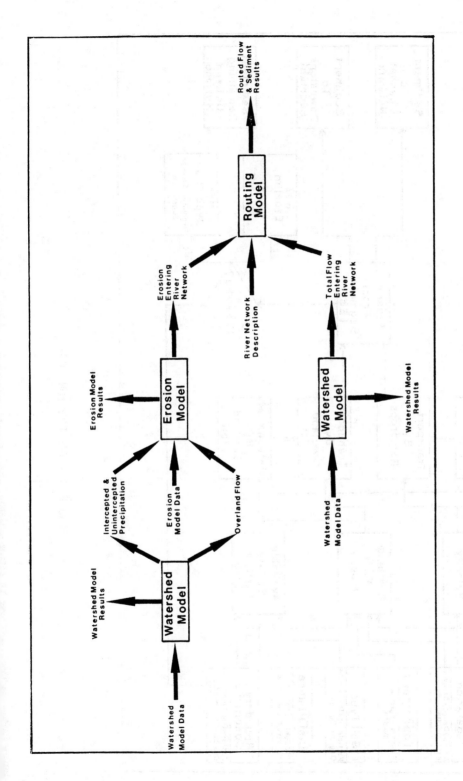

FIGURE 1 - RIVER BASIN MODEL

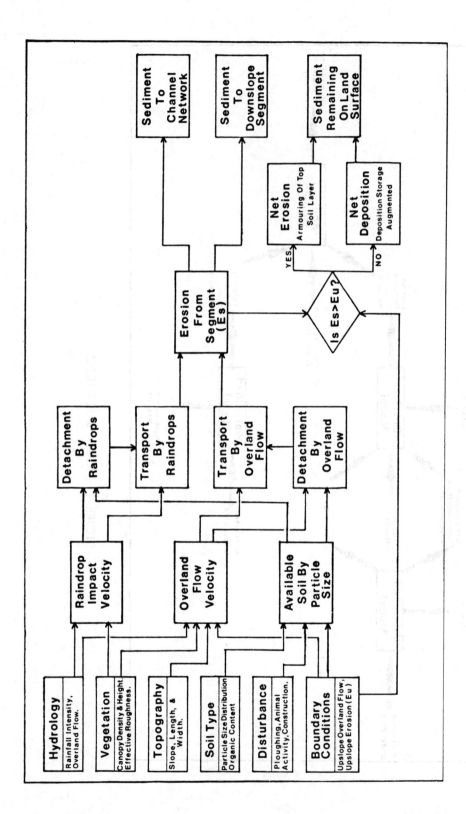

FIGURE 2 - THE EROSION MODEL

The sediment and flow routing element of the model uses a branching network to represent the channel system of mixed trapezoidal reaches and reservoirs. Kinematic flow routing techniques are used in channel reaches and level pool routing in reservoirs. As in the erosion model, Bagnold's approach (Bagnold 1966) is used to represent sediment transport and the mass balance equations are related to the range of particle sizes found both in suspension and in the river bed.

SEDIMENT RESPONSE CURVE FOR A RIVER BASIN

The need to numerically represent the various dominant factors con-tributing to sediment erosion from land surfaces and sediment yield at different points in a river channel network has long been recognised and various notable contributions have been advanced for better under-standing of the problem. (Wischmeier & Smith 1960, Kirkby 1971, Foster & Meyer 1975). The compromise between the assessment and the data available has been highlighted (Fleming   & Al Kadhimi ,   1982). The Universal Soil Loss equation identified the general factors in-fluencing soil erosion as soil erodability (i.e. size, cohesion, organic content etc.), rainfall energy slope, slope length, cropping management (i.e. vegetation) and conservation planning (i.e. disturbance or physical alteration to the land surface). The approach was aimed at quantifying a _relative_ value of maximum erosion potential, for a field plot. Sediment yield elsewhere in the basin would have to take into account delivery ratios of sediment and the variation of field plot conditions over the whole basin. The method does not provide the engineer or planner with the dynamic integrated response of the River Basin for the seasonal variation of basin conditions. Often the main source of sediment at the outlet of a river is from a small source area. Understanding the range of possible dynamic sediment responses and identifying sensitive zones within a river basin is a pre-requisite to good planning and design of the water and sediment resources.

The concept of using a dynamic simulation model of the hydrology and sediment processes to develop a sediment response curve for a river basin is a marriage of the Universal Soil Loss equation concept with

127

deterministic model concept. (Fleming 1981). The sediment response curve
concept is shown in figure 3 and relates a soil erosion index with
variable sediment yield from the basin. The soil erosion index can
include any one or all of the factors dominating the sediment response
of the basin and is derived from a knowledge of the relative weight of
each factor on the sediment response. The response curve can be
developed for the sediment yield at the outlet or for the sediment
erosion at the land source and the two curves compared.

The basic steps involved in preparing the curve are as follows:-

  (i)   Calibrate the Watershed model to the recorded response of
        the basin.
  (ii) Run the Erosion model and Routing model for the range in
        physical data representing the soil and water characteristics
        of the basin.
  (iii) Choose the median set of conditions which represents the
        median recorded erosion or sediment yield for the basin.
  (iv) Assess the relative weighting of the dominant factors such
        as vegetation, slope, soil particle size etc. and plot the
        response curve.

With such a family of curves together with good land surface data,
the influence of developing one part of a river basin in relation to
another can be determined

SEDIMENT YIELD FROM ALPINE BASINS

Two river basins are considered in this paper; the Avisio and the Brenta,
both are located in the Italian Alps as shown on figure 4. On these water
sheds an accurate hydrological investigation has been supported by the
'Dipartimento dell'Ambiente Naturale e Difesa del Suolo della   Provincia
Autonoma di Trento' for the Avisio watershed and by the 'Nucleo   Operati
vo di Padova del Magistrato alle Acque' for the Brenta watershed.

Detailed studies were undertaken into the effect of a large   number   of
factors on the water and sediment yield of the two basins   (Fattorelli &
Fleming, 1982; Baroncini, Fattorelli & Fleming, 1983).   In   this   paper
only a few are considered and these include the land surface slope,   the
vegetation cover and the range in soil particle sizes.

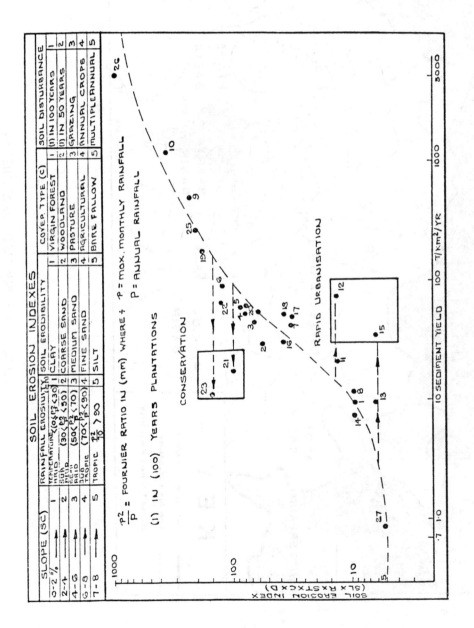

FIGURE 3 – SEDIMENT RESPONSE CURVE (FLEMING et al 1982)

129

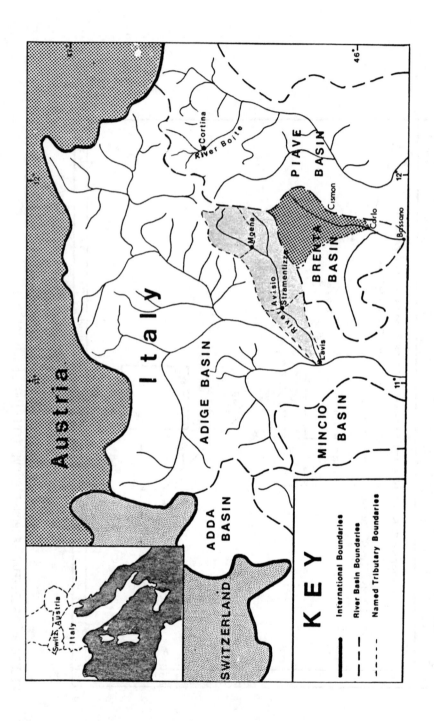

FIGURE 4 – RIVER BASIN LOCATIONS

A general indication of the different physical characteristics of slope, vegetation cover and soil particle size as sampled for the two basins is given in Tables 1 and 2 and figure 5 respectively. The Avisio was treated as two segments; 1. all areas above the 2000 m contour and 2. all areas below the 2000 m in contour.

The Brenta was subdivided into three segments with the upper above 1800 m, the middle between 900-1800 m and the lower below 900 m.

TABLE 1   PHYSICAL LAND-SURFACE CHARACTERISTICS OF THE AVISIO
           & BRENTA BASINS

River Basin	Segment	Area (km²)	Average Flow (m)	Average Slope (%)	Average Slope Length (m)
AVISIO TO	1	202	2355	0.37	672
STRAMENTIZZO	2	509	1619	0.29	418
CISMON TO	1	151	2091	0.61	998
BRENTA	2	348	1352	0.46	541
	3	141	617	0.34	358

TABLE 2   VEGETATION CHARACTERISTICS OF THE AVISIO AND BRENTA BASINS

River Basin	Vegetation Class	Area km²	%
Avisio to	Forest 1	359.10	50.46
Stramentizzo	Bare Area	87.21	12.25
	Impermeable Area	13.12	1.85
	Range Area	202.93	28.51
	Agricultural Area	49.34	6.93
Cismon to	Coniferous Forest (dense)	379.0	59.03
Brenta	Deciduous " (light)	39.9	6.2
	Bare Area	46.0	7.1
	Range Area	106.0	16.51
	Agricultural Area	71.1	11.07

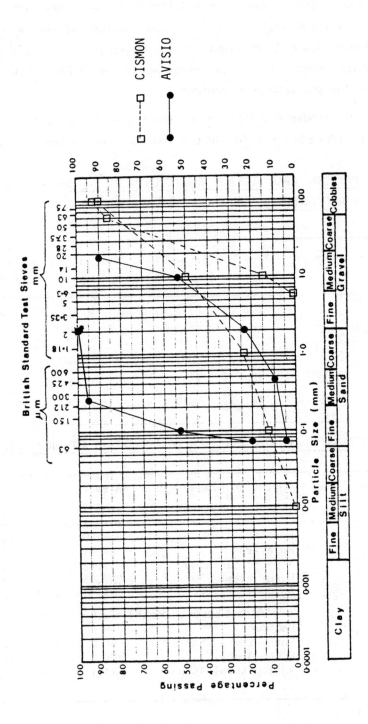

FIGURE 5 - SOIL SIZES FOR THE AVISIO AND CISMON BASINS.

The two catchments have relatively similar sizes and vegetation cover, but the sampled sediment size distributions show that the Cismon to Brenta sediments are coarser. This may be due to unrepresentative sampling or a distinct difference in sediment types. The Cismon to Brenta has considerably greater steepness than the Avisio.

Recorded data on sediment yield is available only as a result of sedimentation surveys carried out in the reservoirs located in these catchments. Table 3 shows the data obtained.

TABLE 3   SEDIMENT YIELD DATA AVISIO AND BRENTA BASINS

LOCATION	Period Covered	Sedimentation m³/Km²/yr	Approx. Sediment Yield T/Km²/yr
Avisio to Stramentizzo	1951-1970	130	111
Cismon to Noana Dam	1959-1973	1320	1056
Cismon to Senaiga Dam	1955-1970	424	339
Cismon to Ponte Serra	1909-1919	376	300

From a comparison of the physical characteristics of the sub basins generally the sediment yields would have been expected to lie within the same range. However the range is seen from table 3 to be considerable, particularly since they represent average rates over a long time period.

A detailed simulation of the sediment response of the Avisio to Stramentizzo was undertaken (Suleiman,   1983) with a similar study of the Cismon to Brenta (Rufai,   1984). The purpose of each study was to examine the relative sensitivity of sediment response under variable physical

133

conditions.

Calibration of the watershed model for each basin was achieved with·typical results shown in figures 6 and 7. The erosion model was then run for the Avisio to test out the sensitivity of the slope, vegetation cover and the sediment particle size on the sediment yield. Table 4 shows a summary of the results obtained. The erosion model was initially calibrated to simulate relative sediment yields at the rate measured at Stramentizzo reservoir, using the physical conditions currently found in this basin.

The sensitivity of slope, vegetation, and sediment particle size are also shown in figures 8 , 9 and 10 respectively. For the range in values used for each of the physical para- meters it is seen that there is a significant variation in the sediment yields produced at the outlet of the basin. Further it is noted that the range is large enough to explain the range in measured sediment yields recorded in both the Avisio and the Cismon.

A weight factor was then computed for the results of each sensitivity test, where the weight factor represented the ratio of sediment yield for a given parameter over the sediment yield of the calibration value of that parameter. The calibration value represents the current measured value of a particular physical state i.e. vegetation, etc. A sediment erosion index is then formed, by the product of the relative weight factors. Figure 11 shows the sediment response curve for the Avisio river. The influence of vegetation was found to be most sensitive with the slope and sediment size relatively less sensitive, but still sig- nificant in their effect on sediment response.

In the study of the Cismon basin the erosion model was used to study the variation in sediment erosion from each segment zone i.e. upland, middle and lower and to compare this with the sediment yield from the basin. A number of alternative trials were undertaken in a similar manner as described

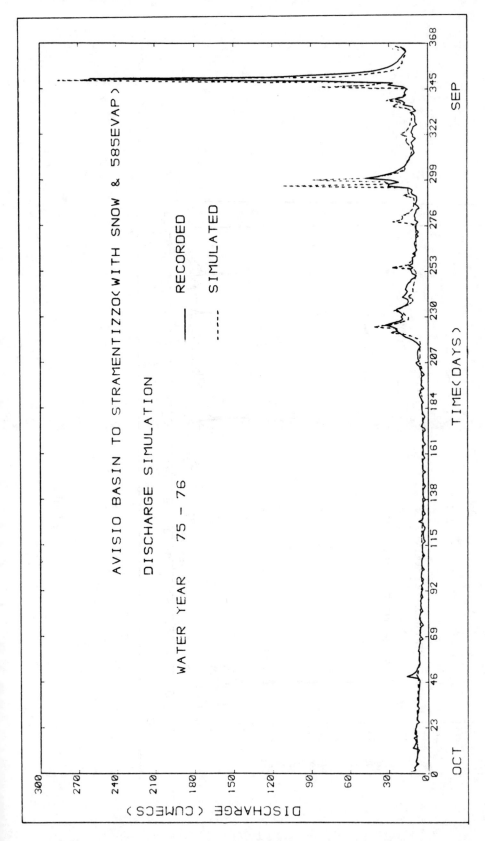

FIGURE 6

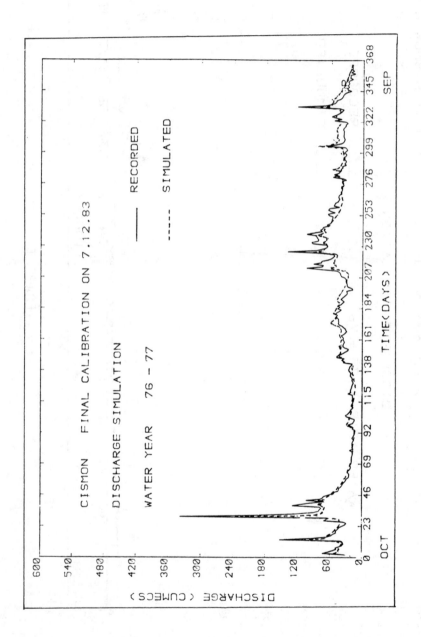

FIGURE 7

TABLE 4   -   SUMMARY OF RESULTS OF SENSITIVITY TRIALS WITH EROSION MODEL

VEGETATION COVER (VEGETAZIONE)			SOIL SIZE (SUOLO)			SLOPE (PENDENZA)	
% Area	t km^{-2}	Rv	D$_{50}$ (mm)	t km^{-2}	Rg	%	t km^{-2}
20-30	720	2.86	0-2	221	0.89	0-10	16
			0.55	248	1.00		
30-40	630	2.54	2-4	116	0.47	10-20	48
40-50	590	2.38	4-6	80	0.32	20-30	136
						0.33	248
50-60	510	2.06	6-8	68	0.27	30-40	280
60-70	430	1.73	8-10	62	0.25	40-50	500
70-80	360	1.45					
80-90	300	1.21					
90-100	255	1.03					
97	248	1.00					

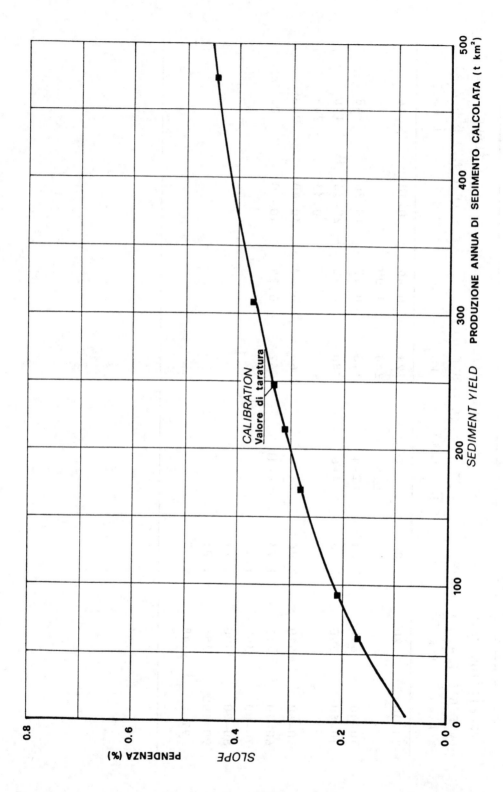

FIGURE 8 - THE INFLUENCE OF SLOPE

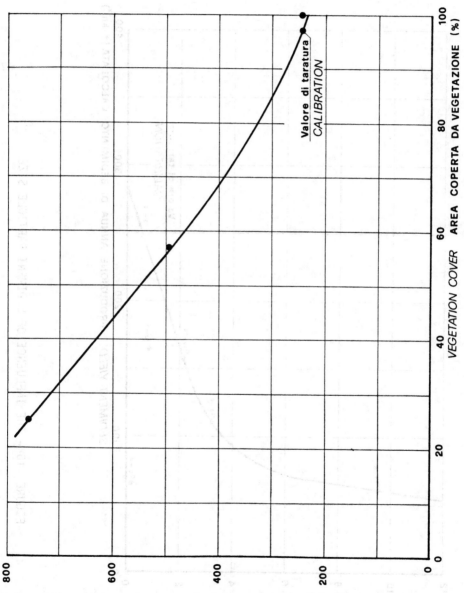

FIGURE 9 – THE INFLUENCE OF VEGETATION

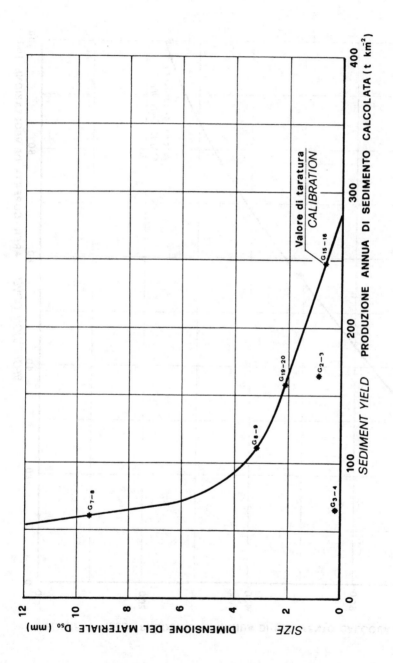

FIGURE 10 - THE INFLUENCE OF SEDIMENT PARTICLE SIZE

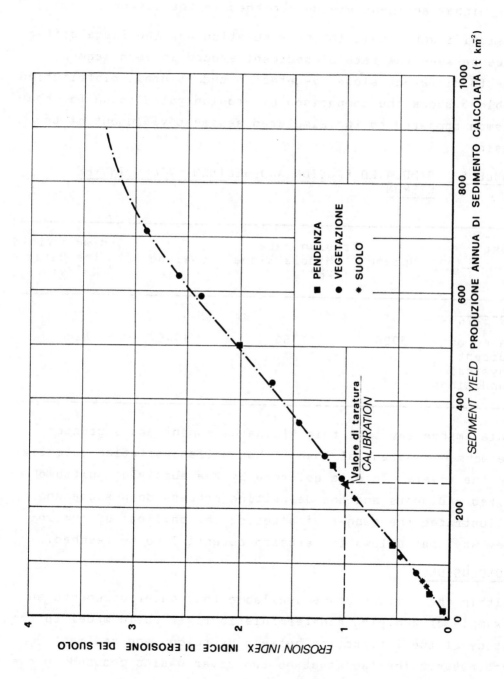

FIGURE 11 - SEDIMENT RESPONSE CURVE-AVISIO

above for the Avisio and a similar sensitivity was found to exist. However calibration of the erosion model had to include an adjustment to the parameters controlling the detachment and transportation functions in order to simulate the higher sediment yields recorded in the Cismon.

A significant result in the simulation was the large difference between the rate of sediment eroded in each segment due to different slope, vegetation and rainfall distribution. Table 5 shows the comparison of erosion rates found for the Cismon compared to the simulated sediment yield out of the basin.

TABLE 5   SIMULATED EROSION AND SEDIMENT YIELD IN THE CISMON

| Test | Erosion rate | | | Sediment Yield to the River $T/Km^2/yr$ |
	Upland	Middle Slope $T/Km^2/yr$	Lowland	
1975-76 Based on current physical conditions	3906	1386	1310	624

This emphasises that this Alpine Catchment has a greater detachment potential than it has a transportation potential to the river. This is governed by the supply of suitable sized sediments and the deposition process down slope and illustrates the danger of altering the physical system in any way that allows the erosion potential to be reached.

CONCLUSIONS

Within the limited space available this paper documents an example of applying a deterministic river basin model to the study of the interaction between hydrology and sediment processes. The fact that no two river basins respond in the same way is further underlined, but a method is presented which enables better understanding of why there is such a range in response between different river basins and within the same river basin. The paper further highlights the gross inadequacy of existing measurements of sediment yield for describing the sediment problems of a basin. At the same time it demonstrates a method that allows a comparative assessment to be made which has some use in future planning and design of river basin water and sediment resources.

The next stage in this research approach will be to undertake a detailed soil survey of the two basins to obtain a better comparison of soil types. The extension of the analysis of sensitivity into the river channel processes and reservoir sedimentation will also help to advance understanding on the influence they have on sediment response.

ACKNOWLEDGEMENTS

This study was made possible due to the co-operation of a number of individuals and organisations. These include research students at Strathclyde University including Mr. R. McKenzie, Mr. A. Sulieman and Mr. G. Rufai. Further they include research staff at Padova University including Mr. G. Dalla Fontana, Mr. F. Ca' Zorzi and Mr. L. Marchi.

References:

- BAGNOLD R.A. (1966): "An approach to the Sediment transport problem from general physics". USGS Prof-Paper 4221.

- BARONCINI E., FATTORELLI S. & FLEMING G. (1983): "Simulazione idrologica del bacino del fiume Brenta mediante modello continuo". Rapporto informativo n. 4 della Cattedra di Idrologia Forestale della Università di Padova.

- CHEN C. (1976): "Flow resistance in broad shallow grassed channels" Proc. ASCE HY3 March 1976. Paper 1194 p. 307.

- CRAWFORD N.H. & LINSLEY R.K. (1966): "Digital simulation in Hydrology: Stanford Watershed Model IV". T.R. 39. Stanford Univ., California.

- FARMER E.E. (1973): "Relative detachability of soil particles   by rainfall". Proc. Soil Science of America, 37. pp. 629-633.

- FATTORELLI S. & FLEMING G. (1982): Indagine sui fattori influenzanti il regime idrologico del bacino dell'Avisio". Rapporto informativo n. 1 della Cattedra di Idrologia Forestale dell'Università di Padova.

- FLEMING G. (1969): Design Curves for Suspended Load Estimation" . Proc. Inst. of Civil Eng. 43, pp. 1-9.

- FLEMING G. (1972): "Sediment Erosion-Transport-Deposition   Simulation: State of the Art". Proc. USDA Sediment Yield Workshop,Oxford, Miss., USA. (ARS-3-40). pp. 274-285.

- FLEMING G. & WALKER R.A. (1980): "Sediment Transport as part of Total Catchment modelling". Published in: "Assessment of Erosion" , Editor Dr. Boodt & Gabriels, Wiley 1980.

- FLEMING G. (1981): "The Sediment Problem Related to Engineering" . Proc. of S.E. Asian Regional Symposium, on "Problems of Soil Erosion and Sedimentation". Editors Tingsandiali & Eggers, Asian Inst. of Tech. Bangkok.

- FLEMING G. & KADHIMI A. (1982): "Sediment modelling and Data Sources - a compromise in assessment". Proc. IASH, Exeter Symp., July Publ. No. 137.

- FLEMING G. (1983): River Basin Model for Water and Sediment Resource Assessment". User Guide, Vol. 2 " The Sediment erosion transport deposition model and flow routing.

- FLEMING G. & MCKENZIE R. (1984): "River Basin Model for Water and Sediment Resource Assessment". User Guide Vol. 1 "The Watershed Model, Strathclyde Univ., Glasgow.

- FOSIER G.R. & MEYER L.D. (1975): "Mathematical simulation of upland erosion by fundamental erosion mechanics". U.S.D.A. ARS-5-40, pp. 190-207.

- GLEN J.W. (1982) editor: "Hydrological aspects of alpine and high mountain areas". Proc. I.A.S.H. Symp., Exeter. Publ. No. 138.

- KIRBY M.J. (1971): "Hillslope process-response model based on continuity equation". Inst. of British Geographers. Special Publ. No. 3, pp. 15-20.

- MUTCHLER & HANSEN (1970): "Splash and Waterdrop terminal velocity". Science 169, pp. 1311-1312.

- RUFAI G. (1984): "Influence of River Basin Response on Reservoir Sedimentation". Unpublished M. Sc. Thesis, University of Strathclyde, Dept. of Civil Eng., Glasgow, U.K.

- SULEIMAN A. (1983):"Computer simulation of the effects if land use on soil erosion". M. Sc. Thesis, Univ. of Strathclyde, Dept. of Civil Eng., Glasgow U.K.

- WISCHEMEIER W.H. & SMITH D.D. (1960): "A Universal soil loss equation to guide conservation from Planning Part 7". Int. Congress of Soil Science, Wisconsin.

SIMPLE MODEL FOR ASSESSING ANNUAL SOIL EROSION ON HILLSLOPES

R.P.C. Morgan, D.D.V. Morgan and H.J. Finney

Silsoe College, Silsoe, Bedford MK45 4DT, U.K.

ABSTRACT

A model is presented for predicting annual soil loss from field-sized areas on hillslopes. The erosion process is separated into a water phase and a sediment phase. In the sediment phase erosion is taken to be the result of the detachment of soil particles by raindrop impact and their transport by runoff. Splash detachment is related to rainfall energy and rainfall interception by the crop. Runoff transport capacity depends upon the volume of runoff, slope steepness and crop management. Rainfall energy and runoff volume are estimated in the water phase. The predicted rate of soil loss is compared with a top soil renewal rate to determine changes in the depth of soil over time. Model validation was carried out using data from published studies of soil loss from 67 sites in twelve countries.

INTRODUCTION

A new generation of hillslope erosion models, founded on the scheme presented by Meyer and Wischmeier (1969), is being developed for use as predictive and management tools in non-point source pollution studies (Foster, 1982). These models simulate the erosion process with a stronger physical base than the Universal Soil Loss Equation (Wischmeier and Smith, 1978) but, taking CREAMS as an example (Knisel, 1980), still rely on it for empirical modelling of soil erodibility and conservation practice. They are generally daily or event models which require a lot of input data, not all of which are easily obtainable. Since annual erosion rates are determined by summation of the daily values, the models are time consuming to use and require considerable computer storage, especially when operated for periods of twenty years or more. Although the models provide a reasonably accurate representation of the erosion system, they do not allow an easy understanding of the effects of individual parameters on the system. Thus, in their development, simplicity has been sacrificed. Whilst this may be acceptable for detailed management studies, its loss is not helpful for rapid reconnaissance assessments of erosion.

147

With the concentration of research on sophisticated event models, the development of simple models has lagged behind and the recent advances in our understanding of soil erosion processes has not been applied to them. This paper describes an attempt to amend this deficiency by bringing together the results of recent research by geomorphologists and agricultural engineers to produce a simple model for predicting annual soil loss. Preliminary studies related to soil erosion prediction in south east Asia (Morgan, 1981) showed that developing such a model was feasible. An early version was formulated and successfully tested for Malaysia (Morgan, Hatch and Sulaiman, 1982). The present version was developed and documented by Morgan, Morgan and Finney (1982).

DESIGN STRATEGY

The model was designed to conform with the following requirements.

1)   Applicability to field-sized areas on hillslopes. This restricts the model to the processes of raindrop erosion, overland flow and rill flow and removes the complications posed by including gully erosion and large channel flows. This means that the model cannot be used to predict sediment yield from drainage basins.

2)   Prediction rather than explanation. Nevertheless, the model attempts to provide a reasonable simulation of the soil erosion system including the effects of soil conservation practice.

3)   Prediction of annual erosion direct rather than through the summation of soil losses in individual events. As a result the model is necessarily empirical.

4)   Use of existing published material. Thus, the operating functions are selected from the literature according to their predictive ability, simplicity and ease of determining their input parameters. No attempt is made to develop and calibrate operating functions specially for the model.

MODEL STRUCTURE

The model separates the soil erosion process into a water phase and a sediment phase (Figure 1). It requires fifteen input parameters (Table I) and six operating functions (Table II).

The sediment phase is a simplification of the soil loss model described by Meyer and Wischmeier (1969). It considers soil erosion to result from the detachment of soil particles by raindrop impact and the transport of those particles by overland flow. The processes of splash transport and detachment by runoff are ignored. Thus the sediment phase comprises two predictive equations, one for the rate of splash detachment and one for the transport capacity of overland flow. The respective inputs to these equations of rainfall energy and runoff volume are determined from the water phase.

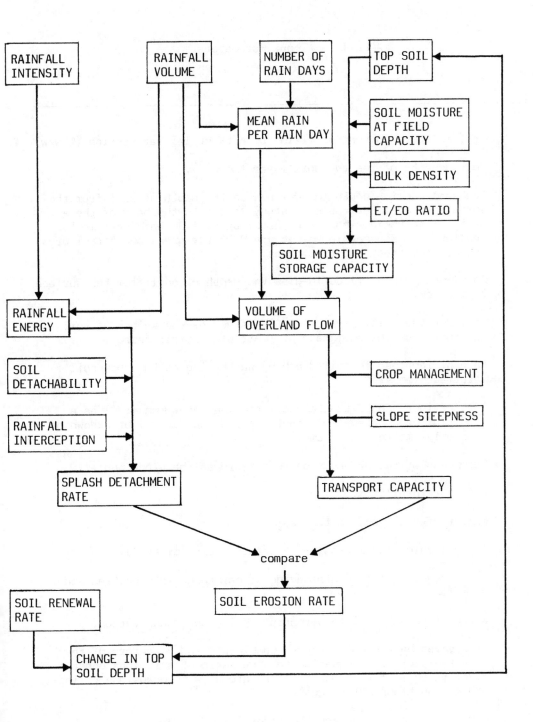

Figure 1 - Flow Chart of the Model

TABLE I - Input Parameters

MS   Soil moisture content at field capacity or 1/3 bar tension (% w/w).

BD   Bulk density of the top soil layer (Mg m^{-3}).

RD   Top soil rooting depth (m) defined as the depth of soil from the
          surface to an impermeable or stony layer; to the base of the A
          horizon; to the dominant root base; or to 1.0m whichever is the
          shallowest. Reasonable values are 0.05 for grass and cereal crops
          and 0.1 for trees and tree crops.

SD   Total soil depth (m) defined as the depth of soil from the surface
          to bedrock.

K   Soil detachability index (g J^{-1}) defined as the weight of soil
          detached from the soil mass per unit of rainfall energy.

W   Rate of increase in soil depth by weathering at the rock-soil
          interface (mm yr^{-1}).

V   Rate of increase of the top soil rooting layer (mm yr^{-1}) as a
          result of crop management practices and the natural breakdown of
          vegetative matter into humus.

S   Steepness of the ground slope expressed as the slope angle.

R   Annual rainfall (mm).

R_n Number of rain days in the year.

I   Typical value for intensity of erosive rain (mm hr^{-1}).

P   Percentage rainfall contributing to permanent interception and
          stemflow.

E_t/E_o Ratio of actual (E_t) to potential (E_o) evapotranspiration.

C   Crop cover management factor. Combines C and P factors of the
          Universal Soil Loss Equation to give ratio of soil loss under a
          given management to that from bare ground with downslope tillage,
          other conditions being equal.

N   Number of years for which the model is to operate.

TABLE II - Operating Functions

Water phase

$$E = R (11.9 + 8.7 \log_{10} I) \qquad \ldots (1)$$

$$Q = R \exp (-R_c/R_o) \qquad \ldots (2)$$

where

$$R_c = 1000 \ MS.BD.RD \ (E_t/E_o)^{0.5} \qquad \ldots (2a)$$

$$R_o = R/R_n \qquad \ldots (2b)$$

Sediment phase

$$F = K (E \ e^{-aP})^b . 10^{-3} \qquad \ldots (3)$$

$$G = C \ Q^d \sin S. \ 10^{-3} \qquad \ldots (4)$$

E = kinetic energy of rainfall $(J \ m^{-2})$.

Q = volume of overland flow (mm).

F = rate of splash detachment $(kg \ m^{-2})$.

G = transport capacity of overland flow $(kg \ m^{-2})$.

Values of exponents: $a = 0.05$; $b = 1.0$; $d = 2.0$

Annual rainfall (R) is the basic input parameter to the water phase.
Rainfall energy (E) is determined by calculating the energy for a typical
intensity (I) of erosive rains, using the relationship developed by
Wischmeier and Smith (1978) and multiplying the energy term by the annual
rainfall amount (De Ploey, 1972). Guide values for rainfall intensity are
11.0 mm hr^{-1} for temperate climates, 25.0 mm hr^{-1} for tropical climates and
30 mm hr^{-1} for strongly seasonal climates such as the Mediterranean type.
Where local information is available on typical intensities of erosive rain,
however, it should be used in preference.

The annual volume of overland flow (Q) is predicted from the annual rainfall
using an equation presented by Kirkby (1976). This assumes that runoff
occurs whenever the daily rainfall total exceeds a critical value which
represents the soil moisture storage capacity (R_c) of the soil-landuse
complex and that daily rainfall amounts approximate an exponential frequency
distribution. The parameter R_c is determined following Withers and Vipond
(1974) from the soil parameters of moisture storage at field capacity (MS),
bulk density (BD) and top soil rooting depth (RD) after allowing for the
effects of crop cover. The latter are accounted for, following Kirkby
(1976), through evapotranspiration, using the ratio of actual evapotranspiration (E_t) to potential evapotranspiration (E_o). Typical values for MS
and BD are given in Table III but values measured in the field should be
used where possible. The top soil rooting depth (RD) should be interpreted
as the depth to which the bulk of the root mat extends. Guide values are
0.10 m for cereals and vegetables and 0.15 m for tree crops. Annual values
for E_t/E_o are determined as time-weighted averages of values for the landuse
types occurring during the year, e.g. six months of spring-sown barley
(E_t/E_o = 0.60, Table IV) and a six-month winter fallow of bare soil
(E_t/E_o = 0.05) results in a value of 0.33. Changes in crop conditions
lasting less than two months are ignored. Thus, six months of spring-sown
barley followed by one month of bare soil and five months of winter beans
results in E_t/E_o ratios of 0.60 for seven months and 0.65 for five months,
giving a weighted average of 0.62.

TABLE III - Input Values for Selected Soil Types

Soil	MS	BD	K
Clay	0.45	1.1	0.02
Clay loam	0.40	1.3	0.4
Silty clay	0.30	-	-
Sandy loam	0.28	1.2	0.3
Silt loam	0.25	1.3	-
Loam	0.20	1.3	-
Fine sand	0.15	1.4	0.2
Sand	0.08	1.5	0.7

Sources: MS - Brady (1974), Withers and Vipond (1974); BD - Hall (1945),
Brady (1974); K - Quansah (1981).

TABLE IV - Typical Input Values for Plant Parameters

	P	E_t/E_o	C
wet rice		1.35	0.1-0.2
wheat	43%	0.59-0.61	0.1-0.2 (winter sown)
			0.2-0.4 (spring sown)
maize	25%	0.67-0.70	0.2
barley	30%	0.56-0.60	0.1-0.2
millet/sorghum		0.62	0.4-0.9
cassava/yam			0.2-0.8
potato	12%	0.70-0.80	0.2-0.3
beans	20-25%	0.62-0.69	0.2-0.4
groundnut	25%	0.50-0.87	0.2-0.8
cabbage/Brussels sprouts	17%	0.45-0.70	
banana		0.70-0.77	
tea		0.85-1.00	0.1-0.3
coffee		0.50-1.00	0.1-0.3
cocoa		1.00	0.1-0.3
sugar cane		0.68-0.80	
sugar beet	12-22%	0.73-0.75	0.2-0.3
rubber	20-30%	0.90	0.2
oil palm	30%	1.20	0.1-0.3
cotton		0.63-0.69	0.3-0.7
cultivated grass		0.85-0.87	0.004-0.01
prairie/savanna grass	25-40%	0.80-0.95	0.01-0.10
forest/woodland	25-35%	0.90-1.00	0.001-0.002 (with
(coniferous & tropical)			undergrowth)
	15-25%		0.001-0.004 (no)
(temperate broad-leaved)			undergrowth)
bare soil	0%	0.05	1.00

Note: C values should be adjusted by the following P values if mechanical soil conservation measures are practised:

contouring: multiply by 0.6
contour strip cropping: multiply by 0.35
terracing: multiply by 0.15

Sources: summarised in Morgan, Morgan and Finney (1982).

The sediment phase is divided into two components: splash detachment and runoff transport. Splash detachment is modelled as a function of rainfall energy (E) using the well-established power relationship with a value of 1.0 for exponent b(Meyer, 1981) and interpreting the intercept term k as an index of soil detachability. Typical values of k (Table III) have been determined experimentally (Quansah, 1981). The relationship is modified to allow for the effects of rainfall interception of the crop (P) by reducing the rainfall energy exponentially with increasing interception, using a value of 0.05 for exponent a (Laflen and Colvin, 1981). Annual values of interception are determined from time-weighted averages for the values of the landuse types (Table IV) occurring during the year.

The transport capacity of overland flow is determined from an equation developed by Kirkby (1976) and depends upon the volume of overland flow, the slope steepness and the effect of crop cover. The relationship between transport capacity and the first power of the sine of the slope angle conforms with that derived theoretically in a sediment transport equation for overland flow (Morgan, 1980). The value of 2.0 for the runoff exponent is similar to the 1.7 power found in most sediment transport equations and has field support, using annual data, in the studies of Mou and Xiong (1980). The crop cover management factor (C) uses the C-factor values of the Universal Soil Loss Equation (Wischmeier and Smith, 1978). Annual values can be determined from the simplified information presented in Table IV using a time-weighted average of the values for the separate landuse types occurring during the year.

The effects of soil conservation practices can be accounted for within the separate phases of the model. For example, the introduction of agronomic measures of erosion control is allowed for by changes in evapotranspiration, interception and crop management which respectively affects the volume of runoff, the rate of splash detachment and the transport capacity of overland flow.

The model compares the predictions of the rate of splash detachment and the transport capacity of overland flow and assigns the lower of these two values as the rate of soil loss, thereby indicating whether detachment or transport is the limiting factor.

The predicted rate of soil loss is compared with an estimate of the rate of weathering (W) at the soil-bedrock interface to calculate the loss or gain in soil depth (SD). Information on weathering rates is very limited but they are generally about 0.01 to 0.02 mm yr^{-1}. The difference between the rate of soil loss and the rate of top soil renewal (V) allows a similar calculation to be made for the change in top soil rooting depth (RD). Estimates of top soil renewal rates can be based on the guidelines outlined by McCormack and Young (1981) with modifications to allow for different agricultural practices. Recommended rates for soil depths less than 0.25 m are 0.20 mm yr^{-1} for forest, grassland or where crop residues are returned to the soil; 0.15 mm yr^{-1} for standard husbandry and 0.10 mm yr^{-1} for poor husbandry. The new values of SD and RD are used as inputs to the following year of similation. In this way, the effects of a continued reduction in top soil rooting depth are simulated through positive feedback, resulting in reduced soil moisture storage, increased runoff and greater erosion. The process continues until first, the top soil and finally the subsoil disappear.

SENSITIVITY ANALYSIS

The extent to which predictions of soil loss by the model are affected by small changes in the values of the input data was examined using partial differentiation (Morgan, Morgan and Finney, 1982). The results (Table V) show that sensitivity is greatest to changes in annual rainfall and the soil parameters when erosion is transport-limited and to changes in rainfall interception and annual rainfall when erosion is detachment-limited. Values for these parameters thus need to be assessed with the greatest accuracy. Generally, annual rainfall data are readily available but good information on soils and interception is limited. No matter how good the estimates of these parameters and the resulting predictions of soil loss are, however, the success of the model for long-term simulations of erosion is also dependent upon the estimates for top soil renewal rate and weathering rate and good quality data on these are extremely hard to obtain.

TABLE V - Sensitivity Analysis

1% change in	% change in	
	transport capacity	detachment rate
R	$2(1 + R_c/R_o)$	1
MS; BD; RD; R_n	$-2 \ (R_c/R_o)$	
E_t/E_o	$-R_c/R_o$	
C; sin S	1	
K		1
I		$(3.1 + 2.3 \ \log_{10}I)^{-1}$
P (1% absolute change)		-5

VALIDATION

Validation of the model was carried out using data from erosion plot studies published in various journal articles, reports and monographs. Data for 67 sites in 12 countries were obtained covering Tanzania, Ivory Coast, Senegal, Thailand, Italy, Belgium, Germany, Taiwan, Zimbabwe, China, United Kingdom and Malaysia (Morgan and Finney, 1982).

The model generally predicted erosion as being limited by the transport capacity of the runoff but where runoff rates were very high, erosion was detachment-limited. This occurred on bare soil in the Ivory Coast and Belgium, under clean-cultivated citrus in Taiwan and under cropped land in the loess of China. The soil loss predictions were worst at very low rates of soil loss (< 0.1 kg m^{-2}yr^{-1}), when they were often an order of magnitude or more out, as in most of the sites in the United Kingdom, and at very high rates of soil loss (> 20 kg m^{-2}yr^{-1}), as in China. Runoff predictions were often poor where mechanical soil conservation measures such as ridging and terracing were used, as on the orchard sites in Taiwan and the coffee plantations in Tanzania, or where mulching was adopted, as under banana in the Ivory Coast and Taiwan, citrus in Taiwan and maize in Malaysia. Under these conditions the model frequently, but not always, overpredicted. This can be explained by the failure of the model to allow for surface depression storage of rainfall in the water phase.

In devising a method for assessing the goodness of fit of the predictions, account was taken of the role of the model in reconnaissance survey. Predicted values (Y) were compared with the measured values (X) using reduced major axis lines (Kermack and Haldane, 1950; Till, 1973). These were selected in preference to regression lines because of the likelihood of errors in the measured data as well as in the predicted values. The following relationships were obtained:

$$Y = 19.776 + 0.775X \quad \text{for runoff}, \quad r = 0.735, \ n = 56$$
$$Y = 0.472 + 0.503X \quad \text{for soil loss}, \quad r = 0.583, \ n = 67$$

The lower value of the correlation coefficient (r) for soil loss is partly explained by the failure of the model to predict sufficiently closely two extremely high soil loss rates in China. If these two cases of 23.4 and 43.9 kg m^{-2}yr^{-1} are omitted, the relationship becomes:

$$Y = -0.090 + 0.896X \quad \text{for soil loss}, \ r = 0.671, \ n = 65$$

where the slope of the regression line is not significantly different from unity ($P > 0.05$).

This type of analysis gives equal weight to differences between predicted and measured values regardless of their magnitude whereas, in reconnaissance surveys in particular, much greater differences may be acceptable at very high and very low values of soil loss. If the erosion rate is low enough not to present a problem, it may be sufficient for practical purposes to predict so even if the prediction itself is not very accurate. If erosion rates are catastrophic, it may be sufficient for the model to indicate this general condition without giving a close prediction of the soil loss rate. An alternative evaluation procedure was therefore applied which viewed predictions as successful if (1) the annual predicted and observed values were both less than 0.1 kg m^{-2} or otherwise (2) the ratio of the predicted value to either a single observed value or to the mid-point of a range of observed values was between 0.5 and 2.0. No threshold was applied to the runoff predictions which were assessed only by the second criterion.

Judged against these criteria, the model successfully predicted runoff for 33 out of 56 test sites and soil loss for 47 out of 67 sites, giving respective success rates of 59 and 70 per cent. If only those sites, 31 in all, are considered where data on soil properties were taken from field measurements instead of guide values, the success rates are 57 and 90 per cent.

CONCLUSIONS

It is feasible to produce from published sources a simple model to predict annual soil erosion from hillslopes which takes account of recent advances in our understanding of erosion processes. The model has proved satisfactory in validation trials involving 67 sites in twelve countries. Except for very low and very high rates of erosion, the model gives realistic predictions over a wide range of conditions. The model simulates the generation of runoff and sediment in a manner which, qualitatively at least, represents what happens in practice. The major factors which influence runoff and erosion processes are included in a way in which the user can understand their effects and therefore decide which factors to change to ameliorate an erosion problem. The model is therefore potentially useful for reconnaissance assessments of erosion and for designing strategies for erosion control. Applications of the model are beyond the scope of this paper but it has been successfully used to simulate soil erosion under shifting cultivation (Morgan, Morgan and Finney, 1984), tree crops, commercial timber extraction and continuous cropping in Malaysia (Morgan, Finney and Morgan, 1982) and in assessments of recreational soil erosion (Morgan, 1983).

ACKNOWLEDGEMENTS

The assistance of my colleagues, Miss H.J. Finney and Mr. D.D.V. Morgan, is greatly appreciated. The present form of the model was developed with the aid of research grants from the UK Natural Environment Research Council and the International Institute for Applied Systems Analysis, Laxenburg, Austria.

REFERENCES

Brady, N.C. 1974. The nature and properties of soils. Macmillan, New York.

De Ploey, J. 1972. A quantitative comparison between rainfall erosion capacity in a tropical and a middle-latitude region. Geographia Polonica, 23, 141-150.

Foster, G.R. 1982. Modeling the erosion process. In Haan, C.T., Johnson, H.P. and Brakensiek, D.L. (eds), Hydrologic modeling of small watersheds. American Society of Agricultural Engineers Monograph No. 5, 297-380.

Hall, A.D. 1945. The soil. John Murray, London.

Kermack, K.A. and Haldane, J.B.S. 1950. Organic correlation and allometry. Biometrika, 37, 30-41.

Kirkby, M.J. 1976. Hydrological slope models: the influence of climate. In Derbyshire, E. (ed), Geomorphology and climate. Wiley, London, 247-267.

Knisel, W.G. 1980. CREAMS: a field scale model for chemicals, runoff and erosion from agricultural management systems. USDA Conservation Research Report No. 26.

Laflen, J.M. and Colvin, T.S. 1981. Effect of crop residue on soil loss from continuous row cropping. Transactions, American Society of Agricultural Engineers, 24, 605-609.

McCormack, D.E. and Young, K.K. 1981. Technical and societal implications of soil loss tolerance. In Morgan, R.P.C. (ed), Soil conservation: problems and prospects. Wiley, Chichester, 365-376.

Meyer, L.D. 1981. How rain intensity affects interrill erosion. Transactions, American Society of Agricultural Engineers, 24, 1472-1475.

Meyer, L.D. and Wischmeier, W.H. 1969. Mathematical simulation of the processes of soil erosion by water. Transactions, American Society of Agricultural Engineers, 12, 754-758, 762.

Morgan, R.P.C. 1980. Field studies of sediment transport by overland flow. Earth Surface Processes, 5, 307-316.

Morgan, R.P.C. 1981. The role of the plant cover in controlling soil erosion. In Tingsanchali, T. and Eggers, H. (eds), South-east Asian regional symposium on problems of soil erosion and sedimentation. Asian Institute of Technology, Bangkok, 255-265.

Morgan, R.P.C. 1983. The impact of recreation on mountain soils: towards a predictive model for soil erosion. Conference on the ecological impacts of outdoor recreation on mountain areas in Europe and North America, Recreation Ecology Research Group, Ambleside, Cumbria.

Morgan, R.P.C. and Finney, H.J. 1982. Stability of agricultural ecosystems: validation of a simple model for soil erosion assessment. International Institute for Applied Systems Analysis Collaborative Paper CP-82-76.

Morgan, R.P.C., Finney, H.J. and Morgan, D.D.V. 1982. Stability of agricultural ecosystems: application of a simple model for soil erosion assessment. International Institute for Applied Systems Analysis Collaborative Paper CP-82-90.

Morgan, R.P.C., Hatch, T. and Sulaiman, W. 1982. A simple procedure for assessing soil erosion risk: a case study for Malaysia. Zeitschrift für Geomorphologie Supplementband 44, 69-89.

Morgan, R.P.C., Morgan, D.D.V. and Finney, H.J. 1982. Stability of agricultural ecosystems: documentation of a simple model for soil erosion assessment. International Institute for Applied Systems Analysis Collaborative Paper CP-82-59.

Morgan, R.P.C., Morgan, D.D.V. and Finney, H.J. 1984. A predictive model for the assessment of soil erosion risk. International Conference for the Diamond Jubilee of the National Institute of Agricultural Engineering, AG ENG 84, Cambridge.

Mou, J. and Xiong, G. 1980. Prediction of sediment yield and evaluation of silt detention by measures of soil conservation in small watersheds of north Shaanxi. Preprint, International symposium on river sedimentation, Chinese Society of Hydraulic Engineering, Beijing. In Chinese with English summary.

Quansah, C. 1981. The effect of soil type, slope, rain intensity and their interactions on splash detachment and transport. Journal of Soil Science, 32, 215-224.

Till, R. 1973. The use of linear regression in geomorphology. Area, 5, 303-308.

Wischmeier, W.H. and Smith, D.D. 1978. Predicting rainfall erosion losses. A guide to conservation planning. USDA Agricultural Handbook, No. 537.

Withers, B. and Vipond, S. 1974. Irrigation: design and practice. Batsford, London.

SOME ASPECTS OF SOIL EROSION MODELLING

D. TORRI, Centro per lo Studio della Genesi, Classificazione e Cartografia del suolo. CNR, Piazzale delle Cascine 15 – FIRENZE, ITALY.

M. SFALANGA, Istituto Sperimentale per lo Studio e la Difesa del Suolo. Piazza D'Azeglio, 30 – FIRENZE, ITALY.

ABSTRACT

Soil erosion is often accelerated by agricultural activities. As sediment acts both as pollution factor and pollutant carrier, soil loss can be considered a non-point pollution process. An experiment on splash detachment and runoff transport shows that interrill erosion depends on the runoff transport capacity or on the detachment rate following which one is the limiting agent. Mathematical description of these processes indicates that statistical equations are inadequate to describe interrill erosion. A residual variability of the data may be attributed to dishomogeneity at the soil surface (aggregate and clod distribution, crusts, etc.). This variability can induce errors in the estimate of soil loss. Consequently, it should be predicted as it can assume a relevant role when the damages due to pollution depend on critical thresholds.

INTRODUCTION

Soil erosion by rain is a natural phenomenon; it can be accelerated by human activity such as agriculture. It has some negative effects on the environment, which can be categorized as fertility loss and pollution.

Fertility loss depends on the fact that erosion usually takes place on the most superficial soil layer, which is the best structured and the richest in nutrients. The fraction of the detached material which is transported by superficial runoff to the channel system may cause excessive silting when deposited. Moreover, large quantities of sediments generally cause disequilibria in the aquatic environment. Sediment is usually rich in chemicals due to nutrients, herbicides, etc., present in the soil. Consequently, sediment also contributes to chemical pollution.

The study and control of the erosive processes are primarily relevant where agriculture is associated to a high risk of erosion or where the soil is rich in the clay, silty-clay fraction, that is very effective in trapping chemicals.

The prevision of pollution and the control of pollutant factors can be achieved through models of different kinds (such as statistical, deterministic, etc.). Their prevision should be performed on a single rainstorm basis as pollution is a discontinuous phenomenon which is often linked to critical values not to be surpassed.

EROSION STUDIES IN ITALY

Italy is characterized by a high erosion potential (hilly and montaneous agriculture) and by soils rich in clay (20% ca. of the agricultural territory). Aggressiveness of the climate and excessive antropization make erosion control pertinent to pollution. Unluckily, climatic differences make erosion difficult to be studied. Roster [1], divided Italy into five climatic areas; a simple analysis of the season - to - season variation of precipitation is enough to subdivide Italy into three zones (one peak of precipitation in winter, one peak in summer, two peaks in spring and autumn).

Those climatic differences might have influenced the results of the studies on soil erosion performed in Italy as data were usually analysed on a single rainstorm basis while using statistical techniques.

The erosivity index proposed by Wischmeier and Smith [2] was found to be both well correlated [3,4] and uncorrelated [5,6] to erosion. Aggressiveness indices depending on runoff

haracteristics only [3,7] or on both rain and runoff [6,8,9,10]
ere also proposed. Only once was an additive index (rain +
unoff) compared to erosion [3].

ew data allowed a complete comparison between measured erosion
nd estimations made through the Wischmeier and Smith's equation
11]: a slope effect described through a convex parabola was
uccessfully investigated in Sicily [4] while an over-stimation
f soil loss was observed for clayey soils [12].

he above mentioned results, more completely summarized by Chisci
t al. [13] only approximatively indicate the erosion hazard in
nvironments very similar to those in which the equations were
eveloped. Moreover, statistical equations are usually not
hysically based as pointed out by Kirkby [14]. Consequentely
hey generally fail when used to predict erosion values close to
xtremes of the tested range. An example of how even a single
spect of erosion cannot be easily described is presented in the
ollowing paragraph.

N EXPERIMENT ON SPLASH DETACHMENT AND RUNOFF TRANSPORT

: Scientific background

When a drop hits the soil surface it splashes detaching particles
and aggregates. The mechanism of splash detachment has been
studied by many researchers [15 to 37]. The main features through
which a drop detaches particles [15,16] are as follows:

a – a drop hits a soil particle releasing a part of its momentum
to it;
b – a drop, during the impact, generates a corona of lateral jets
of water. The shear stress produced across the solid-liquid
contact area determines the detachment of particles.

The fact that a film of water develops at the soil surface can
cause a reduction of the detached material. In fact it resists
the expansion of the jets and reduces their speeds [17].
Moreover, the drop impact is partially dissipated into the water
film [18]. On the other hand, an increase of the pore water
pressure within the aggregates might increase detachment [19,20].
Some researchers [21] observed an initial increase followed by a
decrease in detachment as the height (h) of the water film
increases. The value of h at which the detachment reaches its
maximum was estimated to be between 0.14 and 0.20 times the
diameter of the hitting drop. On the contrary, other researchers
observed a continuous decrease of detachment with increasing h
[22,23,24,25].

As proposed by different authors [26 up to 34], indices of
detachment power of rain are the kinetic energy or the momentum
or the instantaneous intensity of rain or factors as m^a
v^b where m is the mass of the rain drop, v its speed and a

and b empirical exponents. In addition, slope has showed a positive effect on detachment [28,31,35]. Also the angle between the trajectory of rain-drops and the slope might effect detachment [36,37].

The detached material can be transported by the runoff or by saltation due to the drop impacts. The runoff transport role depends on the equilibrium between the runoff transport capacity (TC) and the detachment rate (DR). If DR is greater than TC the transported material cannot exceed TC. When the opposite situation takes place, the transported material cannot exceed DR. The passage from one situation to other can be abrupt [28] or gradual [31].

ii: Materials and methods

The experiment was planned in such a way to have:

1 - detachment rate due to raindrop impact only;
2 - raindrop splash transport excluded from the measurement of soil loss.

TAB. I : Textural and aggregate distribution (in %) of the soil samples (vertic xerochrepts).

| SIZE | TEXTURE | AGGREGATE | |
| | | initial* | splashed-out** |
μ	%	%	%
4000	-	100.0	100.0
2000	100.0	93.3	99.6
1000	99.6	83.8	97.6
500	99.5	76.1	94.3
250	99.2	67.5	89.0
125	98.7	60.3	83.6
63	96.2	54.0	78.3
30	94.1		
15	89.4		
5	78.7		
2	60.4		
1	48.2		

* - Distribution of aggregate after to 24[h] of saturation by capillary rise.
** - Average distribution of aggregate splashed-out during the tests.

The reduction of the runoff detachment to a negligible value (Condition 1) was achieved using a cohesive soil (Tab.1). It has also been controlled if high runoff rates, in absences of rain, produced erosion. Condition 2 was achieved shielding the runoff collector.

Soil samples were 10 cm deep and 50 cm wide; slope and lengths
were variable: from 0.5 to 2 m and from 5% to 30% respectively.
The samples were prepared using air dried soil, passed through a
4.0 mm sieve [38]. The initial water content was 4.5% (air dried)
or 38% (48h of saturation by capillary rise).

The caracteristics of the simulated rains were as follows:
intensity: 15, 30, 60, 110 mm h^{-1};
median drop diameter: 1.9 mm;
kinetic energy per unit of mass of rain: 24.1 Joule kg^{-1}.
Drop sizes and kinetic energy per unit of mass of rain were kept
constant in all the runs.

Additional runoff (clear water) was supplied from upslope during
some runs. The runoff speed and the height of the water film were
calculated using the programme proposed by Savat [39].

iii: Experimental results

The data taken into account correspond to constant rate of runoff
and soil loss. Using steady state data the variability due to
initial breakdown of aggregates [19,20] is reduced.

The ratios A/i, where A is the measured soil loss in g min^{-1}
m^{-2} and i is the rain intensity in mm min^{-1}, are drawn
versus the height of the film of water in Fig.1a. The observed
behaviour agrees fairly well with the one described by Mutchler
and Young [21] even if there is a subdivision due to slope. The
experimental data allow two interpretations:

1 - Differences in detachment rate depend on the effect of the
film of water;
2 - Soil loss is controlled by the transport capacity until the
maximum is reached, then by detachment rate.
According to 1 -, the detachment rate can be described by the
following equation:

$$DR = 3800 \ i \ \sin^{0.32} \gamma \ h^{1.9} \exp(-6.8 \ h) \tag{1}$$

where:
DR = detachment rate (g min^{-1} m^{-2})
h = average height of the water film along the plot (mm)
γ = slope angle.

This hypothesis is not completely satisfying as it clashes against
the data produced by Ghadiri and Payne [22]. Moreover, there is
no detachment at zero slope.

A function depending only on runoff characteristics is required
to support hypothesis 2. This function, which is an estimation
of the runoff transport capacity, must verify the following
condition:

165

$$A/TC = \text{constant when } h < h_{cr} \qquad (2)$$

where:
h_{cr} = value of h at which the ratio A/i is maximum.

A function approximating condition (2) is as follows:

$$TC = 120 w q v^{1/2} \qquad (3)$$

where:
TC = transport capacity (g min^{-1})
w = width of the plot (m)
q = runoff discharge rate per unit of width (cm^2 s^{-1})
v = runoff speed at the bottom of the plot (cm s^{-1}).

The ratio A/TC versus h is drawn in Fig.1b. It is possible to observe a certain constancy when h is smaller than h_{cr} while the slope does not separate data anymore. The decrease, which follows, indicates that the detachment is already the limiting factor. This result supports hypothesis 2.

Park et al. [15] used an exponential function to describe the effect of the height of the film of water on the detachment. Using the same kind of function and taking into account the slope angle - even if data are not enough to state any relation for sure - the detachment rate can be expressed as follows:

$$DR = 160 (\sin^{0.77} \gamma + 0 22) | \exp(-3.07 h) \qquad (4)$$

The exponent of the slope and the additive term fairly well agree with the values suggested by Khaleel et al. [31].
An estimate of erosion can be performed using equation (3) when TC<DR and equation (4) when DR<TC (Fig.2). The data show a scattering which is larger than the estimate of the maximum error due to the measuring apparatus. Actually, data should not scatter more than 10% from the 45 sloping straight line. On the contrary the 50% of the data scatter more than the 20%. This indicates that there are sources of variability not included in equations (3) and (4). As moisture content, shear strength, cohesion, bulk density showed correlation with the residual variability other sources of variation must be taken into account such as differences in aggregate distribution at the soil surface [20].

CONCLUSION

The experiments on splash detachment and runoff transport showed that:

1 - two equations are needed to explain soil loss;
2 - detachment rate shows an exponential decrease with the height of the water film;

3 - a residual variability exists which may depend on surface dishomogeneity of aggregate distribution.

Item 1 and 2 indicate that a statistical equation is inadequate to describe interrill erosion. Item 1 clearly states that two equations are needed while item 2 limits severely the use of statistical techniques. In fact, the detachment rate cannot be approximated by a single equation over its entire range of variation because of the exponential. Moreover, rain intensity, soil characteristics, length and slope of the interrill are also implicitly present in the detachment equation as they can predict the height of the water film. The effects of the mentioned factors cannot be easily separated as the usual regression equations require.

The residual variability due to superficial dishomogeneity (aggregates, clods, crusts, etc.) might last in natural conditions affecting the estimates of soil loss. It should be, consequentely, predicted to define the probability levels of the erosion estimates which is relevant when pollutant contents must be kept under critical values.

REFERENCES

1. ROSTER, G., "Climatologia dell'Italia", Unione Tipografico - Editrice Torinese, 1909.
2. WISCHMEIER, W.H. and SMITH, D.D., "Rainfall and its Relationship to Soil Loss", Trans. Am. Geoph. Union, 39, 2, 285-291, 1958.
3. ZANCHI, C., "Previsione dell'erosione e della concentrazione delle torbide in funzione di alcune caratteristiche fisiche della pioggia e del ruscellamento", Annali Ist. Sperim. Studio Difesa Suolo 217-230, IX, 1978.
4. LI DESTRI NICOSIA, O., "Indagine sperimentale sui fattori dell' erosione idrica superficiale", Congr. Int. Problemi Idraulici nell'assetto territoriale della montagna, Milano, Maggio 1981.
5. CHISCI, G. and ZANCHI, C., "The Influence of Different Tillage Systems and Different Crops on Soil Losses on Hilly Silty-clayey Soil", "Conservation 80", Int. Conf. Soil Conservation, Silsoe-Bedford, U.K., John Wiley Sons, Chichester U.K., 211-218, 1981.
6. BOSCHI, V. and CHISCI, G., "Influenza delle colture e delle sistemazioni superficiali sui deflussi e l'erosione in terreni argillosi di collina", Genio Rurale XLI, 4, 7-16,1978.
7. VAN ASCH, T.W.J., "Water Erosion on Slopes in Some Land Unit in a Mediterranean Area" in "Rainfall Simulation, Runoff and Soil Erosion", J.De Ploey ed., Catena Supplement, 4, 129-140, Braunschweig, 1983.
8. RAGLIONE, M., SFALANGA, M. and TORRI, D.,"Misura dell'erosione in un ambiente argilloso della Calabria", Annali Ist. Sperim. Studio Difesa Suolo, XI, 159-182, Firenze, 1980.
9. CARONI, E. and TROPEANO, D., "Rate of Erosion Processes on

Experimental Areas in the Marchizza Basin (Northwestern Italy)", Int. Symposium on "Erosion and Sediment Transport", IAHS-AISH, No.133, Firenze, 1981.

10. TROPEANO, D., "Soil Erosion in Vineyards in the Tertiary Piedmontese Basin (Northwestern Italy). Studies on Experimental Areas", in "Rainfall Simulation Runoff and Soil Erosion", J.De Ploey ed., Catena Supplement, 4, Braunschweig, 1983.

11. WISCHMEIER, W.H. and SMITH, D.D., "Predicting Rainfall Erosion Losses from Cropland East of the Rocky Mountains", USDA, ARS, Agr. Handbook No.282.

12. ZANCHI, C., "Influenza del diverso carico di pascolamento sul ruscellamento superficiale, sul drenaggio e sulle asportazioni di suolo: esperienze pluriennali nel Centro Sperimentale di Fagna (Firenze)", Annali Ist. Sperim. Studio Difesa Suolo, 193-216, XII, 1981.

13. CHISCI, G., GIORDANO, A., INDELICATO, S., LI DESTRI NICOSIA, O., SFALANGA, M. and TORRI, D., "Acquisizione per la previsione dell'erosione idrica sui versanti", Convegno Conclusivo P.F. Conservazione del Suolo, 188-202, Roma, Giugno 1982.

14. KIRKBY, M.J., "Modelling Water Erosion Processes", in "Soil Erosion", Kirkby M.J. and Morgan R.P.C. eds., John Wiley Sons, Chichester, U.K., 183-216, 1980.

15. PARK, S.W., MITCHELL, J.K. and BUBENZER, G.D., "Splash Erosion Modelling: Physical Analysis", Trans. of the ASAE, 25,356-361, 1982.

16. HUANG, C., BRADFORD, J.M. and CUSHMAN, J.H., "A Numerical Study of Raindrop Impact Phenomena: The Elastic Deformation Case", Soil Sci. Soc. Am. J., 47, 855-861, 1983.

17. HARLOW, F.H. and SHANNON, J.P., "The Splash of a Liquid Drop", J. of Appl. Phys., 38, 10, 3855-3866, 1967.

18. PALMER , R.S., "The Influence of a Thin Water Layer on Waterdrop Impact Forces", I.A.S.H. publ. 65, 141-148, 1963.

19. FARRES, P., "The Role of Time and Aggregate Size in the Crusting Processes", Earth Surface Processes, 3, 243-254,1978.

20. LUK SHIU-HUNG, "Effect of aggregate size and microtopography on rainwash and rainsplash erosion", Z. Geomorph. N.F. 27, 3, 283-295, Berlin-Stuttgart Sept. 1983.

21. MUTCHLER, C.K. and YOUNG, R.A., "Soil Detachment by Raindrops", in "Present and Perspective Technology for Predicting Sediment Yields and Sources", Proceedings of the Sediment Yield Workshop, Oxford, Mississippi, USDA-ARS-40, 113-117, 1975.

22. GHADIRI, H. and PAYNE, D., "Raindrop Impact and Soil Splash", in "Soil Physical Properties and Crop Production in Tropics", Lal R., Greenland D.J. Eds., John Wiley Sons, Chichester,U.K., 95-104, 1979.

23. DE PLOEY, J., "Crusting and time-dependent rainwash mechanisms on loamy soils", in "Conservation 80", Int. Conf. Soil Conservation, Silsoe-Bedford, U.K., John Wiley Sons, Chichester, 139-154, 1981.

24. POESEN, J., "Rainwash Experiments on the Erodibility of Loose Sediments", Earth Surface Processes and Landforms, 6, 285-307,

1981.

25. POESEN, J. and SAVAT, J., "Detachment and Transportation of Loose Sediments by Raindrop Splash. Part II. Detachability and Transportability measurements", Catena, 8, 19-41,Braunschweig, 1981.

26. ELLISON, W.D., "Soil Erosion Studies", 1, Agr. Eng. 28, 4, 145-146, 1947.

27. FREE, G.R., "Erosion Characteristics of Rainfall", Agric. Engng., 41, 7, 447-449, 1960.

28. MEYER, L.D. and WISCHMEIER, W.H., "Mathematical Simulation of the Process of Soil Erosion by Water", Trans. ASAE 12, 6, 754-758, 1969.

29. BUBENZER, G.D. and JONES, B.A., "Drop Size and Impact Velocity Effects on the Detachment of Soils under Simulated Rainfall", Trans. ASAE, 14, 4, 625-628, 1971.

30. ELWELL, H.A. and STOCKING, M.A., "Rainfall Parameters for Soil Loss Estimation in a Subtropical Climate", J. Agric. Engng. Res., 18, 169-177, 1973.

31. FOSTER, G.R. and MEYER, L.D., "Mathematical Simulation of Upland Erosion by Fundamental Erosion Mechanics", in "Present and Perspective Technology for Predicting Sediment Yields and Sources", Proc. of Sediment Yield Workshop, Oxford, Mississippi, USDA-ARS-40, 190-206, 1975.

32. KHALEEL, R., FOSTER, G.R., REDDY, K.R., OVERCASH, M.R. and WESTERMAN, P.W., "A Non-point Source Model for Land Areas Receiving Animal Wastes: III. A Conceptual Model for Sediment and Manure Transport", Trans. ASAE, 22, 6, 1353-1361, 1979.

33. VAN ASCH, T.W.J. and EPEMA, G.F., "The Power of Detachment and the Erosivity of Low Intensity Rains", Pedologia XXXIII, 1, 17-27, Ghent 1983.

34. AL-DURRAH, M.M. and BRADFORD, J.M., "The Mechanism of Rainsplash on Soil Surfaces", Soil Sci. Soc. Am. J. 46, 1086-1090, 1982.

35. KIRKBY, A.V.T. and KIRKBY, M.J., "Surface Wash at the Semi-arid Break in Slope", Z. fur Geomorph. Suppl. Bd. 21, 151-176, 1974.

36. MOEYERSONS, J., "Measurements of Splash-saltation fluxes under Oblique Rain", in "Rainfall Simulation, Runoff and Soil Erosion", J.De Ploey ed., Catena Supplement, 4, 19-32, Braunschweig, 1983.

37. POESEN, J., "Field Measurements of Splash Erosion to Validate a Splash Transport Model", Int. Symposium (I.G.U.). The Role of Geomorphological Field Experiments in Land and Water Management, Bucaresti, Romania, Aug. 25th-Sept. 3rd 1983 , in press.

38. TORRI, D. and SFALANGA, M., "Stima dell'erodibilita' dei suoli mediante simulazione di pioggia in laboratorio. Nota II: Preparazione dei campioni di suolo", Annali Ist. Sperim. Studio Difesa Suolo, XI, 141-157, 1980.

39. SAVAT, J., "Resistence to Flow in Rough Supercritical Sheet Flow", Earth Surface Processes, 5, 103-122, 1980.

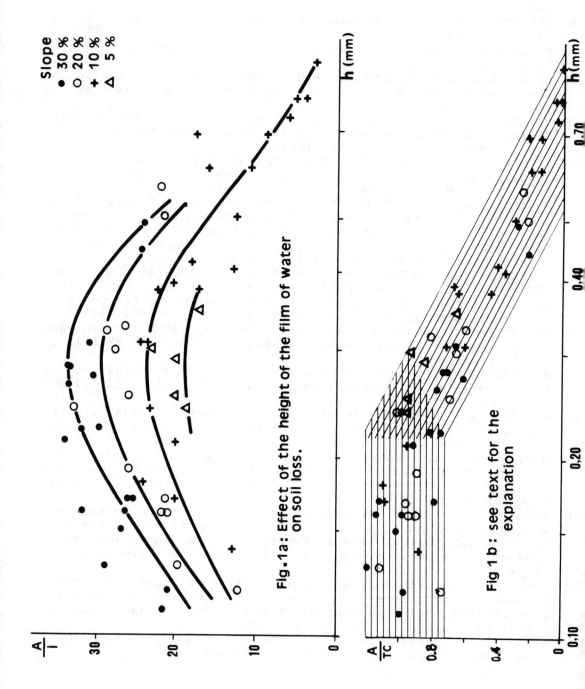

Fig.1a: Effect of the height of the film of water on soil loss.

Slope
- ● 30 %
- ○ 20 %
- + 10 %
- △ 5 %

Fig 1 b : see text for the explanation

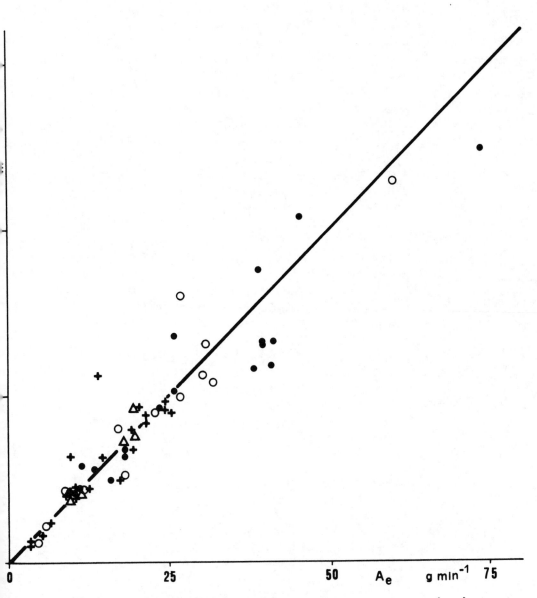

Fig 2: measured soil loss (A_m) versus estimated soil loss (A_e)

THE APPLICATION OF AN AGRICULTURAL WATER BALANCE AND EROSION MODEL IN ENVIRONMENTAL SCIENCE

A USER PERSPECTIVE

T. E. Hakonson,[1] G. R. Foster,[2] L. J. Lane,[1] and J. W. Nyhan[1]

[1]Environmental Science Group
Los Alamos National Laboratory
P.O. Box 1663
Los Alamos, NM 87545

[2]USDA/Agricultural Research Service
National Soil Erosion Laboratory
Purdue University
W. Lafayette, IN 47907

INTRODUCTION

The science of ecology made major advances in the development of theory, models, and supporting data in the late 1940's. That progress, in large part, can be attributed to atmospheric testing of nuclear weapons and the generous sponsorship of the United States Atomic Energy Commission. Concern about the fate of fallout radionuclides in the environment led to a plethora of studies on the distribution and food chain transport of radionuclides such as ^{137}Cs, ^{90}Sr, and ^{131}I. Based upon that early radioecological work, and on the work of the International Biological Program, it became clear that ecological processes, which have evolved over millions of years to incorporate and distributed materials in the environment, dictated the eventual fate of fallout radionuclides. Because many radionuclides are chemical analogs of naturally occurring elements (i.e., Cs is an analog of K, Sr is an analog of Ca), knowledge of ecosystem processes was perceived as necessary for understanding the behavior of many, if not all, anthropogenic chemicals introduced into the environment.

As the science of ecology progressed, it became clear that the basic principles and interrelationships on material flow in ecosystems could be applied to understanding the consequence of (i.e., predicting) and potentially resolving environmental issues arising from many man-caused environmental stresses. Reclamation of disturbed lands, development of better practices for fertilizer and herbicide application to agricultural areas, and disposal of hazardous and radioactive waste have all benefited from application of ecosystem concepts to the problem.

In the early 1970's, the US Department of Energy (DOE, the Atomic Energy Commission's successor), began a major research program to understand the behavior and consequences of long-lived actinide elements in the environment. The environmental concern over actinides arose because these materials are associated with the nuclear fuel cycle, they are generally very long lived (i.e., ^{239}Pu has a 24,000 year physical half-life), and they are associated with waste streams generated by the nuclear industry.

Radioecological studies at Los Alamos (1, 2, 3, 4), as well as at many other locations (3, 4), resulted in two rather significant findings. First, greater than 99% of the actinide elements released to the environment deposit in soil and sediment. Secondly, the actinides are tightly bound to soil and sediment. Thus, processes that transport soil and sediment also transport these radionuclides. Studies at Los Alamos have shown that the hydrologic erosion of the soil is a major factor in the translational movement of plutonium deposited on the ground surface (5, 6) and, as a consequence of rain splash of soil, also greatly influences transport of plutonium to plants (7, 8) including vegetable crops (9).

This paper discusses the use of CREAMS (Chemicals, Runoff, and Erosion from Agricultural Management Systems) (10), in developing environmental research programs at Los Alamos and in designing and monitoring the performance of shallow land burial (SLB) sites for low-level radioactive waste (LLW). Discussion is also presented on research needs and ongoing studies involving Los Alamos to supply some of those needs.

ENVIRONMENTAL ISSUES CONCERNING LOW-LEVEL RAIOACTIVE WASTE DISPOSAL

Shallow land burial has been used as a waste disposal technique since the beginning of man. From recorded history, we know that as early as 6000 BC, Neolithic and pre-Elamite civilizations, in what is now Iran, used SLB for disposal of waste (11). In more recent times, as a consequence of expanding populations, industry, and development of energy for resources, concern has arisen about the adequacy of SLB for containing the potentially hazardous waste by-products generated by these activities.

In the United States low-level radioactive waste, such as generated by the nuclear power industry, hospitals, universities, and nuclear research and development facilities, is typically buried in shallow earth excavations of variable size but generally averaging 15-m wide by 15-m deep by about 200-m long (Figure 1). Trenches are filled with waste consisting of a heterogeneous mixture of materials, including laboratory trash, reactor parts, and dismantled buildings. A trench cap, of about 1-2 m thickness, is applied as a final covering to complete isolation of the buried waste from the biosphere.

Over the past 40 years, operating experience at SLB sites for LLW, demonstrates that current practices work fairly well in isolating buried radionuclides although virtually every one of the six commercial and five DOE sites has not proved 100% effective in confining the wastes to the trench environs (12, 13, 14). Of the six commerical LLW sites, three are currently operational; the closure of the three commercial sites is at least partly attributable to unanticipated problems with subsurface water and solute movement. Contamination of groundwater is of particular concern because it is so vital to man and his activities and is not readily subject to corrective measures for removal of pollutants. At the moment, no new low-level sites are being licensed by the US Nuclear Regulatory Commission partly because pending regulations (15) controlling siting, design, monitoring, and closeout of SLB sites have not been finalized.

Two important aspects of the pending regulation that affect SLB for LLW are that the waste must be buried in the unsaturated zone and that the performance of the site must be modeled. The first requirement is relatively easy to satisfy by not selecting sites located in or near groundwater aquifers. The

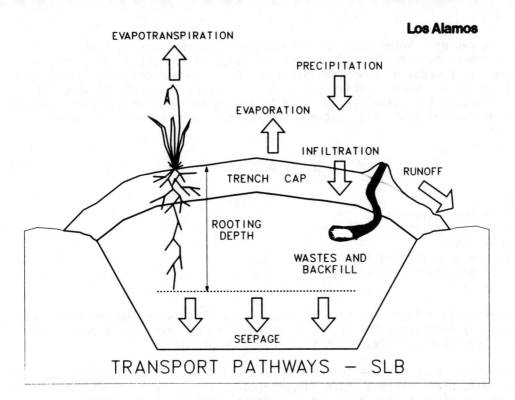

EVAPOTRANSPIRATION

PRECIPITATION

EVAPORATION

INFILTRATION

RUNOFF

TRENCH CAP

ROOTING
DEPTH

WASTES AND
BACKFILL

SEEPAGE

TRANSPORT PATHWAYS — SLB

FIGURE 1.   Hydrologic processes effecting shallow land burial sites.

second requirement is very difficult to satisfy because the models and data
are, as yet, inadequate for predicting site performance, particularly as
influenced by water and solute transport in the unsaturated zone.

If we examine the ecosystem processes that influence site performance with
potential impact on dose to man (Figure 1), we note that water and soil dynamics, as influenced by physical and biological factors, account for most of the
performance-related problems (12, 13). For example, erosion associated with
the runoff from a trench cap can breach the cap and expose waste to the biosphere. Consequently, erosion rates on the cap must be within tolerances that
leave the cap intact over the 100-200 year life of the LLW disposal facility.
Likewise, water that infiltrates into the trench cap can accumulate in the
trench (bathtub effect) and/or percolate in association with solutes into
groundwater. Percolation also enhances subsidence of the trench cap as a
result of decomposition of bulky waste in the trench. Finally, both plants
and animals, in addition to playing an important role in water balance, can
penetrate into the waste and transport radionuclides to the ground surface as
a result of root uptake and/or burrowing activities.

A WATER BALANCE APPROACH FOR SLB

The conceptualization of water, soil, and biological processes that affect SLB
integrity (Figure 1), reveals the interdependence of the physical and biological components of the trench cap (14). Precipitation incident on the site is
subject to losses from runoff, infiltration into the soil, and interception by

the plant canopy. Water that infiltrates into the soil can be lost back to the atmosphere by evaporation (E), transpiration (T) by the plant cover, or as the combined process of evapotranspiration (ET). Water remaining in the soil can be stored or, when it moves below the root zone, percolate or seep (L) into or through the waste trench. The following formulation describes some of the relationships that exist between the various components of the water balance:

$$\frac{dS}{dt} = P-Q-ET-L \tag{1}$$

where
    S = soil moisture
    P = precipitation
    Q = runoff
    ET = evapotranspiration
    L = percolation or seepage
    t = time

The experession relates the rate of change in soil moisture in the trench cap to input (P) and output (Q, ET, L) in units of volume per unit area per unit time, or equivalently, depth per time (e.g., mm per day).

Soil moisture stored in the trench cap is a function of water holding capacity of the soil, plant rooting depth and antecedent and current values for the terms on the right hand side of Eq. 1. Precipitation (P) is a function of the waste site locale and is highly variable in time and space. Runoff (Q) is a function of precipitation, soil characteristics, vegetation cover, soil moisture, and surface management practice, including slope and slope length. Evapotranspiration (ET) is a function of climatic variables, including precipitation, temperature, solar radiation, soil properties, vegetation type, and soil moisture. Percolation is a function of soil moisture and soil properties. Soil erosion and sediment transport are strongly related to precipitation and runoff, and, indirectly, to other terms in Eq. 1. Because plant and animal intrusion into and through a trench cap influences water balance, they also influence infiltration rates and erosion. Although the effects of burrowing animals are not directly represented in Eq. 1, they could be accounted for by the terms influencing soil moisture and erosion.

A water balance approach to resolving SLB issues offers the following advantages:

- it accounts for most of the hydrologic and biological factors that influence site integrity,

- water balance models can be used to screen various modifications in cap design for effect on erosion, percolation, and etc., and

- it can be used to estimate upper boundary conditions for subsurface water flow.

SIMULATING WATER BALANCE

Hydrologic and erosion processes are highly variable in time and space. As such it is not practical to measure them under all possible combinations of soils, climate, topography, biological conditions, and land use.

Consequently, mathematical models are needed to predict those processes under a wide range of conditions.

In response to similar needs in agriculture, the US Department of Agriculture (USDA) developed a reasonably simple computer simulation model called CREAMS (10, 17, 18, 19), which included water balance, erosion/sediment transport, and chemistry components. The model was intended to be useful in agricultural scenarios, without calibration or collecting of extensive site specific data to estimate parameter values, by taking advantage of extensive data sets (10) collected over several decades by USDA and others.

The CREAMS model has been widely used for agricultural applications (20) and recently has been used as a tool in waste management studies (21, 22). Although the model has been applied to SLB sites, it was developed for cropland situations and does not account for some of the physical and biological processes that are specific to non-agricultural ecosystems.

The CREAMS model predicts both water balance and erosion. The water balance components include two options, a daily rainfall model based on the US Soil Conservation Service runoff equation and an infiltration model using rainfall intensity data (23). The soil profile, to the plant rooting depth, is represented by up to seven layers (which could be a multilayered trench cap) each with a given thickness and water storage capacity. The evapotranspiration calculations, which are based on Ritchie's method (24), include soil evaporation and plant transpiration based on mean monthly air temperature, mean monthly solar radiation, and seasonal leaf area index. Flow through the rooting zone is computed using a soil water storage-routing routine and percolation is estimated when soil moisture exceeds field capacity. These calculations maintain a water balance as described in Eq. 1.

Using storm inputs from the hydrology component, the erosion/sediment yield component computes soil detachment, sediment transport and deposition by routing sediment through overland flow and in concentrated flow (19). Gross erosion and sediment yield are computed by sediment size classes, including soil aggregates. A more detailed description of the CREAMS model is presented in Conservation Research Report No. 26 (10), including results of model testing and evaluation, sensitivity analysis and a users manual for preparing model input.

APPLICATION OF CREAMS TO SLB

Evaluating Trench Cap Designs - The following two examples demonstrate the use of CREAMS in SLB to illustrate that water and soil dynamics in and on a trench cap can be modeled and that the ability to predict water balance and erosion in a highly disturbed trench cap can be used to optimize design in order to minimize or prevent unacceptable levels of erosion and/or percolation.

A paper entitled "Use of a State-of-the-Art Model in Generic Designs of Shallow Land Repositories for Low-Level Wastes" (22), describes the use of CREAMS to evaluate the effectiveness of various trench cap configurations in limiting erosion and percolation by varying trench cap soil type, soil depth, vegetative cover, slope steepness, and slope length. Selected results from that paper are presented below.

Model parameters for the simulation study were selected from conditions representative of Los Alamos, New Mexico, a semi-arid location in north-central New Mexico. Los Alamos receives an annual average precipitation input of 46 cm. Mean monthly temperature, solar radiation, and daily rainfall were selected for the 20 year period spanning 1951 to 1970. Trench cap soil parameters were selected from measurements on Hackroy soil (25) and a sandy backfill composed of crushed tuff configured with a uniform slope of 22 m and a slope steepness of 5%. Cover conditions included a non-vegetated soil, a sparse (20%) range grass cover, and a dense (40%) alfalfa cover.

Some significant results of the model simulation were that vegetation plays a key role in controlling both runoff, erosion, and percolation compared with a non-vegetated surface (Figure 2). Although the plant cover increased infiltration into the trench cap by reducing runoff (a 6-fold decrease for the alfalfa cover over that from the bare trench cap), the transpirational losses were sufficiently high to reduce percolation by a factor of at least five over that estimated for the bare cap surface. The great significance of the plant cover in controlling water balance and erosion on the trench cap will be examined in greater detail later in this paper.

Adding a clay layer within the trench cap effectively eliminated percolation compared with the soil/backfill cap design (Figure 3). However, the clay barrier reduced percolation at the expense of increasing runoff by almost 65% because of the higher antecedent moisture in the 15 cm of topsoil.

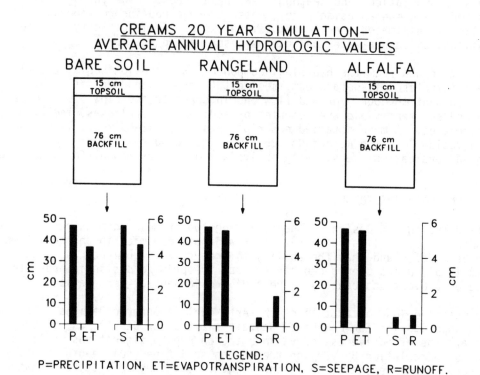

CREAMS 20 YEAR SIMULATION—
AVERAGE ANNUAL HYDROLOGIC VALUES

LEGEND:
P=PRECIPITATION, ET=EVAPOTRANSPIRATION, S=SEEPAGE, R=RUNOFF.

FIGURE 2. Predicted Average annual hydrologic values for a soil over sandy backfill trench cap at Los Alamos, N.M., 1951-1970. (from ref. 22).

Table I. Simulated 20 year average annual water balance for Maxey Flats, Kentucky during 1959–1978 as a function of trench cap management practice.

Practice	Rooting Depth (cm)	Runoff[a] (cm)	ET (cm)	Percolation (cm)	Erosion[b] T/ha
Bare soil	30[c]	49	65	3.8	457[d]
Grass, unmowed	60	23	93	2.1	12
Grass, mowed (3 times/yr)	60	24	93	2.0	15
Grass, mowed (every 3 weeks)	60	34	82	1.3	27
White Pine	90	11	105	0.61	0.0
White Pine	180	9.7	108	0.0	0.0

[a]20 year average precipitation = 117 cm
[b]Sheet and rill erosion only
[c]Depth to which evaporation occurs
[d]Erosion rates greater than about 10T/ha are considered to be excessive in cultivated cropland for maintaining crop productivity.

Table II. Average annual precipitation and percolation (cm water) as a function of cover management practice on SLB trench designs at Maxey Flats, Kentucky.

Year	Annual Precipitation	White Pine	Mowed[a] Grass	Unmowed Grass
1959	112	5.6	5.9	6.9
1960	106	0.0	3.6	1.0
1961	121	0.0	0.0	0.8
1962	125	1.5	3.6	4.2
1963	89	0.0	0.0	0.5
1964	100	0.0	0.0	0.0
1965	117	0.0	0.5	2.8
1966	112	0.0	0.0	0.0
1967	114	0.0	0.0	0.0
1968	111	0.0	0.0	0.0
1969	85	0.0	0.0	0.0
1970	122	0.0	0.0	0.0
1971	114	0.0	0.0	1.4
1972	145	0.03	1.0	2.1
1973	112	0.0	1.1	2.9
1974	144	4.1	5.9	7.9
1975	150	0.8	4.0	5.8
1976	97	0.0	0.9	2.8
1977	111	0.0	0.0	0.0
1978	155	0.0	0.03	2.4
TOTAL	2312	12.0	16.5	41.5

[a]grass cover mowed every 3 weeks

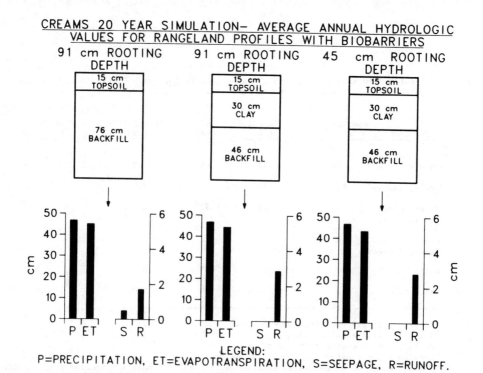

CREAMS 20 YEAR SIMULATION— AVERAGE ANNUAL HYDROLOGIC
VALUES FOR RANGELAND PROFILES WITH BIOBARRIERS

LEGEND:
P=PRECIPITATION, ET=EVAPOTRANSPIRATION, S=SEEPAGE, R=RUNOFF.

FIGURE 3.   Predicted average annual hydrologic values for a topsoil over clay over
backfill trench cap at Los Alamos, New Mexico, 1951-1970 (from ref.
22).

These results demonstrate the highly interactive nature of the ecosystem
processes operating on an SLB site. To further emphasize that fact, studies
at Los Alamos (16) have shown that while a clay moisture barrier may prevent
percolation, the integrity of a saturated bentonite clay barrier (subject to
swelling and shrinking) can be rapidly destroyed by invading plant roots,
which abstract the moisture from the clay, causing it to shrink and crack.

In addition to evaluating the hydrologic response from multi-layered trench
caps, CREAMS is useful in optimizing configurations of specific cap materials.
For example, an important variable in the design of an SLB trench cap is the
thickness of the cap material. Optimizing water storage capacity of the cap
where it can be pumped back to the atmosphere by evapotranspiration, provides
a potentially effective means of preventing percolation.

The effect of increased trench cap thickness on various components of the
water balance for both vegetated and the bare soil conditions is illustrated
in Figure 4. If we focus on seepage or percolation as a function of increasing cap thickness, we see that increasing thickness had little effect under
bare soil conditions, but as thickness increased to about 1 m, seepage below
the vegetated surface reached a minimum dictated by a plant rooting depth of
1 meter. Further increases in cap thickness had little effect on seepage
because the plant roots could not exploit the deeper regions. Increasing

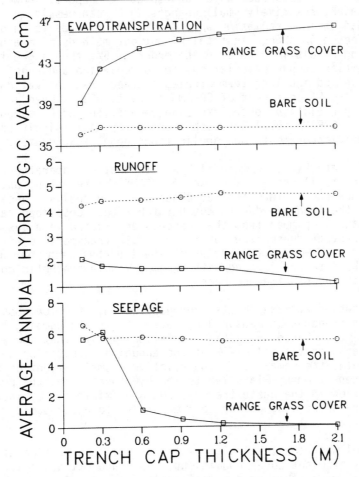

FIGURE 4. Predicted average annual hydrologic values as a function of sandy-loam trench cap thickness at Los Alamos, N.M., 1951-1970.

the cap thickness had little effect on runoff regardless of cover treatment although the very strong influence of vegetation in reducing runoff (compared to the non-vegetated surface) is apparent in Figure 4.

Evaluating Plant Cover Effects - The strong influence of the plant cover in controlling percolation and erosion led to the use of CREAMS to further explore the influence on vegetation type on water balance. Since plants use water at different rates, as controlled by species specific factors including phenology, species or species mixes could be chosen that maintain soil water as low as possible in order to store and eventually transpire precipitation arriving when ET is low (i.e., winter).

An increase of 5% in transpiration, over the year might reduce or eliminate percolation. Because vegetation species use water at different rates through the year (26), careful selection of the plant cover, to optimize trans-pirational losses through the year, may provide an inexpensive, long-term control of subsurface water and solute transport. However, as will be discussed later, relatively small changes in ET can result in large and significant changes in percolation and runoff, which leads to a dilemma. If we need to effect a 5% change in total ET but our methods of measuring and computing ET can result in errors on the order of 5%, then a great deal of uncertainty remains in our calculations for percolation and runoff. Clearly, carefully controlled and long term studies are required to validate ET models in general and the ET component of CREAMS in particular. Even so, the possibility of percolation control through vegetation and ET management as predicted by the CREAMS model has such enormous economic significance that continuing model improvements and applications appears warranted.

Maxey Flats in Kentucky, a commercial LLW site that was operated from 1963 to 1972, was chosen to illustrate the use of CREAMS for selecting optimum plant covers for trench caps. The site was chosen for the analysis because water accumulated in the trenches due to the bathtub effect described earlier. The accumulated water is pumped from the trenches and routed to a gas-fired evaporator to prevent subsurface water and solute transport to offsite areas. About 2.3×10^6 ℓ of water accumulated in the trenches each year until 1982 when most of the site was covered with an impermeable synthetic covering to prevent percolation into the trenches.

Annual precipitation at Maxey Flats averages 121 cm, of which about 3 cm percolated into trenches as measured by water levels and pumping volumes at the site. Estimates based on CREAMS simulations verify that the amount of percolation accounted for only 2-3% of the annual precipitation while runoff and ET distributed the remainder. The relatively small amount of percolation into the trenches at Maxey Flats led to the hypothesis that the problem of water accumulation in the waste trenches could be minimized by increasing runoff, evapotranspiration, and/or soil moisture storage capacity in the trench cover.

The CREAMS model was used to simulate water balance and erosion at Maxey Flats under a variety of plant cover conditions. Soil and climatological data for the CREAMS model were taken from existing site data and from Morehead, Kentucky, a nearby community. The native soil at Maxey Flats is a silt clay-loam that has very poor hydrologic characteristics because of the mechanical mixing of the soil on reapplication as a trench cap. Estimates of evapotranspiration as a function of plant species were made from the literature (27, 28, 29). Slope length was established as 70 m with a convex slope of 2-12% (2% on the peak of the trench cap, 12% on the flanks).

Average annual hydrologic values based on the CREAMS simulation for the 20 year climatologic record (1959-1978) are summarized in Tables I and II. Erosion rates from the vegetated trench cap averaged at least 30 times less than on the bare soil surface regardless of vegetation species used to cover the cap. While the plant cover reduced the amount of runoff over bare soil conditions, it did so at the expense of increased infiltration. However, by adding transpiration as a component of the water balance, the overall effect of the plant cover was to decrease percolation. The size of the decrease in percolation over bare soil conditions appeared to be a strong function of the plant species and cover management practices. For example, frequent mowing of the pasture resulted in less percolation than infrequent mowing or no mowing at all because more precipitation was lost to other sources including runoff. While ET was larger for the grass under the unmowed and infrequently mowed practice, it was not large enough to use the additional water that infiltrated into the cap. Moreover, the evergreen trees appeared to provide greater protection against percolation than the grass cover because of the higher transpiration rates throughout the year and, particularly, during the winter.

Seasonal averages of percolation over the 20 year period (Figure 5) identified late winter as the most critical period for the occurrence of percolation at Maxey Flats. All plant species and management practices prevented percolation during the summer months when evapotranspiration was occurring (Figure 6). However, the only species contributing to transpiration losses of water during the late winter, when the grass species were senescent, was the pine.

The ability to examine the consequences of various design and management practices in SLB sites on long-term hydrologic averages is useful. However, the large year to year variability in precipitation must be considered to design SLB systems that perform under climatic extremes. The data in Table II, on annual average percolation through the trench cap over the 20 year period, shows that the problem of percolation is not an annual occurrence for any of the cover or management practices examined, but is tied closely to fluctuations in annual precipitation. Measurable percolation was predicted in 13 years of the 20 year period for the unmowed grass cover while it was predicted

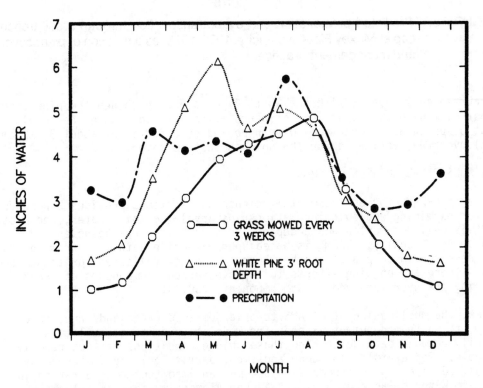

FIGURE 5.   Predicted percolation through a 90 cm thick soil trench cap at Maxey Flats, Kentucky, 1959-1978, as a function of plant cover and management practice.

183

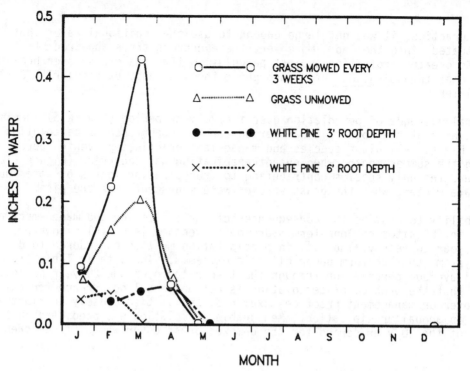

FIGURE 6. Predicted precipitation and evapotranspiration through a soil trench cap at Maxey Flats, Kentucky, 1959-1978, as a function of plant cover and management practice.

to occur in only 5 years with the White Pine cover. Although the pine cover did not eliminate the occurrence of percolation, it did reduce the total amount of percolation over the 20 year period by a factor of about 3 (12 cm vs 42 cm) over that estimated for the unmowed pasture cover.

ADVANTAGES/DISADVANTAGES OF CREAMS

The use of CREAMS for environmental science applications provides a powerful tool for examining ecological relationships involving soil, water, and biota. CREAMS is well-documented and accepted and used by several agencies and groups. Additionally, CREAMS is based upon extensive data sets from agricultural research in croplands and has been tested and validated for these conditions. There are ongoing efforts to improve various components of CREAMS including the hydrology and plant components (30, 31).

However, the application of CREAMS to arid/semiarid rangelands and, specifically, to waste management extends the model beyond its capabilities primarily because data describing those unique conditions are not readily available. For example, in rangelands, but particularly in disturbed systems such as SLB sites, plant succession becomes an important consideration in long-term water balance of a site. At Los Alamos, waste disposal site vegetation changes from initial invader species, such as Russian thistle (Gutierrezia sarothrae) and yellow sweet clover (Melilotus officinalis), to a shrub (Quercus spp, Rhus spp) and evergreen tree (Pinus ponderosa) community within 35 years of site closure. A similar situation exists in terms of changes in soil characteristics as climate and biota contribute to soil weathering. Trench cap soils on Los Alamos waste sites closed in the 1940's have changed from a relatively unproductive sandy material to a silt-loam with improved water retention characteristics and productivity.

Animal interactions, which are not directly represented in the model, are important in altering water balance, erosion and nutrient cycling (32, 33, 34). Studies at Los Alamos (35) suggested that pocket gophers (*Thomomys bottae*), a fossorial rodent that commonly invades disturbed areas, may create significant disturbance of SLB trench caps with potential impact on erosion, percolation, and evapotranspiration (by altering plant density and plant succession) over the 100-200 year life of the site.

More immediate disadvantages of CREAMS for arid/semiarid site use is the lack of parameter estimates for the climatic, edaphic, and biological conditions that exist in these regions. Rainstorms, for example, often occur as intense thundershowers that are highly variable in space and time leading to problems in developing representative precipitation data for the model. A climate generator, reflecting that variability, would greatly facilitate adaption of CREAMS to arid-semiarid rangelands. Additionally, estimates of ET in native species as a function of season are not readily available, primarily because of the difficulty in making such measurements.

RESEARCH TO EXTEND CREAMS TO ARID SITES

Los Alamos is taking an active role in extending CREAMS to arid site SLB by fostering and participating in cooperative research with several groups and agencies. At Los Alamos and Nevada Test Site, USDA-ARS, University of California at Los Angeles, Nevada Applied Ecology Group, DOE, and Los Alamos National Laboratory are conducting joint studies using the USLE erosion plot configuration and the rainfall simulator (36, 37) to develop data and CREAMS model parameters under a variety of conditions ranging from undisturbed desert to semi-arid rangelands to highly disturbed SLB trench cap configurations. Significant results of those studies include:

- emerging data on the importance of soil fauna in mediating erosion and percolation,

- the importance of plant cover in limiting erosion and percolation in SLB,

- the effect of time on the hydrologic and erosional stability of disturbed sites, and

- the dominance of desert pavement (a natural gravel mulch) over vegetation in controlling erosion and percolation in a desert ecosystem.

The first three results are illustrated by the data in Figure 7 showing erosion (normalized to bare soil) from simulated SLB trench caps as a function of biological variables. The purpose of the experiment represented by the data in Figure 7 was to evaluate water balance and erosion as influenced by a 15-20% barley cover, burrowing animals (*Thomomys bottae* - Botts Pocket Gopher), and the interaction of plants and burrowing animals. Three rainfall simulator runs were made during a three month period in 1983 to examine time dependent relationships in the hydrologic response of the various treatments.

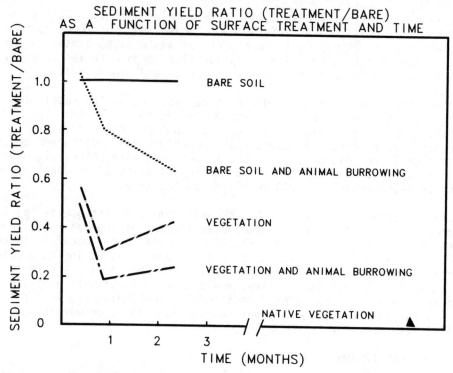

FIGURE 7. Sediment yield ratio (treatment/bare soil) as a function of biological factors and time.

The preliminary data indicated that soil cast to the trench cap surface by pocket gophers steadily decreased erosion to about 60% of bare soil erosion as a consequence of increased infiltration and decreased runoff velocities resulting from the surface soil casts. Erosion from the vegetated plots was 40% of bare soil conditions, after three months, while erosion from the vegetated plots with animals was the lowest at 20% of the bare plot treatment. While plants and animals provide effective control of erosion, they do so at the expense of increasing the infiltration of water into the trench cap.

The data for native vegetation in Figure 7 (solid triangle) is the normalized erosion rate from an undisturbed soil covered by a 15-20% blue gramma (Bouteloua gracilis) grass cover. Erosion rates from the natural plots averaged 2% that of the Barley plots despite the fact that both plots had about the same relative cover. These data suggest that both edaphic and biological successional processes are important in returning disturbed sites to erosional stability. The rates and pathways of disturbed land succession are important research questions that must be answered to improve designs and performance predictions for SLB sites.

Finally, results from ongoing rainfall simulator studies at Nevada Test Site (36) demonstrate the overwhelming importance of desert (or erosion) pavement in controlling runoff and erosion (Table III). In contrast, the sparse vegetation cover plays little direct role in controlling erosion. However, plants do greatly influence antecedent soil moisture, a variable that influences the amount of runoff and erosion.

Table III. Summary of sediment yield data from 12 experimental runoff-erosion plots on the Nevada Test Site. Average sediment yields in g/m^2, two plots per treatment, Spring 1983 (from reference 36).

Location	Treatment	Dry	Wet	Very Wet
Area 11	control, natural	1.3	6.0	12.3
	vegetation removed	1.2	8.2	16.2
	vegetation and erosion pavement removed	82.6	104.0	179.0
Mercury	control, natural	49.5	25.9	29.0
	vegetation removed	61.9	46.2	52.6
	vegetation and erosion pavement removed	555.0	404.0	302.0

SUMMARY AND CONCLUSIONS

The use of CREAMS in arid and semiarid rangelands and for applications to shallow land burial provides a powerful tool for developing management alternatives for land use and waste disposal. Because CREAMS was developed primarily for cultivated agriculture and parameterized in more humid conditions in the eastern half of the United States, it is not directly applicable to arid and semiarid systems without further development and calibration. Model parameters for CREAMS under the agricultural and environmental science applications discussed in this paper for the most part do not exist. Ongoing research by several groups and agencies is intended to rectify these deficiencies.

Perhaps the greatest weakness of CREAMS, or any similar model, is the lack of structure and feedback to account for time dependent changes in physical and biological attributes of a site. Plant and animal succession, and their influence on soils, becomes important over the time scales being considered for SLB performance. Until significant advances are made in our understanding and ability to mathematically describe ecosystem processes, models such as CREAMS will be somewhat limited in scope and utility. However, the need for tools to wisely manage natural resources will continue and represent challenges to the agricultural and environmental scientist to meet those needs.

REFERENCES

1. Hakonson, T. E., Nyhan, J. W. "Ecological Relationships of Plutonium in Southwest Ecosystems," Transuranic Elements in the Environment, W. C. Hanson (ed.), DOE/TIC 22800, US Department of Energy, NTIS, Springfield, Virginia, 1980, pp 403-419.

2. Hakonson, T. E., Nyhan, J. W., Purtymun, W. D. "Accumulation and Transport of Soil Plutonium in Liquid Waste Discharge Areas at Los Alamos," Transuranium Nuclides in the Environment, IAEA-SM-199/99, S. T. I./PUB/410. International Atomic Energy Agency, Vienna, 1976, pp. 175-189.

3. Watters, R. L., Edgington, D. N., Hakonson, T. E., Hanson, W. C., Smith, M. H., Whicker, F. W., Wildung, R. E., "Synthesis of Research Literature." Transuranic Elements in the Environment, W. C. Hanson (ed.), DOE/TIC 22800, US Department of Energy, NTIS, Springfield, Virginia, 1980, pp. 1-44.

4. Watters, R. L., Hakonson, T. E., Lane, L. J., "The Behavior of Actinides in the Environment," Radiochimica Acta, Akademsche Verlagsgesellschaft, Wiesbaden, Vol. 32, 1983, pp. 89-103.

5. Hakonson, T. E., Watters, R. L., Hanson, W. C., "The Transport of Plutonium in Terrestrial Ecosystems," Health Physics, Pergamon Press Ltd., Vol. 40, (January), 1981, pp. 63-69.

6. Lane, L. J., Hakonson, T. E., "Influence of Particle Sorting in Transport of Sediment Associated Contaminants" Waste Management 1982, R. G. Post (ed.), University of Arizona Press, Tucson, Arizona, 1982, pp. 543-557.

7. Dreicer, M., Hakonson, T. E., White, G. C., Whicker, F. W., "Rainsplash as a Mechanism for Soil Contamination of Plant Surfaces," Health Physics, Pergamon Press Ltd., Vo. 46(10), 1984, pp. 177-188.

8. Foster, G. R., White, G. C., Hakonson, T. E., Dreicer, M., "A Model for Splash Retention of Sediment and Soil-Borne Contaminants on Plants," J. Environmental Quality, Madison, Wisconsin, 1984 (in press).

9. White, G. C., Hakonson, T. E., Ahlquist, A. J., "Factors Affecting Radionuclide Availability to Vegetables Grown at Los Alamos." J. Environmental Quality, Madison, Wisconsin, Vol. 10, 1981, pp. 294-299.

10. Knisel, W. G., Jr. (ed.), CREAMS: A Field Scale Model for Chemicals, Runoff, and Erosion from Agricultural Management Systems, USDA-Conservation Research Report No. 26, US Department of Agriculture, Science and Education Administration, May 1980, 640 pp.

11. Langer, W., An Encyclopedia of World History, Houghton Mifflin Co., Boston, Massachusetts, 1968, 1504 pp.

12. Duguid, J. O., "Assessment of DOE Low-Level Radioactive Solid Waste Disposal Storage Activities," Battelle Memorial Institute - 1984, Columbus, Ohio, (November 1977).

13. Jacobs, D. G., Epler, J. S., Rose, R. R., "Identification of Technical Problems Encountered in the Shallow Land Burial of Low-Level Radioactive Wastes," Oak Ridge National Laboratory/SUB-80/136/1, Oak Ridge, Tennessee, (March 1980).

14. Hakonson, T. E., Lane, L. J., Steger, J. G., DePoorter, G. L., "Some Interactive Factors Affecting Trench Cover Integrity On Low-Level Waste Sites," Low-Level Waste Disposal-Site Characterization and Monitoring,

M. G. Yalcintas (ed.), NUREG/CP-0028, CONF-820674, Vol. 2, NTIS, Springfield, Virginia, 1982, pp. 377-400.

15. 10CRF61, Licensing Requirements for Land Disposal of Radioactive Waste,
     (Proposed Rule) Federal Register 46(142), July 1981.

16. Hakonson, T. E., Cline, J. F., Richard, W. H., "Biological Intrusion
     Barriers for Large Volume Waste Disposal Sites," Low-Level Waste
     Disposal Facility Design, Construction and Operating Practices, M. G.
     Yalcintas (ed.), NUREG/CP-0028, CONF-820911, Vol. 3, NTIS, Springfield,
     Virginia, 1983, pp. 289-308.

17. Knisel, W. G., Jr., "Erosion and Sediment Yield Models-An Overview," ASCE
     Watershed Management Symposium, American Society of Civil Engineering,
     July 1980, pp. 141-150.

18. Knisel, W. G., Jr., Foster, G. R., "CREAMS: A System for Evaluating Best
     Management Practices," Economics, Ethics, Ecology: Roots of Production
     Conservation, W. E. Jeske (ed.), Soil Conservation Society of America,
     Ankevy, Iowa, 1981, pp. 174-179.

19. Foster, G. R., Lane, L. J., Nowlin, J. D., Laflen, J. M., Young, R. A.,
     "Estimating Erosion and Sediment Yield on Field-Sized Areas," ASAE Vol.
     24(5), 1981, pp. 1253-1262.

20. Warner, R. C., Dysant, B. C., "Erosion Modeling Approaches for Construc-
     tion Sites," Proc. American Society of Agricultural Engineering Watershed
     Management Symposium, Boise, Idaho, July, 1980.

21. Lane, L. J., Romney, E. M., Hakonson, T. E., "Water Balance Calculations
     and Net Production of Perennial Vegetation in the Northern Mojave
     Desert," J. Range Management, Vol. 137(1), January 1984 pp. 12-18.

22. Nyhan, J. W., Lane, L. J., "Use of a State of the Art Model in Generic
     Designs of Shallow Land Repositories for Low-Level Wastes," Waste Manage-
     ment 1982, R. G. Post (ed.), University of Arizona Press, Tucson,
     Arizona, 1982, pp. 235-244.

23. Smith R. E., Williams, J. R., "Simulation of the Surface Water
     Hydrology," CREAMS: A Field Scale Model for Chemicals, Runoff, and
     Erosion from Agricultural Management Systems, W. G. Knisel (ed.), USDA-
     Conservation Research Report No. 26, May 1980, pp. 13-35.

24. Ritchie, J. T., "A Model for Predicting Evapotranspiration From a Row
     Crop with Incomplete Cover," Water Resources Research, Vol. 8(5), 1972,
     pp. 1204-1213.

25. Nyhan, J. W., Hacker, L. W., Calhoun, T. E., Young, D. L., "Soil Survey
     of Los Alamos County, New Mexico," Los Alamos Scientific Laboratory
     Report LA-6779-MS, June 1978, 102 pp.

26. Saxton, K. E., "Evapotranspiration," Hydrologic Modeling of Small Water-
     sheds, (C. T. Haan, H. P. Johnson, and D. L. Brackensiek, eds.), The
     American Society of Agricultural Engineers, 2950 Niles Road, St. Joseph,
     Michigan, 1982, pp. 229-273.

27. Federer, C. A., "Evapotranspiration (Literature Review, 1971-1974)," Reviews of Geophysics and Space Physics, Vol. 13(3), 1985, pp. 442-445.

28. Ritchie, J. T., Rhoades, E. D., Richardson, C. W., "Calculating Evaporation from Native Grassland Watersheds," Transactions of ASAE, Vol. 19(6), 1976, pp. 1098-1103.

29. Hibbert, A. R., "Forest Treatment Effects on Water Yield," International Symposium on Forest Hydrology, (W. E. Sopper and H. W. Lull, eds.), Pergamon Press, Oxford, pp. 527-543.

30. Wright, J. R. (ed.), Spur- Simulation of Production and Utilization of Rangelend: A Rangelend Model for Management and Research, USDA-ARS Miscellaneous Publication No. 1431, 1983, 120 pp.

31. Renard, K. G., Foster, G. R., "Soil Conservation: Principles of Erosion by Water," Dryland Agriculture, (H. E. Dregne and W. O. Willis, eds.), ASA, CSSA, SSSA Monograph No. 23, 1982, pp. 155-176.

32. Abaturov, B. D., "The Role of Burrowing Animals in the Transport of Mineral Substances in the Soil," Pedobiologia, Vol. 12, 1972, pp. 261-266.

33. Chew, R. W., "The Impact of Small Mammals on Ecosystem Structure and Function," Populations of Small Mammals Under Natural Conditions, Pymatuning Laboratory of Ecology, Special Publication No. 5, May 1976.

34. Ellison, L., "The Pocket Gopher in Relation to Soil Erosion in Mountain Ranges," Ecology, Vol. 27, 1946, pp. 101-114.

35. Hakonson, T. E., Martinez, J. L., "Disturbance of a Low-Level Waste Site Trench Cover by Pocket Gophers," Health Physics, Vol. 42(6), June 1982, pp. 868-871.

36. Bostick, K. V., Simanton, J. R., Lane, L. J., Hakonson, T. E., "Results of Erosion and Contaminant Transport Research on the Nevada Test Site," Nevada Applied Ecology Program Symposium, Las Vegas, Nevada, June 28-30, 1983, (in press).

37. Nyhan, J. W., DePoorter, G. L., Drennon, B. J., Simanton, J. R., Foster, G. R., "Erosion of Earth Covers Used in Shallow Land Burial at Los Alamos," J. Environmental Quality, May-June Issue, 1984, (in press).

VALIDITY AND LIMITATIONS OF DIFFERENT TRANSPORT MODELS WITH PARTICULAR REFERENCE TO SEDIMENT TRANSPORT

Giampaolo Di Silvio^

^ Istituto di Idraulica "G. Poleni" dell'Università di Padova, via Loredan, 20 - 35131 Padova, Italy

ABSTRACT - The consequences of successive simplifications (Fig. 1) made on the transport equations in order to obtain simpler and less detailed models, are examined.

After a review of surface flow models (transport of mass and momentum) with constant density (Fig. 2), the transport models of a neutrally buoyant tracer are discussed (Fig. 3).

Groundwater flow and groundwater tracer transport models are also considered and the approaches for modeling transport processes in stratified flows are mentioned.

Afterwards, a survey of sediment transport models is made, stressing the effects of space and time averaging. Starting from the most general three-dimensional model of suspension, a one-dimensional unsteady model is obtained, aimed at simulating short time/scale processes both with uniform and non-uniform grainsize distribution.

The quasi-steady onedimensional model (averaged over the hydrological cycle) is discussed, considering at first different grainsize classes and then all the classes altogether in the overall transport.

1. INTRODUCTION

The question has often arisen whether modeling is an art or a science.

I remember a discussion, during the 1981 Berkeley conference on "Predictive ability of surface water flow and transport models", between G. Abraham from the Delft Hydraulics Laboratory and J.J. Leendertse from the Rand Corporation of Santa Monica. The presentation of Abraham et al. [1] was a valuable attempt to analyse the process of mathematical modeling, schematized as a logical sequence along which a series of performance controls (verifications) nests appropriate feed-back loops to be followed during model construction. This "scientific" approach was firmly questioned as unrealistic by Leendertse, who saw modeling, instead, as an intuitive and creative activity, and its final product, the model, as the result of a number of fortunate attempts made along an inductive, rather than deductive, process.

I am personally inclined to agree with Leendertse in the sense that the actual pattern by which modelers operate, bears little resemblance to the neat, logical sequence of operations which is often reported in their papers when the model is presented to the scientific community. At the same time, however, I believe that a "logical" effort of rationalizing and systematizing the "creative" process that has led to the model, is always very useful, no matter if made by the modeler himself or by some reviewer. In other words, if we accept the idea of the modeler as an artist, while it is true that the Renaissance painters did not develop the perspective representation as a mathematical exercise of projective geometry, it is true as well that a rigorous understanding of geometric implications of this technique has been essential to their successors, even to modern painters who have overcome the very principles of the perspective.

By this premise, I intend to reassure my confreres the modelers, that my intention is not to question the artistic or creative side of the trade. What I will try to do is to screen the general common denominator in the number of transport models regularly employed in various fields of investigation, and to examine the consequences of the assumptions inherent to each model.

2. SIMPLIFICATION PROCEDURES IN MODELING TRANSPORT PHENOMENA

A frequent source of misunderstanding among modelers themselves or, even more often, among modelers, users and experimenters, is that they may use the same name for different things or call the same thing by different names (typically: 'diffusion constant', 'dispersion coefficient', 'exchange parameter', etc.). Generally confusion arises because of the underlying assumptions of the relative conceptual (mathematical) model which are not clearly stated by the interlocutors; as we shall see later, in fact, the definition of the above mentioned quantities is strictly related to the peculiar structure of the model, that is, to its degree of approximation.

No matter what natural environment (sea, river, aquifer, hillslope, etc.) where transport takes place or what specific tracers transported by water (from planktonic species to soil particles, from pesticides to heat) are considered, the structure of a model is always aimed at simulating the most important mechanisms of transport, keeping in mind the required resolution in space and time of the tracer distribution. In other words, the model structure depends on how detailed the information one looks for should be.

Every transport model, in fact, is the result of a process of simplification performed, in a more or less aware and explicit manner, on two groups of equations: the waterflow equations (portraying the velocity field) and mass balance equations of the tracers (providing the concentration distribution). It is to be noted that water-flow equations are also balance equations, as they express the conservation of water-volume and -momentum.

In general, two kinds of simplification procedure can be applied to the above mentioned equations (Fig. 1).

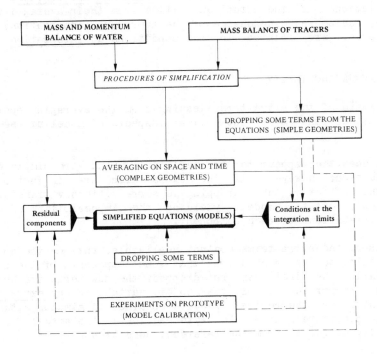

Fig. 1 — Simplification procedures leading to different waterflow and transport models.

The first procedure consists of dropping certain terms from the equations, taking advantage of particular situations of the flow field. Dropping of some terms lead, for example, to steady conditions (absence of time variable), plane or axialsymetric conditions (absence of one space variable), uniform conditions (absence of velocity and concentration gradients in the flow direction), etc. Such a procedure is typical of "fluid mechanics", as simple geometrical and experimental conditions can easily be set in a laboratory.

A second procedure of simplification, instead, consists in decreasing the number of independent variables (i.e. time and space dimensions), by averaging the equations with respect to the variable to be removed. The averaging procedure is most frequent in "hydrology", or hydraulics of natural environments, where the extension and complication of geometries generally do not allow for a detailed and complete description of the phenomena.

After the averaging operation, the resulting equations hold a simpler structure than the original ones, since the number of independent variables is reduced. The same averaging operation, however, is not completely devoid of drawbacks: as it will be better seen in what follows, the non-linear terms of the original equations yield some residual components, embodying the deviations of the actual quantities from their averaged values; other terms, also a consequence of the averaging (integration) over a certain variable, are related to the conditions at integration limits.

3. WHAT DOES AVERAGING IMPLY?

A correct estimate of the extra-terms issuing from the averaging operations represents one of the most most delicate aspects of modeling (see again Fig. 1).

In many instances they appear to be negligible, and are simply dropped from the equation; however, if the extra-terms are in fact somewhat important, the effect of their dropping is surreptitiously transferred to other components of the model, yielding the impossibility of correct physical interpretations of each component.

In other cases, the extra-terms are not negligible, but, on the contrary, they may result to be much larger than some components of the original equations. While the last can be dropped, the new terms are sometimes expressed as a proper funtion of the model variables, by extrapolating the experimental or theoretical results obtained in simpler situations; note that this procedure is tobe taken with caution if the actual flow field is substantially different from the reference condition.

Finally, another way of expressing the extra-terms yielding by the average process, is through model calibration, that is, by resorting to specific measurements on the prototype.

In general, the more simplified (averaged) the model, the more difficult the evaluation of extra-terms without referring to specific measurements on the prototype; in this case, the numerical values obtained through calibration are limited to the particular situation subject to measurements and should not be extended in principle to other circumstances.

At any rate, experimental data required for calibration (and verification) of a transport model, are strictly dependent on the model structure. In other words, according to the averages performed on the equations and according to the treatment of resulting extra-terms, one needs to measure a certain type of physical quantities with certain modalities (frequency in space and time) in order to quantify correctly each "constant", "coefficient" or "parameter" mentioned above. As many dissappointed modelers have learned, a nice collection of experimental data may turn out to be insufficient or even useless, because not dedicated to the requirements of their model.

4. SURFACE FLOW MODELS WITH CONSTANT DENSITY

Let us see, specifically, which types of model can be obtained [4] by succeeding averaging operations performed on the fundamental equations of waterflow (that is, to the transport equations of water volume and momentum) when applied to water bodies without stratification (Fig. 2).

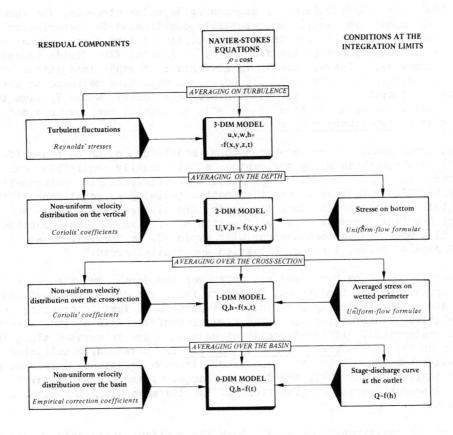

Fig. 2 — Waterflow models (with constant density)yielded by subsequent space— and time—averaging of Navier—Stokes equations.

Water-flow equations with constant density (namely Navier-Stokes and continuity equations) have been subjected for a long time to a number of different simplifications; the most popular and accepted mathematical models employed by the hydraulic engineers during the last 200 years, in fact, may be regarded as more or less simplified forms of the Navier-Stokes equations, although originally developed in a completely independent way.

A fundamental time-averaging operation, aimed at eliminating turbulent fluctuations, lead to the so-called Reynolds equations, where the velocity components u, v, and w (purged by the fluctuations) and the free-surface elevation h appear as a function of the three space coordinates and time (three-dimensional model). Because of the non-linearity of the Navier-Stokes equations, turbulent fluctuations are not completely suppressed in the averaged equations, where they appear in the form of residual components ("virtual" or "Reynolds' stresses"). How those stresses are expressed as a function of the averaged quantities, represents one of the most intriguing problems of fluid mechanics, unfortunately far from being solved in a general form, i.e., for any flow fields.

Apart from the difficulties of expressing Reynolds stresses, the three-dimensional model is still unnecessarily complicated for description of natural environments. A fundamental simplification of three-dimensional equations, definitely acceptable for nearly-horizontal flow fields (coastal areas, estuaries, rivers, etc.), is attained by further integration over the flow depth. In this way the dependent variables decrease to three horizontal components of the depth-averaged velocity, U and V, plus the water level, h, as a function of the sole horizontal components x and y, and of time t (two-dimensional model).

Residual components of the two-dimensional model (depending on non-uniform velocity distributions along the depth), albeit usually negligible, can be estimated by the so-called Coriolis coefficients in uniform-flow conditions. Shear stresses at the bottom (arising as conditions at the lower limit of integration along the depth) are also generally expressed by any uniform-flow formula, as a function of U, V and h. Two-dimensional water flow models have been successfully employed, in the last decades, in many studies of lakes, lagoons, estuaries and coastal areas.

Space averaging can be performed not only over the depth, but also over the entire cross section of a stream. Often, in fact, the distribution of water level and water velocity over the cross section is less interesting that the relative distribution along the direction of the main stream (rivers, canals, estuaries). If two-dimensional equations are integrated along the cross section, the one-dimensional model is obtained, constituted by the well-known De St.Venant equations. In such equations, developed by direct inspection in 1871, the flow discharge, Q, and the water level, h, are the dependent variables (functions of the distance along the stream, x, and the time, t).

Also for the one-dimensional model, both the residual components (Coriolis coefficients) and the conditions at integration limits (average shear stress on wetted perimeter) are set approximately equal to those of corresponding uniform flow conditions. It is to be noted that when the cross section of the stream is not compact (for instance, when the main channel is flanked by flood plains or tidal flats), the residual components (Coriolis coefficients) are far from being negligible; in this case the transversal velocity distribution is to be computed upon partition of the cross section in a number of subsections. The one-dimensional model, either in steady or unsteady conditions, still represents the most suitable approach for a large number of problems of hydraulic engineering.

Although a reliable and handy tool, currently used in the professional practice, one-dimensional models can be further simplified by averaging. The extreme space-averaging operation is certainly that performed over the entire flow-field; this operation leads to zero-dimensional models, also said "models with concentrated parameters" or "black-box models". In those models momentum equation is reduced to a simple algebraic equation (usually a discharge rating curve in a certain section of the flow field, sometimes expressed as a relationship between water discharge and amount of water stored upstream of this section).

The zero-dimensional model has more frequently found its applications in the field of hydrology. Particular applications of the zero-dimensional model are the so-called "hydrological" methods for flood propagation (e.g. the Muskingum method), as well as most of the methods of runoff generation in natural watersheds and artificial drainage systems. Space-averaging may be either performed over the entire basin altogether or over a number of subbasins (multibox model or model with distributed parameters); those subbasins, arranged in cascade or as a network, may, in turn, be connected by components simulating resistance and inertial effects. This approach lead to remarkable simplifications in the treatment of very complicated hydraulic systems, whenever the flow equations can be linearized [16].

The critical point of zero-dimensional models is the requested relationship between level (or storage) and discharge, representing the final relic of the momentum balance equation. Only in few instances (deep reservoir controlled by a gate; regular channel) this relationship can be legitimately predicted from the space-averaging of the one-dimensional equation, by neglecting the residual components. In most cases, however, especially for natural watersheds, the relationships between storage and discharge can be only obtained through calibration; as the simple, often linear, empirical relationships singled out for a certain basin absorbs all the residual components, one can well understand why the ratio between storage and discharge (i.e. "concentration" time) does not appear to be consistent in time and why this quantity is so loosely related to morphological characteristics of the basin. In spite of that, because of its ability to synthesize complicate processes, the one-dimensional model in its various forms is still an unreplaceable tool of investigation in hydrology.

5. TRANSPORT MODELS OF A NEUTRALLY BUOYANT TRACER

A neutrally buoyant tracer (i.e. having the same density of water) does not affect the velocity distribution of the waterstream in which the tracer is dissolved; the transport equation of the tracer can then be solved independently from the waterflow equations, as soon as the velocity field has been determined. On the other hand, the degree of approximation of the tracer model should be comparable to that of the waterflow model; it does not make sense, in fact, to look for high resolution of tracer concentration if a related detail in simulating velocity fields would not be assumed.

Fig. 3 depicts the transport models resulting from time- and space-averages mentioned in the previous paragraph, as well as the extra-terms arising for subsequent averaging operations. The residual components, appearing on the left of the figure, represent an additional transport with respect to the average transport (product of average velocity by average concentration). This additional transport is generally expressed by a diffusion equation, formally identical to the Fick-law molecular diffusion. The proportionality coefficient between transport and concentration gradients (i.e. the "diffusion" or "dispersion" coefficient) depends, of course, on the averaging operation performed on the equation; that is, it depends upon the particular model on which attention is focused.

For example, in the three-dimensional model issuing from averaging over the turbulence, the residual components represent the additional transport along the three coordinates due to the turbulent fluctuations, and the corresponding turbulent diffusion coefficients are computed by assuming that the tracer is transported by the same mechanisms carrying momentum (Reynolds analogy).

The residual components of the two-dimensional (unsteady) model, represent the additional transport due to non-uniform vertical distributions of velocity and concentration ("shear effect"). The dispersion coefficients in this case (computed by Taylor, Elder and others [7] , [17] by extrapolating uniform flow conditions) have nothing in common with and are numerically much larger than the turbulent coefficients mentioned above.

As for the one-dimensional (unsteady) model, the residual components represent the transport due to the non-uniform velocity and concentration distribution over the cross-section. The dispersion coefficient in this case may be computed via Elder's formula (shear effect) only if the channel is very regular. The dispersion coefficient results instead much larger if the channel has a cross-section variable in the longitudinal direction or if there is an alternate flow ("trapping and pumping effect" in tidal estuaries). For a convenient simulation of this mechanism, it is convenient to put explicitly into account the transversal water exchange between main channel and lateral subsections [4] , rather than making an uncertain estimate of a dispersion coefficient.

The series of space-averaging is terminated by the zero-dimensional (unsteady) model, where the transport equation is averaged over the entire flow field. In this model, the basin is schematized as a fully mixed reactor, and the effects of non-uniform distribution over the basin is usually taken into account by a conventional definition of the "average" concentration, or by subdividing the basin in a appropriate [18] number of tanks.

Along with "unsteady" models, providing the time-history of concentration

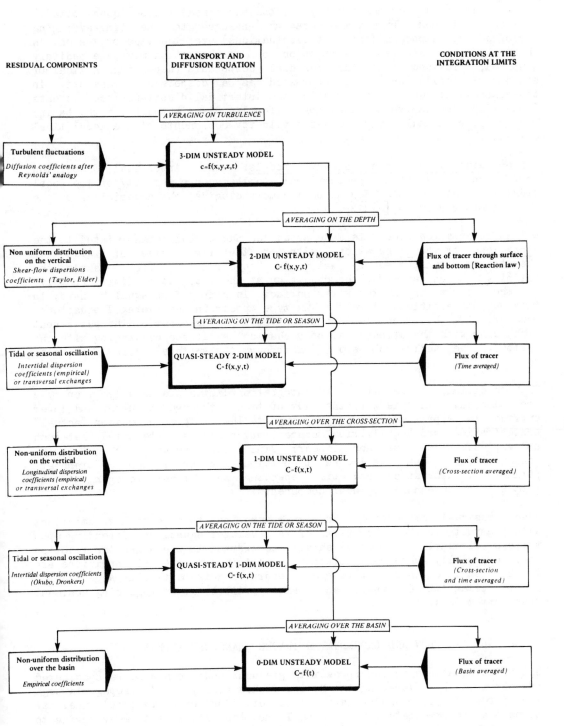

RESIDUAL COMPONENTS

**TRANSPORT AND
DIFFUSION EQUATION**

**CONDITIONS AT THE
INTEGRATION LIMITS**

AVERAGING ON TURBULENCE

Turbulent fluctuations

*Diffusion coefficients after
Reynolds' analogy*

3-DIM UNSTEADY MODEL
c=f(x,y,z,t)

AVERAGING ON THE DEPTH

**Non uniform distribution
on the vertical**

*Shear-flow dispersions
coefficients (Taylor, Elder)*

2-DIM UNSTEADY MODEL
C= f(x,y,t)

**Flux of tracer through surface
and bottom (Reaction law)**

AVERAGING ON THE TIDE OR SEASON

Tidal or seasonal oscillation

*Intertidal dispersion
coefficients (empirical)
or transversal exchanges*

QUASI-STEADY 2-DIM MODEL
C= f(x,y,t)

Flux of tracer
(Time averaged)

AVERAGING OVER THE CROSS-SECTION

**Non-uniform distribution
on the vertical**

*Longitudinal dispersion
coefficients (empirical)
or transversal exchanges*

1-DIM UNSTEADY MODEL
C=f(x,t)

Flux of tracer
(Cross-section averaged)

AVERAGING ON THE TIDE OR SEASON

Tidal or seasonal oscillation

*Intertidal dispersion coefficients
(Okubo, Dronkers)*

QUASI-STEADY 1-DIM MODEL
C= f(x,t)

Flux of tracer
*(Cross-section
and time averaged)*

AVERAGING OVER THE BASIN

**Non-uniform distribution
over the basin**

Empirical coefficients

0-DIM UNSTEADY MODEL
C= f(t)

Flux of tracer
(Basin averaged)

Fig. 3 — Transport models yielded by subsequent space— and time—averaging of the tracer
balance equation.

199

during the tidal or seasonal cycle, one may consider also "quasi-steady" models obtained by a time-averaging (analogue to the time-averaging performed to suppress turbulent fluctuations) over the tide or season. In this way, the spatial distribution of concentration may be expeditely computed, provided that the residual components (depending on tidal or seasonal oscillations) are expressed as a dispersion transport. In two-dimensional quasi-steady models, the intertidal dispersion coefficients (generally non-isotropic and non-uniformly distributed over the basin) result proportional to the squared velocity components of the local tidal current [3].

In one-dimensional quasi-steady models, the intertidal dispersion coefficients are still proportional to the square velocity of the local tidal current but also depend on the morphological characteristics of the channel cross-section [6] [13].

A last observation should be made about the additional terms related to the conditions at the integration limits (on the right side of Fig. 3). If the tracer transported by water is "conservative" (for example, the salinity in an estuary), the condition at the integration limits (namely, at the bottom and at the water surface) is simply flux equal to zero; in this way, no additional term of this type appear in the averaged equations. However, if the tracer is "not conservative" (for example, the water heat exchanging with the atmosphere, or a chemical substance exchanging with the bottom), the additional term (flux through the surface) is far from being zero.

The conditions to be set at the integration limits are usually given by a "reaction law": in this way, the flux of tracer through the upper and lower surface is expressed as proportional to the difference between "actual" concentration and "equilibrium" concentration. It is to be noted that such an expression may be non linear: since both actual and equilibrium concentrations, as well as proportionality coefficient ("transfer" parameter), are space and time depending, any further averaging of this term will produce more residual components.

Those residual components, in their turn, can be more or less important with respect to the other terms of the averaged equations, depending upon various circumstances: care should be exerted in averaging the reaction law, especially when data obtained in certain experimental conditions are to be fed into the model. We shall return on this particular aspect later on in paragraph 8, where an example will be given regarding sediment transport models.

6. GROUNDWATER FLOW AND GROUNDWATER TRACER TRANSPORT MODELS

Waterflow equations in incompressible granular media can also be considered as obtained by averaging the Navier-Stokes equations. As seepage through the pores of a granular material is very slow (creeping flow), inertial terms can be immediately dropped while no need exists of time-averaging to eliminate turbulence fluctuations. Instead, space integration (averaging) is to be made over a small volume of water (yet including a sufficiently large number of grains) in order to suppress deviations and tortuosity of the actual velocity distribution through the pores. Tangential and normal stresses transferred onto the water volume by the granular skeleton (namely the conditions at the integration limits introduced by the averaging) are altogether expressed as a body force by the Darcy law.

The three-dimensional unsteady groundwater flow model obtained in this way can be subsequently subjected to further simplifications as we have seen in the previous paragraphs: averaging over the depth or over the cross section, respectively leads to two-dimensional or onedimensional model. Dupuit's approximation (hydrostatic pressure and uniform velocity distribution over the depth or over the cross sections) corresponds to assume negligible residual components yielded by the respective averages.

Zero-dimensional groundwater flow models are obtained by averaging over the entire flow field. Again, like for surface waterflow models, the momentum equation is radically reduced to a simple (often linear) relationship between storage and discharge, while the mass-balance partial differential equation becomes the well-known ordinary differential equation of a reservoir. Zero-dimensional models (either as a single reservoir or as a network of reservoirs) are frequently used for simple simulations in groundwater hydrology.

In a completely analogous way as made above for the Navier-Stokes equations, the transport and diffusion equation of a tracer can be also averaged over a small volume of water (containing a sufficiently large number of grains), in order to obtain the three-dimensional tracer dispersion model in groundwater.

If the tracer is conservative, the only residual components to be taken into account are those related to the deviations, from the volume-averaged values, of the local water velocity and tracer concentration in the pores. These residual components are put into account as a dispersive transport, being the dispersion coefficient function of the grain and flow characteristics.

If the tracer transported by water is not conservative (for example, a chemical substance [16] that exchanges with the surrounding solid grains), the averaging operation yields another term corresponding to the flux of tracer through the solid surface. Sometimes this term is definitely much larger than the dispersion so that the last can be neglected. Three-dimensional equations of trancer transport by groundwater may be subjected to the cascade of space-and timeaveraging, already seen in the preceeding paragraphs. It is to be noted, in this respect, that the tracer exchange with solid grains being usually expressed by a "reaction law" which is often non linear, attention should be paid, in the following space-and time-averages, to the relative importance of the residual components.

7. TRANSPORT MODELS WITH VARIABLE DENSITY

Whenever the tracer presence in the water yields significant density gradients (for example, when stratification occurs, both in surface and underground flow, due to incomplete mixing of fresh and salt water), waterflow and tracer transport equations cannot be solved independently. Water density, in fact, becomes a function of the local tracer concentration, _via_ another equation (state equation) which is also to be put into account. As a consequence, besides velocity components and tracer concentration, two more space-and time-dependent variables are to be simoultaneously considered, that is, pressure (instead of water level) and water density.

In principle, these equations can be subjected to the very same averaging operations seen before; but in fact, because of the presence of stratifications, new problems and difficulties arise. The three-dimensional model, already very hard to be treated with the hypothesis of constant density, becomes even more difficult because the effect on the "internal" gravity of turbulent stresses is still not very clear.

In any case, whenever the stratification is of paramount importance, the distribution of the physical quantities along the vertical axis shall be privileged. Although few examples of three-dimensional models applied to stratified water environment [11] are available, it has been tried to contain the difficulties by resorting to two-dimensional models on the vertical plane. In this sense, models of stratified fjords and estuaries have been proposed [15], where the usual averaging is made in the transversal direction to the main water flow (unsteady two-dimensional vertical model). The residual components depending on non-uniform distribution along the transversal direction, are often negligible, at least in narrow and deep flow fields; on the other hand, the presence of bays or inlets branching from the main channels, can be accounted by a network of such models.

When stratification plays a minor role, the ordinary averaging over the depth (two-dimensional horizontal model) or over the cross section (one-dimensional horizontal model) may be performed. In this case, however, the stratification tends to increase the longitudinal dispersion coefficient with respect to the flow with homogeneous density [7].

Another type of simplification is sometimes introduced to study the particular type of stratified flows called "jets" and "plumes" (e.g. warm water discharged from cooling plants or fresh water flowing from not-tidal river mouths). At least up to a certain distance from the outlet (the so-called "near-field"), jets and plumes can be studied by a substantially one-dimensional approach, assuming suitable similitude laws for the horizontal and vertical profiles of water velocity and density (integral models). Integral models require of course the empirical estimate of a number of parameters, which conglomerate the effects of averaging [7].

8. MODELS OF SEDIMENT TRANSPORT. AN EXAMPLE OF SUBSEQUENT AVERAGING IN SPACE AND TIME

A particular type of tracer is represented by sediments transported in suspension by water. Whenever sediment concentration is not very large, the velocity field is practically unaffected by the presence of the transported grains, so that waterflow equations and tracer transport equation can be solved independently.

The situation, in this hypothesis, is the same as in par. 5, except that the sediment transport equation differs slightly from that of a neutrally buoyant tracer: in fact, as the density of sediment is larger than that of water, local and instantaneous velocities of stream and grains generally do not coincide. Following the time-averaging aimed at eliminating turbulence fluctuations, this difference is simply represented by a non-zero vertical velocity of the "tracer" (grain fall velocity, depending on the grain size), larger than the flow vertical component (equal to zero or anyhow negligible). At this point, the subsequent simplifications (time- and space-averaging) to which the sediment transport equation may be submitted, are analogous to those portrayed in Fig. 3, with much the same problems to be solved: how to express the residual components (dispersion coefficients) and the conditions at the integration limits (flux of the tracer through the surface and bottom).

In the following, as an example of validity and limitations of different transport models, we shall write down and discuss the equations of sediment transport emerging at each stage of the simplification process.

Three-dimensional unsteady model

Subsequent to the time-averaging on turbulence, the three-dimensional equation of sediment transport is given by:

$$\frac{\partial c}{\partial t} + \frac{\partial cu}{\partial x} + \frac{\partial cv}{\partial y} + \frac{\partial cw}{\partial z} + \frac{\partial}{\partial x}\varepsilon_x\frac{\partial c}{\partial x} + \frac{\partial}{\partial y}\varepsilon_y\frac{\partial c}{\partial y} + \frac{\partial}{\partial z}\varepsilon_z\frac{\partial c}{\partial z} = 0 \tag{1}$$

where $c(x,y,z,t)$ is the sediment concentration; u, v, $w(x,y,z,t)$ are the velocity components of sediment (purged by turbulent fluctuations); while ε_x, ε_y, $\varepsilon_z(x,y,z,t)$ are the turbulent diffusion coefficients in the three directions. As mentioned above, u and v can be assumed as the flow velocity components while $-w$ is the grain fall velocity. Turbulent diffusion coefficients, like for any other tracer, are usually taken equal to the turbulent diffusion coefficients of momentum; in quasi-horizontal flows, terms containing ε_x and ε_y (corresponding to the horizontal turbulent diffusion) are negligible compared to the term containing the vertical gradient of concentration $\partial c/\partial z$.

Equation (1) can be solved by numerical methods, as soon as the water velocity field is known and boundary conditions for the sediment are prescribed. For uniform waterflow conditions along a wide rectangular channel, this has been done by several authors [17], [9], [5], who have made different hypotheses on the conditions to be set at the bottom. A reasonable hypothesis [5] seems to be that the upward sediment flux D through the bottom surface (net entrainment) is given by a reaction equation:

$$D = w \, (c_{aR} - c_a) \tag{2}$$

where w is the grain fall velocity, c_{aR} is the "equilibrium" sediment concentration near to the bottom and c_a is the actual concentration near to the bottom of the sediment transported in suspension by the stream. Equation (2), in fact, states that the net entrainment is given by the sum of an upward flux wc_{aR} (detatchment capacity), compensated by a downward flux wc_a (deposition); net detatchment becomes zero in "equilibrium" conditions, that is when the actual concentration near to the bottom is equal to the equilibrium concentration. Equilibrium conditions (uniform flow for both water and sediment) are provided by the same eq. (1) where time variation and longitudinal gradients are set equal to zero:

$$\frac{\partial cw}{\partial z} + \frac{\partial}{\partial z} \, \varepsilon_z \, \frac{\partial c}{\partial z} = 0 \tag{3}$$

The well-known analytical solutions of eq. (3), as obtained by Lane [10] or Rouse [15], allows the computation of the "equilibrium concentration" near the bottom c_{aR}, as a function of the water flow, the particle fall velocity and the "transport capacity" of the stream (namely the rate of sediment transported in equilibrium conditions, as given by total transport formulae available in literature).

One-dimensional unsteady model

The numerical solution of eq. (1), however, is still cumbersome for practical problems. By averaging (1) over the depth one attains to the two-dimensional horizontal model, still rather complicate. By averaging eq. (1) over the channel cross-section, one attains to the more practical one-dimensional model corresponding to the following equation in x and t:

$$\frac{\partial \hat{C}A}{\partial t} + \frac{\partial CQ}{\partial x} = \bar{D} b \tag{4}$$

where A is the cross-section area perpendicular to the x-axis; $\hat{C} = (1/A)\int c \, dA$ is the cross-section-averaged concentration; $C = (1/Q)\int cu \, dA$; $\bar{D}$ the net entrainment from the bottom, given by eq. (2), averaged over the width b.

Let $R_R = c_{aR}/C_R$ and $R = c_a/C$ be the ratios between near-to-the-bottom concentration and transport-averaged concentration, respectively in equilibrium and non-equilibrium conditions, and let those ratios remain quite constant over the cross-section. If $\hat{C} \cong C$ in the first term of eq. (4) (i.e., legitimately neglect the longitudinal dispersion of sediments with respect to the other terms), eq. (4) may be written:

$$\frac{\partial CA}{\partial t} + \frac{\partial CQ}{\partial x} = wbR_R \left(C_R - \frac{R}{R_R} C \right) \tag{5}$$

If one neglects right away the first term of eq. (5) (that is, the storage variation of sediments in the streamflow), one may write:

$$\frac{\partial T}{\partial x} = \frac{wbR_R}{Q} \left(T_C - \frac{R}{R_R} T \right) \tag{5'}$$

where $T = CQ$ is the actual rate of transport and $T_c = C_R Q$ is the the transport capacity, that is the rate of transport in equilibrium conditions.

Eq. (5') is similar to non-equilibrium equations for solid transport obtained by other authors. By introducing the "detatchment capacity" of the stream, i.e. the sediment entrainment by clear water:

$$D_c = \frac{wbR_R}{Q} T_c,$$

one obtains the equation proposed by Foster and Meyer [8] for upland areas, except for the presence of the ratios R and R_R; equations coincide for $R = R_R = 1$, that is assuming a uniform vertical distribution of concentration along the depth (which is possibly correct in sediment transport by overland flow along hillslopes of bare soil, since raindrop splash substantially increases the turbulence).

Eq. (6) is also similar to non-equilibrium equation proposed by Tsubaki and Saito [19] and, in a more general form, by Nakagawa and Tsujimoto [12] for bed-load transport. In both cases equations tend to coincide to (6) by introducing as "average step lenght" of jumping particles the expression:

$$\lambda = \frac{Q}{wbR},$$

also depending on the ratio R, that is on the vertical distribution of particle concentration.

However it is apparent that, in general, the residual terms issuing from averaging over the cross-section (basically, over the depth) and represented by the ratios R and R_R, are not negligeable. For suspended transport in channels R_R (in equilibrium condition) can be computed by resorting to the mentioned solutions of eq. (3). The analogous ratio R in nonequilibrium conditions can be assumed approximately given by [5]:

$$R = R_R \frac{C_R + C}{2C} \tag{6}$$

or by:

$$R = R_R \sqrt{\frac{C_R}{C}}. \tag{7}$$

Both expressions (6) and (7) indicate that sediment concentration is more uniformly distributed than in equilibrium conditions $(R < R_R)$ when the stream is overloaded by sediments $(C > C_R)$ and viceversa; expressions (6) and (7) give the same results when $C \to C_R$.

By introducing eq. (6) in eq. (5), the last can be written:

$$\frac{\partial CA}{\partial t} + \frac{\partial CQ}{\partial x} = \frac{wbR_R}{2} (C_R - C) \tag{8}$$

Integration of eq. (8) provides the transport concentration $C(x,t)$ along the stream and during the time for prescribed boundary conditions. Equilibrium parameters R_R and $C_R = T_c/Q$, depending on grain diameter and local hydraulic characteristics, are computed by the already mentioned formulae obtained in uniform flow conditions.

One-dimensional unsteady model for non-uniform size sediment.

Equation (8) implies that the sediment forming the bottom and the sediment transported in suspension by the streamflow does have the same uniform grain size. If the transported sediment is formed by different grain-size classes, eq. (8) for the j-th class should be written:

$$\frac{\partial c_j A}{\partial t} + \frac{\partial c_j Q}{\partial x} = \frac{w_j bR_{Rj}}{2} (r_j c_{Rj} - c_j) \tag{9}$$

when r_j is the percentage of the j-th class of sediment present in the upper layer of the bed. As the right hand side of eq. (9) represents the net flux of sediment belonging to the j-th class exchanged with the bottom, it may be expressed as well as the contribution of j-th class of sediment to the rate of decrease of the bottom elevation:

$$- \frac{\partial z_j}{\partial t} = \frac{w_j bR_{Rj}}{2} (r_j c_{Rj} - c_j) \tag{10}$$

Finally, the balance of the same j-th class within the upper layer of the bed can be written as:

$$\frac{\partial z_j}{\partial t} = \delta \frac{\partial r_j}{\partial t} + r_j \frac{\partial z}{\partial t} \tag{11}$$

where δ represents the thickness of the upper layer where sediment mixing takes place and $\partial z/\partial t = \sum(\partial z_i/\partial t)$ is the total increase- or decrease-rate of the bottom elevation, due to all the grain size classes.

Equations (9), (10) and (11), integrated in space and time, provide [2] for each grain-size class : the transport concentration, C_j, the percentage present in the bed, r_j, and the bottom variation, $\Delta z \triangleq \sum \Delta z_j$.

One-dimensional quasi-steady model.

The unsteady one-dimensional model for non-uniform size sediments, corresponding to equations (9), (10) and (11), should be used to simulate morphological processes in natural or artificial watercourses, having relatively short time- and space-scales (for example, the localized and temporary modifications occuring during the passage of a flood wave; the filling-up of an artificial trench dug in the riverbed, etc.).

For morphological processes taking years or decades to be significant (as, for instance, the general aggradation or degradation of a river), a detailed time-history (say at one-hour steps) is both unnecessary and terribly expensive. A more practical model, in these cases, is obtained by averaging eqs. (9), (10) and (11) over a period of time (typically one year), long enough to eliminate the fluctuations of the hydrological cycle.

By combining eqs. (9) and (10) and by averaging over the year, one obtains:

$$\frac{\partial \overline{c_j Q}}{\partial x} = - \frac{\partial \overline{z}_j}{\partial t} \tag{11}$$

which states that any averaged (annual) variation of the bottom is related to the spatial variation of the averaged (annual) sediment discharge. Now, let us try to evaluate this averaged sediment discharge. By averaging eq. (10) one obtains:

$$\overline{C_j Q} = \overline{r}_j \ \overline{C_{Rj} Q} + \frac{\partial \overline{Z}}{\partial t} \overline{\left(\frac{2Q}{w_j b R_{Rj}}\right)} + \overline{r_j'(C_{Rj} Q)'} + \overline{\left(\frac{\partial Z_j}{\partial t}\right)' \left(\frac{2Q}{w_j b R_{Rj}}\right)'} \quad ; \tag{12}$$

(i)   (ii)   (iii)   (iv)

by usual notation, in this formula the averaged value of a function is designated by the symbol (-), while the deviation of the same function from the averaged value is designated by the symbol ('). Equation (12) states that the averaged (annual) value of sediment transport through a certain section, belonging the j-th class, is given by the sum of four terms (i), (ii), (iii) and (iv). The first term (i), definitely the most important, represents the product of the averaged (annual) transport capacity $\overline{T}_{cj}$ = $\overline{C_{Rj} Q}$ (computed as if the bottom on that section was composed only by the R j-th class), multiplied by the averaged percentage r_j of the j-th class on the bottom. The second term (ii), depends on the averaged aggradation or degradation rate and is always very small (zero in "equilibrated" streams). The third (iii) and fourth (iv) terms depend on the vicissitudes of the solid transport during the hydrological cycle and, in principle, may be present also for equilibrated stream. While the fourth (iv) term can be probably neglected, the third (iii) term seems to have a clear dispersive character and could be expressed as an extra transport proportional to the negative gradient $-\partial \bar{r}_j/\partial x$. The annual transport of the j-th class, in general, can be written as:

$$\overline{C_j Q} = \bar{r}_j \cdot \overline{C_{Rj} Q} + K \frac{\partial \bar{r}_j}{\partial x} \tag{14}$$

where the dispersion constant K essentially depends on the hydrological regime of the watercourse (K = 0 for a constant discharge).

It is interesting to observe that only the presence of the dispersion term explains the non-uniform longitudinal distribution of bottom sediment which is normally found in "equilibrated" natural watercourses (for example, where the water velocity increases, also a larger percentage of coarser material is found). In fact, would the dispersion term become zero (K = 0 in eq. 14), both the averaged percentage $\bar{r}_j$ and the averaged transport capacity of each fraction $\overline{QC}_{Rj}$ should remain strictly constant. Indeed, for an equilibrated stream eq. (11) gives $\partial \overline{C_j Q}/\partial x = 0$; that is, from eq. (14):

$$\frac{\partial \bar{r}_j}{\partial x} = - \frac{\bar{r}_j}{\overline{C_{Rj} Q}} \frac{\partial \overline{C_j Q}}{\partial x}; \tag{15}$$

and, being $\sum \bar{r}_j = 1$, eq. (15) is only satisfied if $\partial \bar{r}_j/\partial x = 0$ and $\partial \overline{C_j Q}/\partial x = 0$.

A correct estimate of the dispersion constant K in eq. (14) seems therefore to be necessary if the long term evolution of the sediment size distribution along a river, $r_j(x,t)$, is to be predicted.

However, if interest is not cast in the composition but just in the overall quantity of the sediment, another average can be made over the grain size distribution. By summing up eq. (14) for all the grain size classes, and assuming that K is constant for all the classes, the total transport is obtained as a function of the averaged transport capacity and the averaged local composition of the bottom:

$$\sum \overline{C_j Q} = \sum \bar{r}_j \cdot \overline{C_{Rj} Q} \tag{16}$$

The long-term evolution of the bottom elevation is given by summing up eq. (11) for all grain classes and introducing eq. (16):

$$\frac{\partial \bar{z}}{\partial t} = - \frac{\partial}{\partial x} \left(\sum \bar{r}_j \cdot \overline{C_{Rj} Q} \right) \tag{17}$$

Computation of the annual sediment transport of a stream.

The annual continuity equation (17) can be solved as soon as the averaged (annual) local composition of the bottom, $\bar{r}_j$, is measured and the averaged (annual) transport capacity, $\overline{C_{Rj}Q}$, is computed as a function of the hydraulic characteristics of the stream.

The "instantaneous" transport capacity for each grain size, $T_{cj} = C_{Rj}Q$, is provided by any experimental transport formula of literature, as a function of the flow characteristics. Within a wide range of grain diameter and water velocities, however, all transport formulae can be written in the form:

$$T_{cj} = \alpha \frac{Q^m i^n}{b^p d^q} \qquad (18)$$

where Q is the instantaneous water discharge, i is the energy slope, b is the channel width and d_j is the grain size of the j-th class; the coefficient α and the exponents $m \cong 2$, $n \cong 2$, $p \cong 1$ and $q \cong 1$, basically depend on the range of grain diameter and waterflow velocity occuring in the stream.

If one assumes that along the hydrological cycle all the parameters remain reasonably constant, exception made for water discharge Q, the annual sediment transport of the stream for the j-th class can be computed as:

$$\overline{r_j \cdot C_{Rj}Q} = \overline{r}_j \cdot \int_{year} T_{cj} \, dt = \frac{\alpha \, \overline{r}_j \, i^n}{m \, d_j^q \, b^p} \left(Q_o^{m-1} V_o \right) \qquad (19)$$

where Q_o and V_o respectively represent the maximum flood discharge of the year and the total yearly runoff. By assuming again that the parameters remain constant for all the grain classes, the total sediment transport of the stream is given by:

$$\sum C_j Q = \sum \overline{r}_j \cdot \overline{C_{Rj}Q} = \sum \frac{\overline{r}_j}{d^q} \cdot \frac{\alpha \, i^n}{m \, b^p} \left(Q_o^{m-1} V_o \right) \qquad (20)$$

According to eq. (20), the total annual transport depends on :

(i) a sedimentological parameter, corresponding to the "equivalent" diameter:

$$\overline{d} = \left[1 / \sum (\overline{r}_j / d_j^q) \right]^{1/q}, \qquad (21)$$

which depends on the grain size bottom composition (averaged over the year);

(ii) a morphometric parameter:

$$\left(i^n / b^p \right), \qquad (22)$$

depending on the local water slope and bottom width;

(iii) a hydrological parameter:

$$\frac{\alpha}{m} (Q^{m-1} \, V_o),$$
(23)

depending on the annual flood discharge and runoff.
Contrary to the "instantaneous" solid transport, hardly correlated to the "instantaneous" waterflow, the averaged (annual) transport is strictly dependent on the annual flood peak and runoff (hydrological factor).

All the parameters α, m, n, p and q can be obtained by calibrating eq. (20) against measurements in the same river or in similar watercourses. The same parameters may also be introduced in eq. (18) for calculating the "instantaneous" transport capacity of each grain-size class, to be used for non-averaged models (short time- and space-scales simulations).

9. CONCLUSIONS

More or less simplified models can be obtained by time- and space-averaging performed on the transport equations.

Since everything has a price, simplifications by averaging are paid not only with a minor detail of the information provided by the model but also with the presence of extra-terms which may or may not be neglected or predicted as a function of the averaged values. This depends on the specific transport process considered and on the particular flow field in which the process occurs. Only when extra-terms can neither be neglected nor predicted, their effects will be taken into account through overall empirical coefficients obtained by calibration of the model, namely by comparison with proper data collected on the prototype.

The choice of the model should hence be made keeping in mind both the required time- and space-resolution of the model output and the available means of treating the extra-terms issuing from the simplifications.

It is to be recalled,however,that different models (more or less simplified) have often been applied to the same environment to investigate phenomena at various scale. Different models have also been built for mutual support: sometimes, a small number of tests run on the more sophisticated models have been directed to check the simpler and more economical ones; in other cases, the less detailed models (checked on the basis of available experimental data), have been used to feed the more detailed ones on inputs (e.g., boundary conditions) not directly provided by the experiments.

The best use of different models, in summary, is made when their respective abilities and drawbacks are fully recognized.

10. REFERENCES

1 Abraham, G.; van Os, A.G. and Verboom, G.K., 1981. Mathematical Modeling of Flows and Transport of Conservative Substances: Requirements for Predictive Ability. Transport Models for Inland and Coastal Waters, H.B. Fischer ed., Academic Press.

2 Armanini, A. and Di Silvio, G., 1982. Sudden Morphological Modifications along a Mountain River Simulated by a Mathematical Model. Thrid Congress of the Asian and pacific Regional Division of I.A.H.R., Bandung (Indonesia).

3 Avanzi, C. e Fiorillo, G., 1983. Modello di dispersione intermareale della Laguna di Venezia. Memoria II-9, Convegno del Magistrato alle Acque di Venezia, Venezia, 10-12 giugno.

4 Di Silvio, G., 1978. Modelli matematici per lo studio della propagazione di onde lunghe e del trasporto di materia nei corsi d'acqua e nelle zone costiere. Metodologie Numeriche per la Soluzione di Equazioni Differenziali della Idrologia e della Idraulica, Patron Editore, Bologna.

5 Di Silvio, G. and Armanini, A., 1981. Influence of the Upstream Boundary Conditions in the Erosion Deposition Processes in Open Channels. Paper n. 22, Subj. A(d), Int. Ass. for Hydraulic Res. (I.A.H.R.), XIX Congress, New Delhi, India.

6 Dronkers, J., 1978. Longitudinal Dispersion in Shallow Well-mixed Estuaries. J. of Coastal Eng.,

7 Fischer, H.; List, E.J.; Koh, R.C.Y.; Imberger, J. and Brooks, N., 1979. Mixing in Inland and Coastal Waters. Academic Press.

8 Foster, G.R. and Meyer, L.D., 1972. A Closed-form Soil Erosion Equation for Upland Areas. In: Sediment Symposium to Honour Prof. H.A. Einstein, ed. by H.W. Shen, Colorado State University, Fort Collins, Colorado.

9 Kerssens, P.M.J, 1974. Adaptation Lenght of the Vertical Suspensed Sand Profiles. Techn. Hogeschool, Delft (in Dutch); mentioned in P.Ph. Jansen et al.: Principles of River Engineering; Pitman, 1979.

10 Lane, E.W. and Kalinske, A.A., 1941. Engineering Calculations of Suspended Sediment. Trans. Am. Geophys. Un., 22.

11 Leendertse, J.J. et al., 1973-77. A Three-dimensional Model for Estuaries and Coastal Seas. Five volumes, The RandCorporation, Santa Monica, Cal.

12 Nakagawa, H. and Tsujimoto, T., 1980. Sand Bed Instability Due to Bed Load Motion. Proc. ASCE, Journ. of Hydr. Div., HY12, 15936.

13 Okubo, A., 1973. Effect of Shoreline Irregularities on Streamwise Dispersion in Estuaries and Other Embayements. Neth. J. of Sea Res., 6, 213.

15 Rouse, H., 1973. Modern Conceptions of the Mechanics of Fluid, Turbulence. Trans. ASCE, vol. 102.

14 Perrels, P.A.J. and Karelse, M., 1981. A Two-dimensional, Laterally Averaged Model for Salt Intrusion in Estuaries. Transport Models for Inland and Coastal Water, H. Fisher ed., Academic Press.

16 Rinaldo, A. and Marani, A., 1984. Runoff and Receiving Water Models for NSP Discharge into the Venice Lagoon. Int. Conf. on Agriculture and Environment, Venezia, 11-12 giugno.

17 Sayre, W.W., 1975. Dispersion of Mass in Open-channel Flow. Hydrology Paper 75, Colorado State University.

19 Tsubaki, T. and Saito, T., 1967. Criteria forSand Waves in Erodible Bed Channels. (in Japanese); mentioned in: Alluvial Bedform Analysis I, by Hayashi, T. and Ozaki, S., Application of Stochastic Processes in Sediment Transport, ed. by H.W. Shen and H. Kikkawa, Water Resources Publications (Littleton, Col.,USA).

18 Shanahan, P. and Harleman, R.F., 1984. Transport in lake water quality modeling. Journal of Environmental Engineering, ASCE, (101)1: 42-57.

SELECTION AND APPLICATION OF MODELS FOR NONPOINT SOURCE POLLUTION AND RESOURCE CONSERVATION

Ralph A. Leonard and Walter G. Knisel, Jr.
USDA-ARS, Southeast Watershed Research Laboratory
P. O. Box 946, Tifton, Georgia, 31793, USA

ABSTRACT

Some guidelines are given for consideration in selecting a model for evaluating nonpoint source pollution and for resource conservation. An example application of the CREAMS model is presented to indicate how the model might be used to analyze pesticide losses in runoff and with sediment for two different soils with pesticides of different characteristics.

INTRODUCTION

Mathematical models to assess nonpoint source pollution and evaluate the effects of management practices are needed to adequately respond to the water quality legislation in the United States during the past 15 years. Action agencies must assess nonpoint source pollution from agricultural areas, identify problem areas, and develop conservation practices to reduce or minimize sediment and chemical losses from fields where potential problems exist. Monitoring every field or farm to measure pollutant movement is impossible, and landowners need to know benefits before they apply conservation practices. Only through the use of models can pollutant movement be assessed and conservation practices be effectively planned.

Passage of the Clean Waters Act, PL 92-500, by the U.S. Congress in 1972 resulted in the need for mathematical models to evaluate nonpoint source pollution from diffuse agricultural areas. This need produced a proliferation of model development. Although hydrology and erosion models were available, few models for chemical transport were available. Models for evaluating nonpoint source pollution have been assembled, oftentimes by "piggy-backing" of erosion and chemical components onto hydrology models for both field- and basin-sized areas. This is generally done without total conceptualization of the interactive processes involved, and therefore components may be used that do not provide the necessary quantification for other components.

Models are formulations of processes and logic as represented by the modeler. The formulations may be representations of simple processes or combinations of processes as the modeler deems necessary to solve the problem. Sometimes parameters may be included that are not physically measurable.

The burden is upon the model user to select from several potential models the one which will best represent the conditions, practices, and desired results for his specific problem. The purpose of this paper is to: (a) describe some criteria for model selection, (b) present an overview of the CREAMS model (28), (c) present details of the pesticide component of CREAMS, and (d) give a representative application of how CREAMS may be used.

REVIEW OF MODELS

Crawford and Donigian (3) developed the pesticide transport and runoff (PTR) model to estimate runoff, erosion, and pesticide losses from field-sized areas. The hydrologic component of the PTR model is the Stanford watershed model (4), and the erosion component was developed by Negev (20). The Stanford watershed model was one of the first computer simulation hydrologic models and was developed for basin-sized areas.

Donigian and Crawford (7) incorporated a plant nutrient component with the basic PTR model to develop the agricultural runoff model (ARM). The hydrology, erosion, and pesticides components are the same as the PTR model. The ARM is also for field-sized areas. Both the PTR and ARM models require data for calibration.

Frere et al. (12) developed an agricultural chemical transport model (ACTMO) to estimate runoff, sediment yield, and plant nutrients from field- and basin-sized areas. The hydrology component is the USDA Hydrograph Lab model (14), which is based on an infiltration concept. The erosion component is based on the rill and interrill erosion concepts and USLE modifications developed by Foster et al. (11). The ACTMO model does not require calibration.

Bruce et al. (2) developed an event model (WASCH) to estimate runoff, erosion, and pesticide losses from field-sized areas for single runoff-producing storms. This model requires calibration to the specific site of consideration.

Beasley et al. (1) developed the ANSWERS model to estimate runoff, erosion, and sediment transport from basin-sized areas. The model does not have a chemical component. It has been used to identify sources of erosion and areas of deposition within the basin.

In 1978, the U. S. Department of Agriculture, Agricultural Research Service, began a national project to develop mathematical models for evaluating non-point source pollution. The first development was the CREAMS model, published in 1980 (28). CREAMS is a relatively simple, computer efficient, physically based model for field-size areas. Scientists from several disciplines worked together and interacted to develop a cohesive model of physical processes. CREAMS does not require observed data for parameter calibration, although observed data are beneficial to adjust the sensitive parameters such that good agreement can be obtained with simulation results. CREAMS is especially useful in making relative comparisons of pollutant loads from alternate management practices.

CRITERIA FOR MODEL SELECTION

Model users should carefully consider attributes of different models relative to the specific problem to determine which can provide the desired results.

Some considerations include model purpose, representation, data requirements and availablity, ease of parameter estimation, and both ease and cost of simulation.

Some models were developed to simulate response for a single, design-type storm. This may be adequate when considering only surface runoff and sediment yield, but a design storm may not be one that results in a high percolation loss of nitrate or high pesticide loss. For nonpoint source pollutants, continuous (or daily) simulation over a relatively long climatic record is recommended to examine risk analysis.

Often, modelers think that physical processes are the same regardless of scale, and a model can be applied equally as well for fields as for basins. While the processes are the same, the sensitivity of processes vary with size of area. For example, the infiltration process and the simulation time increment in the model are much more significant for small (plot size) areas than for large basins. A user should consider the scope of a model and the problem when making a selection. Another consideration is whether or not a model is sensitive to changes in management practices. If not, there is little chance of success in selecting among alternate practices for nonpoint source pollution control.

Models that require calibration to evaluate parameter values are generally calibrated for a specific site and practice. If relationships for the phys-ical processes are not carefully formulated, parameter values can be seriously distorted. Calibration of a model with data for a specific site and manage-ment practice may give erroneous results when the model is applied to a different site or management practice without recalibration. Therefore, minimization of the need for calibration is desirable. A model is most useful when values for its parameters are readily available as functions of easily measured features of the site and practice being evaluated. Generally ob-served data are not available for problem locations, and particularly not for different management practices. If such data are available, then it is not necessary to use a model.

Computer costs may not be important when considering a single model simulation run. However, the model user may be interested in several alternate manage-ment practices so a farmer can select the one that controls pollution and is economically feasible within his given constraints. Repetitive 20-year simulations for five to eight alternatives may be prohibitive in computer cost.

These are only a few of the considerations a user must give when selecting a model. Obviously there are others, but space does not permit examination of a complete list.

CREAMS MODEL STRUCTURE

CREAMS consists of three major components: hydrology, erosion/sedimentation, and chemistry. The hydrology component estimates runoff volume and peak rate, infiltration, evapotranspiration, soil water content, and percolation on a daily basis, or if detailed precipitation data are available, calculates infiltration at histogram breakpoints. The erosion component estimates erosion and sediment yield including particle-size distribution at the edge of the field on a daily basis. The chemistry component includes elements for plant nutrients and pesticides. Stormloads and average concentrations of adsorbed and dissolved chemicals in the runoff, sediment, and percolate fractions are estimated.

The Hydrology Component

This component consists of two options, depending upon availability of rainfall data. Option 1 estimates storm runoff when only daily rainfall data are available. If hourly or breakpoint (time-intensity) rainfall data are available, Option 2 estimates storm runoff by an infiltration-based method.

Option 1 - Williams and LaSeur (31) adapted the Soil Conservation Service (29) curve number method for simulation of daily runoff. The method relates direct runoff to daily rainfall as a function of curve number representing soil type, cover, management practice, and antecedent rainfall. The relationship of runoff, Q, to rainfall, P, is:

$$Q = \frac{(P-0.2S)^2}{P + 0.8S} \tag{1}$$

where S is a retention parameter related to soil moisture and curve number. An equation for water balance is used to estimate soil moisture from:

$$SM_t = SM + P - Q - ET - O \tag{2}$$

where SM is initial soil moisture, SM_t is soil moisture at day t, P is precipitation, Q is runoff, ET is evapotranspiration, and O is percolation below the root zone.

The percolate component uses a storage routing technique to estimate flow through the root zone. The root zone is divided into seven layers--the first layer is 1/36 of the total root zone depth, the second layer 5/36 of the total, and the remaining layers, all equal in thickness, are 1/6 of the root zone depth. The top layer is approximately equivalent to the chemically active surface layer and the layer where interrill erosion occurs. Percolation from a layer occurs when soil moisture exceeds field capacity, and time of percolation depends on saturated hydraulic conductivity.

The peak rate of runoff, q_p, (required in the erosion model) is estimated by the empirical relationship (28) relating size of drainage area, mainstem channel slope, daily runoff volume, and the field length-width ratio.

Option 2 - The infiltration model is based on the Green and Ampt equation (13, 26). The concept assumes some soil water initially in a surface infiltration-control layer; and when rainfall begins, the soil water content in the control layer approaches saturation at which time surface ponding occurs, t_p.

The amount of rain that has infiltrated by the time of ponding is analogous to initial abstraction in the SCS curve number model (Option 1), but is also a function of rainfall intensity. After the time of ponding, water is assumed to move downward as a sharply defined wetting front with a characteristic capillary tension as the principle driving force. The infiltration curve is approximated to give the infiltrated depth ΔF in a time interval, Δt, as:

$$\Delta F = \{4A(GD + F) + (F - A)^2\}^{1/2} + A - F, \tag{3}$$

where $A = K_s t_i/2$, $D = \theta_s - \theta_i$, θ_s is water content at saturation, θ_i is initial water content, G is the effective capillary tension of the soil, and K_s is the effective saturated conductivity. The average infiltration rate is used to determine incremental runoff (rainfall excess) rate during a time interval and summed for the storm to give total runoff. The infiltration-based model has three parameters: G, θ_s, and K_s.

Percolation is estimated as in Option 1, except that a single layer below the infiltration control layer represents the root zone. Percolation is calculated using average profile soil water content above field capacity and the saturated hydraulic conductivity, K_s. Peak rate of runoff is estimated by attenuating the rainfall excess using the kinematic wave model with parameter values to account for nonuniform steepness and roughness along the slope (33).

Evapotranspiration - The evapotranspiration (ET) element of the hydrology component is the same for both options. The ET model, developed by Ritchie (24), calculates soil and plant evaporation separately. Evaporation, based on heat flux, is a function of daily net solar radiation and mean daily temperature, which are interpolated from a Fourier series fitted to mean monthly radiation and temperature (19). Soil evaporation is calculated in two stages. In the first, soil evaporation is limited only by available energy and is equal to potential soil evaporation. In the second, evaporation depends on transmission of water through the soil profile to the surface and time since stage two began. Plant evaporation is computed as a function of soil evaporation and leaf area index. If soil water is limiting, plant evaporation is reduced by a fraction of the available soil water. Evapotranspiration is the sum of plant and soil evaporation but cannot exceed potential soil evaporation.

Erosion Component

Details of the erosion component of CREAMS are presented by the lead scientist (10) in another part of the workshop. Therefore, only a few brief comments will be given here to complete the overall model structure.

The erosion component considers the basic processes of soil detachment, transport, and deposition. The concepts of the model are that sediment load is controlled by the lesser of transport capacity or the amount of sediment available for transport. If sediment load is less than transport capacity, detachment by flow may occur, whereas deposition occurs if sediment load exceeds transport capacity. Raindrop impact is assumed to detach particles regardless of whether or not sediment is being detached or deposited by flow. The model represents a field comprehensively by considering overland flow over complex slope shapes, concentrated channel flow, and small impoundments or ponds. The model estimates the distribution of sediment particles transported and sediment sorting during deposition, and the consequent enrichment of fine sediment and organic matter.

Soil detachment is described by a modification of the USLE (17) for a single storm event (11). Rates of interrill and rill detachment in the overland flow element are calculated. The concentrated flow or channel element of the erosion model assumes that the peak runoff rate is the characteristic discharge for the channel. This then is used to calculate detachment or deposition and transport of sediment. Discharge is assumed to be steady, but spatially varied and uniformly increasing downstream. Detachment can occur when sediment load is less than transport capacity of the flow and shear stress of the flow is greater than the critical shear stress for the soil in the channel.

Water is often impounded in fields, either as normal ponding from a restriction at a fence line, a road culvert, or in an impoundment-type terrace. These restrictions reduce flow velocity, causing coarse-grained primary particles and aggregates to be deposited.

In addition to calculating the sediment transport by sizes, the model computes a sediment enrichment ratio, based on specific surface area of the sediment and organic matter and the specific surface area for the residual soil. Organic matter, clay, and silt are the principle sediment particles transported, which results in high enrichment ratios. Enrichment ratios are important in transport of chemicals associated with sediment.

Chemistry Component

Plant Nutrients - The basic concepts of the nutrient component are that nitrogen and phosphorus attached to soil particles are lost with sediment, soluble nitrogen and phosphorus are lost with surface runoff, and soil nitrate is lost by leaching with percolation, denitrification, or uptake by plants.

The nutrient component assumes that an arbitrary surface layer 10 mm deep is effective in chemical transfer to sediment and runoff. All broadcast fertilizer is added to the active surface layer, whereas only a fraction is added by fertilizer incorporated in the soil; the rest is added to the root zone. Nitrate in the rainfall contributes to the soluble nitrogen in the surface layer.

Soluble nitrogen and phosphorus are assumed to be thoroughly mixed with the soil water in the active surface layer. This includes soluble forms from the soil, surface-applied fertilizers, rainfall, and plant residues. The imperfect extraction of these soluble nutrients by overland flow and infiltration is expressed by an empirical extraction coefficient. The amounts of nitrogen and phosphorus lost with sediment are functions of sediment yield, enrichment ratio, and the chemical concentration of the sediment phase.

When infiltrated rainfall saturates the active surface layer, soluble nitrogen moves into the root zone. Incorporated fertilizer, mineralization of organic matter, and soluble nitrogen in rainfall percolated through the active surface layer increase the nitrate content in the root zone. Uniform mixing of nitrate in soil water in the root zone is assumed. Mineralization is calculated by a first-order rate equation from the amount of potential mineralizable nitrogen, soil water content, and temperature. Optimum rates of mineralization occur at a soil temperature of $35^{\circ}C$. Soil temperature is approximated from air temperature in the hydrology component.

Nitrate is lost from the root zone by plant uptake, leaching, and denitrification. Plant uptake of nitrogen under ideal conditions is described by a normal probability curve. The potential uptake is reduced to an actual value by a ratio of actual plant evaporation to potential plant evaporation. A second option for estimating nitrogen uptake is based on plant growth and the plant nitrogen content.

The amount of nitrate leached is a function of the amount of water percolated out of the root zone as estimated by the hydrology component and the concentration of nitrate in the soil water. Denitrification occurs when the soil water content exceeds field capacity (soil water content at 1/3-bar tension). The rate constant for denitrification is calculated from the soil organic carbon content with a twofold reduction for each 10-degree increase in temperature from 35°C.

Thus the plant nutrient component estimates nitrogen and phosphorus losses in transported sediment, soluble nitrogen and phosphorus in the runoff, and changes in the soil nitrate content due to mineralization, uptake by the crop, leaching by percolation through the root zone, and by denitrification in the root zone for each storm. Concentrations of nitrogen and phosphorus in the runoff and sediment are also computed. Individual storm losses are accumulated for annual summaries which are also used to compute average concentrations.

Pesticides – The pesticide component estimates concentrations of pesticides in runoff (water and sediment) and total mass carried from the field for each storm during the period of interest. The model accommodates up to ten pesticides simultaneously in a simulation period. Foliar-applied pesticides are considered separately from soil-applied pesticides, because dissipation of pesticides from foliage is often more rapid than from soils. The model considers multiple applications of the same chemical such as insecticides. A flow chart of the pesticide component is shown in Figure 1.

As in the plant nutrient component, a 10-mm deep active surface layer is assumed. Movement of pesticides from the surface is a function of runoff, infiltration, and pesticide mobility parameters. Pesticide in the runoff active soil layer is partitioned between the solution phase and the soil phase by the following relationships:

$$(C_w \ Q) + (C_s \ M) = a \ C_p \tag{4}$$

and

$$C_s = K_d \ C_w \tag{5}$$

where C_w is pesticide concentration in runoff water, Q is volume of water per unit volume of surface active layer, C_s is pesticide concentration in soil that interacts with runoff, M is mass of soil per unit volume of active surface layer, a is an empirical extraction ratio specifying the soil to water ratio in the extraction zone, C_p is the concentration of pesticide residue in the soil, and K_d is the coefficient for partitioning the pesticide between sediment and water phases. The pesticide concentration C_s is that concentration in the soil material of the surface layer. Selective deposition as expressed by enrichment ratio enriches this concentration in the sediment leaving the fields. The amount of pesticide attached to the sediment leaving the field is the product of the concentration C_s, sediment yield, and enrichment ratio.

Pesticide washed from foliage by rain increases the residual pesticide concentration in the soil. The amount calculated as available for washoff is

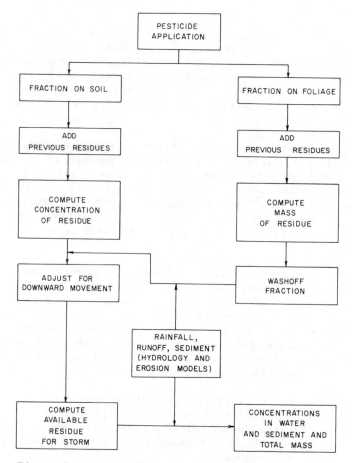

Figure 1. Simplified Schematic Representation
of the Pesticide Submodel in CREAMS

updated between storms by a foliar dissipation function. Pesticide residue in
the surface layer is reduced by extraction in overland flow and infiltrated
rainwater and by degradation which is described by an exponential function
with a half-life parameter.

MODEL APPLICATION

The CREAMS model has been tested with data from research watersheds in several
land resource areas. Comparisons of observed and simulated data over a range
of management systems, results of sensitivity analyses, and a complete user
manual have been published (28).

Simulated values compared favorably with observed data, but CREAMS is not a
predictive (absolute quantity) model. Its main usefulness is for determining
relative differences between alternate management systems and chemicals on a
given field. These differences are the primary basis for selecting best
management practices. The CREAMS model is also useful in planning a complete
resource management system, including fertilizer and pesticide programs.

Alternative fertilization schedules, both time and amount of application, can be examined. CREAMS need not be run on every field; it is feasible to run on selected fields that are representative of major soils and topographies in a land resource area. CREAMS does not require calibration with observed data from the specific field of interest. Most parameter values are physically measurable, and fine tuning them is unnecessary when the purpose is to compare alternative management systems.

Simulations for single design storms or for a single wet, average, or dry year do not provide all the information needed to select appropriate best management practices, especially where pollution from plant nutrients and pesticides is a major concern. Long-term simulations (20 years) are needed to reflect different rainfall patterns (intensities, amounts, and sequences) relative to chemical applications. Several cycles of a crop rotation should be simulated to examine probabilities of exceedances for erosion, plant nutrients, and pesticides. Long-term simulations are also useful in frequency and economic analyses.

General applications of the CREAMS model have been previously discussed (15, 16, 18, 27). Examples of specific applications are assessments of erosion/sediment transport (8, 9, 10, 21), evaluation of specific management practices (5), and assessments of irrigation impacts on water quality (6). Use of the CREAMS model in examining relationships between potential pesticide runoff losses and pesticide and soil properties will be illustrated in this paper.

Four hypothetical herbicides representing a range of properties were envisioned. These were assumed to be applied as post-plant sprays to a very sandy soil and a clay soil at rates of 3 kg/ha each year for 20 consecutive years. Pesticides A and B with K_{oc} = 20 represent non-adsorbed, mobile compounds, whereas C and D with K_{oc} = 20,000 represent compounds that are strongly adsorbed by soil sediments. The parameter, K_{oc}, describes partitioning of pesticides between the soil solution or runoff water and that adsorbed by soil or sediment organic matter. The greater the value of this parameter, the more strongly a pesticide will be adsorbed (23). Pesticides A and C with half-lives of 3 days are non-persistent, whereas B and D with 30-day half-lives are moderately persistent. These combinations of mobility and persistence were selected to span a range encompassing common pesticide classes. A number of pesticides have half-lives longer than 30 days; however, the 30-day value was chosen to avoid any significant carry-over between years of the 20-year simulation period. In practice, corn would not likely be grown for 20 years consecutively, so that each year simulated should be considered as any 1 year of a 20-year period.

Simulations were conducted using a representative 29-ha field of Tifton loamy sand near Tifton, Georgia, USA, and available rainfall/climatic records for the 20-year period, 1955 through 1974. At Tifton the long-term average annual rainfall is approximately 1,163 mm with a maximum monthly amount of 160 mm in July (25). Mean annual temperature is 19.2^{o}C with a mean monthly range from 11.1^{o} in February to 27.2^{o} in August. Solar radiation ranges from an average 263 ly/day in January to 562 ly/day in June. High-intensity, short-duration convective thunderstorms commonly occur in the spring and summer during the times of seedbed preparation, planting, and crop emergence, when the soil erosion potential is high. To compare runoff potentials as affected by soil properties, parameters for a clay soil, Houston black clay, were substituted for those for the Tifton soil in a second series of simulations using identical rainfall/climatic records and field size and configuration. Houston black clay occurs in the Blacklands Prairies region from Alabama to Texas but not in the Tifton area. However, common slopes and expected rainfall patterns are not dissimilar. The surface horizon of Tifton loamy sand is composed of medium to very fine sand (over 70% >50 μm), about 20% silt (2-50 μm), less than 10% non-expanding lattice clay (<2 μm) and 1 to 2% organic matter. In contrast, Houston black clay is about 55% clay, 40% silt, and 5% sand with about 2-3% organic matter. The Houston soil also contains expanding lattice clays. As a result, the Houston soil has low permeability and hydraulic conductivity compared to the Tifton soil. The Tifton soil is classed as hydrologic group B compared to group D for the Houston soil (29). Therefore, much higher runoff volumes may be expected from the Houston soil.

RESULTS AND DISCUSSION

Results of simulations for the 20-year period are summarized in Table I as annual averages. Recall that CREAMS is not a predictive model and results are interpreted to show relative differences among soils and pesticides. For the 20-year period, annual rainfall of 1,220 mm produced 71 mm runoff from the Tifton soil compared to 386 mm from the Houston soil. Sediment yield from the Houston soil was also greater than that from the Tifton soil in approximate proportion to differences in runoff volume. Because of the greater water-holding capacity of the Houston soil and the higher infiltration rate of the Tifton soil, much greater volumes of water percolated through the root zone of the Tifton soil than through the Houston soil; i.e., 336 mm per year compared to only 0.25 mm per year, respectively. As discussed below, these hydrologic differences had pronounced effects on pesticide runoff.

Pesticide losses expressed as percent of the applications ranged from 0.01 to about 10 percent (Table I). These simulated values are consistent with the range of expected values based on actual measurement (30). Total runoff losses of all four pesticides were less from the Tifton soil than from the Houston soil partly because of lower runoff volumes and sediment losses. However, for the mobile pesticides A and B, (Koc = 20) their movement below the soil surface by water infiltrating through the soil surface before initiation of runoff was the primary factor in reducing runoff of these pesticides from the Tifton soil. This effect is illustrated in Figure 2. Figure 2 shows the concentrations of pesticides at the soil surface available for runoff as computed in the model during a 25-day period in 1973 after pesticide application. In estimating the available surface concentration effective in determining runoff concentrations for a given rainfall event, the pesticide concentration at the soil surface at that time is reduced by the amount transported vertically through the surface 0-10 mm layer by infiltrating water. The amount of water infiltrated is assumed to be rainfall less runoff and storage. Since predicted pesticide runoff concentrations are directly proportional to the amount available at the soil surface, this process dramatically reduces runoff potential of mobile pesticides applied to sandy soils with high infiltration rates. As shown in Figure 2, the available surface concentration of pesticides A and B rapidly declined in Tifton soil with rainfall. For example, in 1973, 54 mm of rainfall occurring on the day of application reduced the initial soil surface concentration of pesticides A and B from 20 to 0.14 μg/g. Succeeding rainfall events of 13 and 56 mm on days 2 and 3 removed essentially all of the remaining pesticides from the soil surface. In contrast to this behavior, the relatively immobile pesticides C

Table I. Simulated Runoff, Sediment Yield, and Losses of Hypothetical Pesticides from a Loamy Sand and Clay
Soil. Annual Averages for 20-year Period Using 1955-1974 Rainfall Data from Tifton, Georgia.

	Rainfall	Runoff	Percolation	Sediment Yield	Pesticide Losses Runoff	Pesticide Losses Sediment	Pesticide Sediment Phase	Total Pesticide Losses
	mm	mm	mm	t/ha	g/ha	g/ha	% of losses	% of application
Tifton Loamy Sand	1220	71	336	10.9				
Pesticide A					0.25	0.001	0.4	0.01
Pesticide B					0.35	0.002	0.5	0.01
Pesticide C					4.07	17.880	81.5	0.73
Pesticide D					12.66	62.650	83.2	2.51
Houston Black Clay	1220	386	0.25	55.6				
Pesticide A					158.00	0.750	0.5	5.29
Pesticide B					298.00	1.690	0.6	9.99
Pesticide C					5.74	29.040	83.5	1.16
Pesticide D					21.25	125.900	85.6	4.94

223

and D (Koc = 20,000) remained in the soil surface in both Tifton and Houston soils throughout the 25-day period and declined in concentration as predicted, considering only degradation as described by the pesticide half-life (C = 3 days, D = 30 days). The decline in concentration of pesticides A and B with time in the Houston soil surface reflected both the effects of infiltration and degradation.

When a pesticide has an inherent high mobility in soil, it may also be readily extracted by surface runoff under conditions where infiltration rate is relatively low compared to runoff rate. These conditions prevailed for pesticides A and B in the Houston soil. Average annual runoff losses were about 5 and 10 percent for pesticides A and B, respectively (Table I). Greater losses of B compared to A were incurred because B has a much longer half-life than A.

Pesticides C and D which are strongly adsorbed by soil organic matter, were transported primarily by sediment from both soils (Table I). Since 80 to 85 percent of the total pesticide transport was with sediment, runoff losses from a given site were proportional to sediment yields. However, annual sediment yields as estimated by simple models such as the USLE (32) cannot be used as an index of adsorbed pesticide losses. As an example, annual sediment yields from the Houston soil were

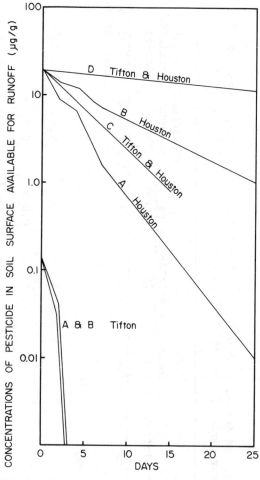

Figure 2. Simulated Concentrations of Pesticides in the Surface Layer (0-10 mm) of Tifton and Houston Soil, 1973

about five times that from the Tifton soil, but annual pesticide losses in sediment from the Houston soil were only about twice that from the Tifton soil. This can be explained by considering the timing of sediment losses with respect to pesticide applications. During the spring growing season, immediately after the pesticide application, much of the rainfall occurred in intense thunderstorms; whereas, in fall and winter rainfall was more of the frontal type and had lower intensity. Runoff and sediment transport from both soils were more similar during the pesticide application season; whereas, under conditions of frontal rainfall in winter months, much greater runoff and sediment transport occurred on the Houston soil. During winter months, most of the applied pesticides had already dissipated and were not present in the transported sediment.

Sediment composition also affects pesticide transport. The CREAMS model computes an enrichment factor for adsorbed chemicals based on changes of specific particle surface of the transported sediment caused by deposition in transport of the larger or more dense particles. Some management practices such as land terracing and conservation tillage are more effective in reducing transport of the coarser sediment such that concentrations of adsorbed chemicals are enriched in sediment compared to those in the residual soil and are increased as the mass of transported sediment decreases (5). In Table I, slight differences between pesticides C and D in percent transported by sediment may be related to differences in sediment enrichment factors later in the growing season which would affect transport of pesticide D, the more persistent pesticide.

Generalizations and comparisons above are based on 20-year averages. A long simulation period is necessary to capture expected year-to-year variability in rainfall patterns and establish meaningful annual averages. Year-to-year variability in simulated pesticide runoff is illustrated for each of the pesticides and soils in Figure 3. The figure shows the number of years in the 20-year simulation period in which losses were within the specified percentage intervals. Within the 20-year period runoff losses of pesticide A from the Houston soil ranged from <0.1 percent (4 years) to >10 percent (2 years). The extent of losses of this non-persistent pesticide is totally dependent on storm occurrence with respect to pesticide application. Runoff losses were relatively high when significant runoff-producing rainfall occurred on or near the application date. However, losses were minimal when runoff did not occur during this time period. The probability of pesticide losses increases with increasing pesticide half-life as shown by comparing losses of pesticide A and pesticide B (Figure 3, Houston soil). The effects of increasing half-life are not evident in comparing losses of A and B from the Tifton soil because leaching from the soil surface is the dominant process affecting these mobile compounds.

Losses of adsorbed pesticides C and D from both the Tifton and the Houston soils increased with increasing pesticide half-life, as expected. Losses of these compounds were dependent on occurrences of sediment-producing events during the time of persistence as determined by pesticide half-life. Therefore, the year-to-year variability in losses (Figure 3) is indicative of the variability in sediment-producing events during the period of pesticide persistence.

It is usually impractical to conduct experiments designed to continuously measure pesticide losses for more than 1 to 3 years. Observations in Figure 3 clearly demonstrate the necessity of long-term assessments to characterize expected year-to-year variability. Also, only a limited number of pesticides, soils, and management variables can be studied experimentally. Use of the CREAMS model allows examination of many combinations of variables that otherwise would be impossible. The length of time simulated is limited only by the length of historical rainfall data available. However, climate generation models (22) can be used to create input data files of any length.

SUMMARY

Mathematical models to assess nonpoint source pollution and evaluate effects of management practices for its control are needed to adequately respond to water quality legislation of the past 5 years. Consequently, the U. S. Department of Agriculture, Agricultural Research Service, has developed CREAMS, a field-scale model for Chemicals, Runoff, and Erosion from

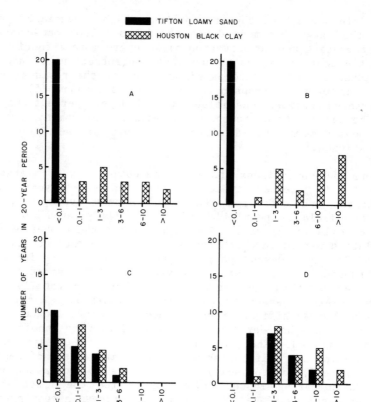

Figure 3. Frequency Distributions of Annual
Pesticide Losses from Tifton and
Houston Soils

Agricultural Management Systems.   The model includes components for hydrology,
erosion/sediment yield, and chemical transport that describe the movement in
runoff of sediment and its characteristics, plant nutrients, and pesticides
from field-sized areas.   It is a continuous simulation model that operates
efficiently to allow consideration of long records (20 years).   The utility of
the model is evaluation of alternate management practices for their impact on
the yield of sediment and chemical pollutants from field-sized areas at
specific sites.   Utility of the model in evaluating runoff of pesticides with
different properties from two very different soils has been demonstrated.   A
number of alternate land management practices (terracing, conservation
tillage, etc.) can also be proposed; and after evaluation of each with CREAMS,
a practice could be chosen from those judged to adequately control sediment
and chemical yield.

REFERENCES

1. Beasley, D. B., Monke, E. J., and Huggins, L. F. ANSWERS: A Model for Watershed Planning. Agricultural Experiment Station Journal Paper No. 7038. Purdue University, West Lafayette, Indiana, 1977.

2. Bruce, R. R., Harper, L. A., Leonard, R. A., Snyder, W. M., and Thomas, A. W. A Model for Runoff of Pesticides from Small Upland Watersheds. Journal of Environmental Quality 4(4), 1975, 541-548.

3. Crawford, N. H., and Donigian, A. S., Jr. Pesticide Transport and Runoff Model for Agricultural Lands. EPA 660/274-013, Office of Research and Development, U. S. Environmental Protection Agency, Washington, DC, EPA 660/274-013, 1973.

4. Crawford, N. H., and Linsley, R. K. The Synthesis of Continuous Stream-flow Hydrographs on a Digital Computer. Technical Report No. 12, Stanford University, Department of Civil Engineering, Technical Report No. 12, Stanford, California, 1962.

5. DelVecchio, J. R., and Knisel, W. G., Jr. Application of Field-Scale Nonpoint Pollution Model. Proceedings, American Society of Civil Engineers, Irrigation and Drainage Specialty Conference, Orlando, Florida, July 20-23, 1982, pp. 227-236.

6. DelVecchio, J. R., Knisel, W. G., and Ferreira, V. A. The Impact of Supplemental Irrigation on Water Quality. Proceedings of American Society of Civil Engineers, Irrigation and Drainage Specialty Conference, Jackson, Wyoming, July 20-23, 1983.

7. Donigian, A. S., Jr., and Crawford, N. H. Modeling Pesticides and Nutrients on Agricultural Lands. Report No. EPA-600/2-76-043, Environmental Protection Technology Series, Office of Research and Development, U. S. Environmental Protection Agency, Washington, DC, 1976. 317 pp.

8. Foster, G. R., and Ferreira, V. A. Deposition in Uniform Grade Terrace Channels. Conference on Crop Production with Conservation in the 80's. Proceedings of American Society of Agricultural Engineers, St. Joseph, Michigan, 1981, pp. 185-197.

9. Foster, G. R., and Lane, L. J. Estimating Sediment Yield from Rangelands with CREAMS. ARM-W-26, Proceedings of Workshop on Estimating Soil Erosion and Sediment Yield from Rangelands. U. S. Department of Agriculture, Agricultural Research Service, 1982, pp. 115-119.

10. Foster, G. R., Lane, L. J., and Knisel, W. G., Jr. Estimating Sediment Yield from Cultivated Areas. Proceedings of American Society of Civil Engineers, Irrigation and Drainage Division, Specialty Conference, July 21-23, 1980, pp. 151-163.

11. Foster, G. R., Meyer, D. L, and Onstad, C. A. A Runoff Erosivity Factor and Variable Slope Length Exponents for Soil Loss Estimates. Transactions of American Society of Agricultural Engineers 20(4), 1977, 683-687.

12. Frere, M. H., Onstad, C. A., and Holtan, H. N.   ACTMO, an Agricultural Chemical Transport Model.   ARS-H-3, U.S. Department of Agriculture, Agricultural Research Service, U. S. Government Printing Office, Washington, DC, 1975.

13. Green, W. A., and Ampt, G. A.   Studies on Soil Physics, I.   The Flow of Air and Water Through Soils.   Journal of Agricultural Science, Vol. 4, 1911, pp. 1-24.

14. Holtan, H. N., and Lopez, N. C.   USDAHL-70 Model of Watershed Hydrology. Technical Bulletin 1435, U. S. Department of Agriculture, 1971.

15. Knisel, W. G., Jr.   A System of Models for Evaluating Nonpoint Source Pollution - An Overview.   Collaborative Paper CP-78-11, International Institute of Applied Systems Analysis, Laxenburg, Austria, 1978.

16. Knisel, W. G.   CREAMS: A Field-Scale Model for Chemicals, Runoff, and Erosion from Agricultural Management Systems.   Vol. 13, Part 4, Proceedings of XII Annual Pittsburgh Conference on Modeling and Simulation, April 22-23, 1982, pp. 1555-1559.

17. Knisel, W. G., Jr., and Foster, G. R.   CREAMS: A System for Evaluating Best Management Practices.   In Economics, Ethics, Ecology: Roots of a Productive Conservation, W. E. Jestse (ed.), Soil Conservation Society of America, Ankeny, Iowa, 1980, pp. 177-194.

18. Knisel, W. G., Jr., Foster, G. R., and Leonard, R. A.   CREAMS: A System for Evaluating Alternate Management Practices.   Proceedings of National Conference on Agricultural Management and Water Quality, Iowa State University, Ames, Iowa, May 26-29, 1981.

19. Kothandaraman, V., and Evans, R. L.   Use of Air-Water Relationships for Predicting Water Temperature.   Report of Investigations No. 69, Illinois State Water Survey, Urbana, Illinois, 1972, 14 pp.

20. Negev, M. A.   Sediment Model on a Digital Computer.   Technical Report No. 76, Stanford University, Department of Civil Engineering, Stanford, California, 1967.

21. Neibling, W. H., and Foster, G. R.   Modeling Shallow-Flow Sediment Transport of Naturally Eroded Sediment.   ASAE Paper No. 83-2160, American Society of Agricultural Engineers, Summer Meeting, Bozeman, Montana, 1983.

22. Nicks, A. D.   Stochastic Generation of the Occurrence, Pattern, and Location of Maximum Amount of Daily Rainfall.   Miscellaneous Publication 1275, Proceedings of Symposium on Statistical Hydrology, U. S. Department of Agriculture, Agricultural Research Service, 1974, pp. 154-171.

23. Rao, P. S. C., and Davidson, J. M.   Retention of Pesticides by Soils. In Retention and Transformation of Selected Pesticides and Phosphorus in Soil-Water Systems: A Critical Review.   EPA/600/3-82-060, Section 4, U. S. Environmental Protection Agency, 1980, pp. 7-39.

24. Ritchie, J. T. A Model for Predicting Evaporation from a Row Crop with Incomplete Cover. Water Resources Research 8(5), 1972, pp. 1204-1213.

25. Sheridan, J. M., Knisel, W. G., Woody, T. K., and Asmussen, L. E. Seasonal Variation in Rainfall and Rainfall-Deficient Periods in the Southern Coastal Plain and Flatwoods Regions of Georgia. Research Bulletin 243, The University of Georgia College of Agriculture Experiment Station, Athens, Georgia, 1979.

26. Smith, R. E., and Parlange, J. Y. A Parameter-Efficient Hydrologic Infiltration Model. Water Resources Research 14(3), 1978, pp. 533-538.

27. Svetlosanov, V., and Knisel, W. G. (eds.). European and United States Case Studies in Application of the CREAMS Model, CP-82-S11, International Institute for Applied Systems Analysis, 1982, 144 pp.

28. U. S. Department of Agriculture, Science and Education Administration. CREAMS - A Field-Scale Model for Chemicals, Runoff, and Erosion from Agricultural Management Systems. Conservation Research Report No. 26, W. G. Knisel (ed.), 1980, 643 pp.

29. U. S. Department of Agriculture, Soil Conservation Service, SCS National Engineering Handbook, Sec. 4, Hydrology, 1972, 548 pp.

30. Wauchope, R. D. The Pesticide Content of Surface Water Draining from Agricultural Fields - A Review. Journal of Environmental Quality, Vol. 7, 1978, pp. 459-472.

31. Williams, J. R., and LaSeur, W. V. Water Yield Model Using SCS Curve Numbers. Journal of American Society of Civil Engineers, Hydraulic Division, 102(HY9), 1976, pp. 1241-1253.

32. Wischmeier, W. H., and Smith, D. D. Predicting Rainfall Erosion Losses-- A Guide to Conservation Planning. Handbook No. 537, U. S. Department of Agriculture, 1978.

33. WU, Y. H. Effects of Roughness and its Spatial Variability on Runoff Hydrographs. Report. No. CED 7778 YHW7, Ph.D. Dissertation, Colorado State University, Civil Engineering Department, Fort Collins, Colorado, 1978.

SIMPLE MODELS TO EVALUATE
NON-POINT POLLUTION SOURCES
AND CONTROLS[4]

B. N. Wilson,[1] B. J. Barfield[2], and R. C. Warner[3]

ABSTRACT

When selecting an algorithm for a non point source model, it is important to strike a balance between complexity and ease of use, so that the predictions can be made with the desired accuracy without requiring that the user become an expert hydrologist. In the development of SEDIMOT II, a design aid model for sediment control structures, an attempt was made to strike such a balance. SEDIMOT II is a distributed parameter model in which runoff and sediment yield are determined for each homogeneous subwatershed and routed downstream to structures and junctions. Runoff is determined with SCS curve number techniques utilizing unit hydrographs with differing shapes depending on land use. Channel routing is accomplished with Muskingum's method. Sediment yield is calculated with either the MUSLE or CREAMS equations. The time distribution is determined by a power relationship between sediment concentration and flow rate. The size distribution is determined by an algorithm which assumes that the larger particles are deposited first. Sediment routing through ponds is determined from either the DEPOSITS plug flow model or the more recent Continuous Stirred Tank Reactor in Series (CSTRS) model. Sediment routing through grass filters is calculated by the GRASSFIL model and through check dams by a combination of a water surface profile model and Camp's overflow rate model.

1. Assistant Professor, Agricultural Engineering Department, Oklahoma State University, Stillwater, Oklahoma 74074.

2. Professor, Agricultural Engineering Department, University of Kentucky, Lexington, Kentucky 40546-0075.

2. Assistant Professor, Agricultural Engineering Department, University of Kentucky, Lexington, Kentucky 40546-0075.

4. The investigation reported in this paper (No. 84-2-207) is in connection with a project of the Kentucky Agricultural Experiment Station and is published with approval of the Director.

INTRODUCTION AND BACKGROUND INFORMATION

Criteria for Selecting a Modeling Technique

Hydrologic modeling techniques vary widely in complexity and approach, ranging from simple regression techniques with few inputs to complex theoretically based equations needing large quantities of input data. These complex model inputs typically require long periods of time to collect for a watershed while inputs for the simple models can be rapidly determined. In order to adequately represent the physical processes occurring on a watershed, it is necessary to resort to the physically based models; however, some applications do not warrant the time requirements of such complex approaches. The trade off between complexity and the requirements of the model application must be considered in selecting a modeling approach.

When selecting a modeling approach, the following criteria should be considered:

1. Proposed use of the model.
2. Potential for adaptation of algorithm to proposed use.
3. Accuracy of the algorithm.
4. Availability of input parameters for the algorithms.
5. Sophistication of potential users.
6. Computational time required for solution.

Based on consideration of these criteria, a set of algorithms can be combined into a model to produce the desired output, with the desired accuracy while minimizing the complexity of the inputs.

Not enough can be said about the need to consider the potential user. All too often, modelers consider only the theoretical aspects of the model and do not evaluate the capabilities of the potential users. The result is often an excellent model which is not adopted by the intended users because its complexity is beyond their grasp. In some cases, a slight compromise in complexity and model sophistication could have resulted in a model being adopted. Such compromise, though typically repugnant to the theoretician, does not always render model computations invalid or unusable.

Approaches to Modeling Non-Point Source Phenomena

Runoff and sedimentation processes on a watershed can be modeled by using either a lumped or distributed parameter approach. The response of an entire watershed is treated as one hydrologic unit in the lumped parameter approach (Betson, et al., 1980; Overton and Crosby, 1979). All input parameters are developed by using area weighting or regression techniques.

With distributed parameter models, hydrologic responses are calculated for homogeneous subwatersheds, each of which are characterized by a single set of parameters. One distributed parameter approach to the problem is to describe the runoff and sediment movement by a set of physically based differential equations which must be solved by numerical techniques (Beasley et al., 1980, Wolfe et al., 1981; Li et al., 1977). With this approach, the spatial variability of the watershed can be described, but the computational time can be excessive. An alternate approach is to subdivide the watershed into homogeneous subunits and treat each subunit with lumped parameter procedures. Outputs from each area are routed to a watershed exit. Examples of this latter procedure include the SCS TR20 model (SCS, 1964), Williams HYMO model (Williams and Hahn, 1973), and the University of Kentucky SEDIMOT II model (Wilson et al., 1984a, 1984b).

Modeling Sediment Control Structures

In general, predictive techniques can be divided into either empirical methods or physically-based models. Developers of empirically based models have not attempted to predict the internal mechanisms that govern the sedimentation processes, but have simply used observed input and output data to obtain predictive relationships. Empirical methods were popular design techniques until the early 1970s, but are now slowly being replaced by physically-based models. These empirical models include the Brune (1953) and Churchill (1948) models for sediment ponds; Ohlander's (1976) model for grass filters; and empirical data for check dams. In the more recent physically-based models, the internal mechanisms that govern the hydraulic and/or sediment response are considered. These mechanisms are typically so complicated that idealization is necessary so that a tractable solution can be found. Included in these models are the DEPOSITS model (Ward et al., 1977), the CSTRS model (Wilson and Barfield, 1984a), the BASIN model (Wilson and Barfield, 1984b), and Modified Overflow rate models (Tapp et al., 1981; Curtis and McCuen, 1977) for predicting pond performance; the GRASFIL model (Hayes et al. 1982); and the Foster model (1982) for grass filters; and the check dam model of Hirschi and Warner (1983).

More recently a more fundamental approach has been taken to modeling flow in reservoirs. This approach involves a solution of the turbulent form of the equations of motion and conservation of mass. Conceptually it is possible to couple the sediment mass continuity equation with the equation of motion and develop a solution. Practically, however, the computational time is excessive. For example, Schamber and Larock's (1981) finite element model for the one dimensional turbulent flow equations without sediment cost over \$500 per run. In addition, the state of the art has not developed to the point where intermittent turbulence typically found in sediment ponds can be modeled.

THE UNIVERSITY OF KENTUCKY SEDIMOT II MODEL

Intended Use for the Model

SEDIMOT II was developed to be used as an operational model for evaluating the effects of land disturbances on the hydrologic regime and to design sediment and runoff control facilities and structures. The intended users are practicing engineers with at least one introductory course in hydrology. It was therefore very important to carefully balance complexity with ease of use in order to assure that the model would be adopted by the engineering community. The following constraints were accepted in developing the model:

1. Algorithms utilized should have sufficient complexity to model the variations observed in the real world.
2. All algorithms should have reasonable validation.

3. An acceptable data base for all model inputs should exist which can be interpreted by users with at least an intermediate background in hydrology.
4. Inputs should be simple enough so that excessive amounts of engineering time are not required for model setup.

Operational Format

SEDIMOT II's operational format utilizes junctions, branches, structures, and subwatersheds, much as the SCS TR20 model. The smallest unit is known as the subwatershed. Runoff and sediment are calculated for each subwatershed and routed to a point of summation known as a structure, as shown in Figure 1. A structure can represent a sediment control facility such as grass filter, sediment pond, or check dam, or it can be simply a summing point known as a null structure.

After runoff and sediment are routed to a structure, they are routed through the structure using one of the algorithms discussed in a subsequent section. The discharge from the structure is then routed downstream along channels and added with lateral inflow until a subsequent structure or junction is reached. In order to be a branch, one or more structures must be designated on the branch. A structure must always be located on the channel at a junction. A junction is a location where one or more branches converge. Runoff and sediment from all branches are summed at the junction and routed downstream to subsequent junctions and structures.

SEDIMOT II can be used to evaluate a watershed up to the following size:

* Junctions: 15
* Branches per junction: 3
* Structures per branch: 4
* Subwatersheds per branch: 29

It is presently available for use on most mainframe computers, the HP3000 and on the IBM PC personal computer.

Description of the Hydrologic Component

Rainfall Runoff Component. The rainfall component in SEDIMOT II converts rainfall depth into a temporal storm pattern utilizing SCS's Type I or Type II synthetic rainfall pattern, or a user defined input breakpoint rainfall pattern. The SCS's synthetic rainfall patterns (SCS, 1973) were developed to describe a 24-hour rainfall event. For storm durations less than 24 hours, the steepest portion of the SCS curves are redistributed in SEDIMOT II by the technique given by Ward et al. (1979).

Rainfall extractions (vegetative interception, depressional storage, infiltration) are evaluated by using the SCS curve number model (SCS, 1972), or,

$$Q = \frac{(P - 0.2S)^2}{P + 0.8S} \text{ for } P > 0.2S \tag{1}$$

where

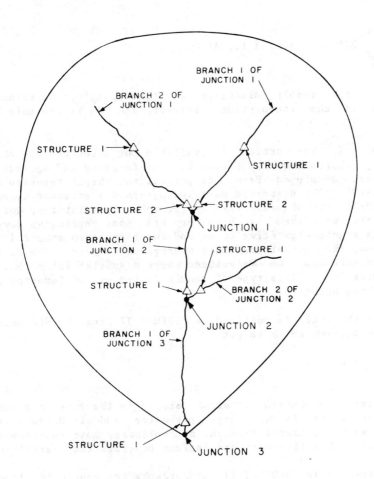

Figure 1.   Illustration of the Operational Format
of the model.

$$S = 1000/CN - 10 \qquad\qquad (2)$$

Q is the cumulative rainfall excess in inches, P is the cumulated rainfall depth in inches, and CN is the SCS curve number. Rainfall excess is calculated at each convolution interval of 0.05 hour (3 minutes). The difference between cumulative rainfall excess of two consecutive time intervals corresponds to the runoff volume for that time increment.

Overland and Subsurface Storm Flow. The overland–subsurface storm flow component of SEDIMOT II is predicted by the convolution of unit hydrographs (see Dooge, 1973) or,

$$q(t) = \sum_{i=0}^{\infty} DUH\ (t - i\Delta t)\ i\ (t)\ \Delta t \qquad\qquad (3)$$

where q(t) is the runoff discharge at time t, $i_e(t_i)$ is rainfall excess intensity, Δt is the convolution interval, and DUH is the unit hydrograph ordinate.

In SEDIMOT II, the option is available for selecting one of the three unit hydrograph shapes shown in Figure 2. The forested and agricultural unit hydrographs were developed from the generalized shapes reported by Overton and Crosby (1979). The disturbed unit hydrograph was selected to approximate the SCS's (1972) curvelinear unit hydrograph. The disturbed unit would be selected if the watershed is bare or has poor vegetative cover. For a watershed with a good vegetative cover and in which storm runoff is dominated by surface flow, the agricultural unit hydrograph is used. The forested unit hydrograph is selected for watersheds where a substantial volume of runoff occurs as quick flow. This type of runoff is typical of forested watersheds in the Appalachian Mountains.

Time to peak (t_p) is defined in SEDIMOT II from the SCS definition of lag time as 0.6 t_c from which we get

$$t_p = 0.6\ t_c + D/2 \qquad\qquad (4)$$

where D is the convolution interval and t_c is the time of concentration. Time of concentration is an input parameter and is defined as the time required for water to travel from the hydraulically most remote point to the watershed outlet. This flow path can include overland and channel segments.

Channel Flow. In SEDIMOT II, hydrographs are routed to structures and between structures by Muskingum's routing procedure (see Wilson et al., 1982), or:

$$0_2 = C_1 I_1 + C_2 I_2 + C_3 0_1 \qquad\qquad (5)$$

where

$$C_1 = \frac{KX + \Delta/2}{K(1 - X) + \Delta/2} \qquad\qquad (6)$$

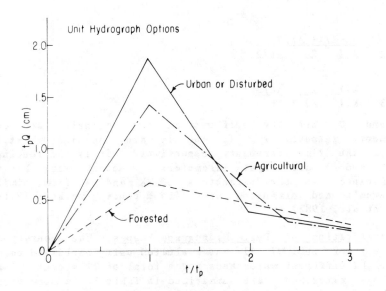

Figure 2. Unit Hydrograph Shapes Used in
SEDIMOT II.

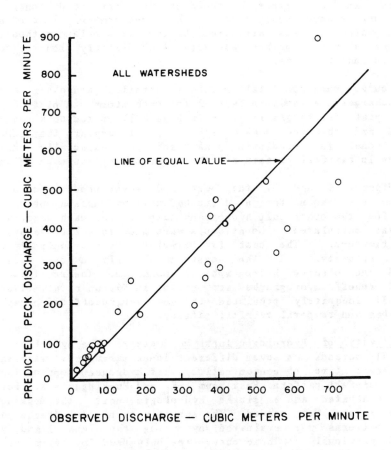

Figure 3. Predicted and Observed Discharges.

$$C_2 = \frac{-KX + \Delta t/2}{K(1 - X) + \Delta t/2} \qquad\qquad (7)$$

$$C_3 = \frac{K(1 - x) + \Delta t/2}{K(1 - X) + \Delta t/2} \qquad\qquad (8)$$

O_1 and O_2 are the outflows at the beginning and end of the time increment, respectively, I_1 and I_2 are the inflows at the beginning and end of the time increment, respectively, Δt is the routing time increment, and K and X are input parameters. Recent work has attached physical significance to the parameters of K and X from similarities between Muskingum's and diffusion wave flood routing methods (Wilson et al., 1982; Dooge et al., 1982).

<u>Verification of Unit Hydrograph Shapes</u>. The accuracy of the hydrologic component of SEDIMOT II was evaluated using published rainfall–runoff data for eight different watersheds for a total of 27 storms. The characteristics of these watersheds are summarized in Table 1. A more detailed description of the drainage basin and storm data can be found in Wilson et al. (1982). The watersheds were selected so that each of the unit hydrograph shapes shown in Figure 2 could be evaluated. All watersheds were considered to be of relatively uniform land use and were evaluated as a single hydrologic unit. Variations in soil types, antecedent moisture conditions, and spatial rainfall depths were incorporated in the curve number. Time of concentration for each watershed was estimated by the SCS's upland method (1972). For each storm a curve number was determined directly from the total rainfall volume, and runoff volume.

The curve number, total rainfall duration, and observed and predicted peak discharges are shown in Table 2 for each storm. A plot of predicted and observed peak discharges is given in Figure 3. Based on a comparison of the predicted and observed peak discharges, it appears that SEDIMOT II is an adequate model for predicting peak discharges, especially considering areal variations in rainfall pattern and the accuracy of instrumentation.

In Figures 4 and 5, the best and worst predictions of the observed hydrograph are shown for a disturbed unit hydrograph option. Results were similar for the other unit hydrograph shapes. For each case, a single curve number was calculated. No attempts were made to calibrate any of the other input parameters. The best fit simulates the runoff hydrograph with excellent accuracy. In the worst case, only a fair agreement between predicted and observed hydrographs was obtained. The remaining observed and predicted runoff hydrographs are given in Wilson et al. (1982). Overall, SEDIMOT II adequately predicted the observed runoff hydrograph for a known curve number and temporal rainfall pattern.

<u>Sensitivity of Hydrologic Input Parameters</u>. The hydrologic component of SEDIMOT II depends on seven different input parameters: watershed area; SCS curve number; time of concentration; unit hydrograph option (see Figure 2); temporal rainfall pattern; Muskingum's K; and Muskingum's X. For a watershed that is evaluated as a single hydrologic unit, the Muskingum's routing coefficients are not needed. The sensitivity of peak discharge to temporal rainfall patterns was evaluated by using the Type I and Type II curves described previously. These curves are only used for storm durations of 24

TABLE 1
Characteristics of watersheds used in hydrologic verification study

Watershed	Land use	Area, ha	Average slope %	Hydraulic length, m	Time of concentration, h
Clays Mill Lexington, KY	Urban	360	2.5	3200	1.45
Landsdowne Lexington, KY	Urban	262	5.0	2438	0.90
Lower Lick Perry County, KY	Forest	15	45.0	521	0.35
Cane Branch Beaver Creek, KY	88% forest 12% mined	172	20.0	2134	1.30
Watershed 10 Coshocton, OH	Grass, crops, woodlands	49	14.0	1161	0.50-0.55
Watershed 166 Coshocton, OH	Grass, crops, woodlands	32	15.0	1144	0.32
Watershed 5 Coshocton, OH	Grass, crops, woodlands	140	10.0	1542	0.64
Watershed W-1 Stillwater, OK	100% grasslands	7	6.2	430	0.19-0.22

TABLE 2
Predicted and observed peak discharge (m^3/MIN)

Watershed	Storm date	Curve number	Rainfall depth, -cm	Storm duration,-h	Peak discharge Observed	Peak discharge Predicted
	4-22-72	85.0	5.6	4.25	656	853
	7-24-73	93.3	2.0	3.00	391	279
	6-05-73	86.7	2.8	3.00	348	265
Clays Mill	5-22-73	77.5	5.5	5.73	540	323
	9-25-71	88.8	2.8	2.00	369	411
	9-26-71	91.4	2.4	1.75	404	384
	6-28-72	84.4	3.1	2.50	357	289
	8-09-72	82.7	4.4	3.75	413	416
	4-22-72	84.8	4.8	4.75	491	697
Lansdowne	9-25-71	86.8	3.7	3.75	510	498
	9-26-71	95.0	2.0	2.00	396	443
	6-28-72	78.2	3.3	2.25	319	182
	4-25-78	52.7	6.7	23.0	0.3	1.5
Lower Lick	5-22-78	57.5	8.3	12.0	3.9	6.8
	5-07-60	80.0	6.1	12.5	100	94
Cane Branch	4-05-62	95.0	6.9	38.5	70	90
	4-24-70	86.8	4.8	10.0	60	89
Coshocton #10	4-25-61	94.0	3.7	5.0	184	169
	8-21-60	63.5	8.6	6.0	76	73
	4-24-70	89.2	4.5	9.5	34	49
Coshocton #166	7-07-69	92.9	2.3	3.0	48	74
	4-24-70	85.4	4.5	10.0	120	178
Coshocton #5	4-25-61	89.8	3.2	7.0	165	252
	8-21-60	69.6	9.9	9.0	574	384
	9-26-67	87.2	6.6	6.0	38	29
Stillwater W-1	4-30-70	92.8	5.4	6.5	41	38
	6-24-67	89.5	8.2	14.0	31	28

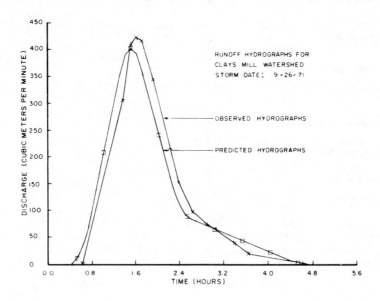

Figure 4.   Best Fit for Disturbed Unit Hydrograph.

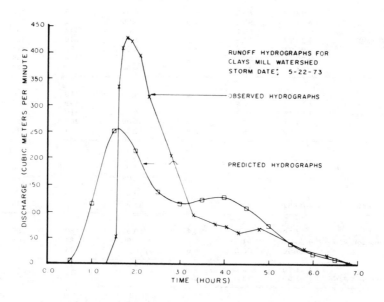

Figure 5.   Worst Fit for Disturbed Unit Hydrograph.

hours or less.   The use of the synthetic rainfall patterns is really a measure of the accuracy of these patterns to predict the peak rainfall intensity.   For design events, the actual storm pattern is unknown; therefore, these results provide the user with an estimate of the errors that would be observed (for the storms examined) if the SCS's type I or Type II curves had been selected.

The results of the sensitivity analysis are shown in Tables 3 and 4. Table 3 contains the percent change in peak discharge for the corresponding change in curve number and time of concentration. Table 4 contains similar results for variations in unit hydrograph shape and temporal rainfall pattern.   In both tables the percent change is defined as,

$$\text{Percent Change} = \frac{q_{ps} - q_{pb}}{q_{pb}} \times 100\% \qquad (8a)$$

where q_{ps} is the predicted peak discharge using the sensitivity parameter and q_{pb} is the predicted peak discharge by using the best estimate of input parameters. The values of q_{pb} are those given in Table 2.

In general, the results shown in Table 3 indicate that the disturbed watersheds (Clays Mill and Lansdowne) are more sensitive to changes in curve number and time of concentration than the other watersheds. In comparison to the agricultural and forested unit hydrographs, the storage-release effect of overland flow is less pronounced in the disturbed watersheds. Hence, changes in the hydrologic parameters have a more direct effect in predicted results. Also shown in Table 3 is the variation in predicted peak discharge with a 25 percent change in time of concentration. For unit hydrograph techniques, time of concentration is an indicator of the importance of surface storage. As time of concentration increases, the attenuation effects of surface storage also increase; consequently, the peak discharge is reduced. For all storms, a 5 percent change in curve number resulted in a larger variation in peak discharge than a 25 percent change in time of concentration. Predicted results are very sensitive to possible errors in curve numbers. This indicates the importance of rainfall abstractions in the rainfall-runoff process.

Table 4 shows the sensitivity of predicted discharge to unit hydrograph shape and temporal rainfall pattern.   There is roughly a 20 percent difference in predicted peak dicharge between the disturbed and agricultural unit hydrograph and approximately a 40 percent difference between the agricultural and forested unit hydrograph. Hence there is about a 60 percent difference in predicted peak discharge between the disturbed and forested unit hydrograph shapes.   In all storms, the Type I rainfall distribution predicted lower peak discharge than the Type II distribution. The Type II rainfall pattern predicted larger intensities than the Type I pattern, and hence, predicted greater peak discharges. For the storms in this study, the Type II pattern appeared to more closely approximate the peak discharge predicted by the observed rainfall pattern.

Description of the Sedimentology Component

SEDIMOT II was designed to allow the systematic evaluation of possible

TABLE 3

Sensitivity of peak discharge to curve number and to time of concentration. Results are given in percent change in predicted peak discharge relative to that predicted with the best estimate of input parameters.

Watershed	Storm date	Curve number +5%	Curve number -5%	Time of concentration +25%	Time of concentration -25%
Clays Mill	4-22-72	22%	-20%	-14%	19%
	7-24-73	104%	-52%	-15%	23%
	6-05-73	62%	-42%	-16%	24%
	5-22-73	39%	-31%	-14%	20%
	9-25-71	56%	-38%	-18%	29%
	9-26-71	70%	-43%	-18%	27%
	6-28-72	54%	-38%	-17%	25%
Lansdowne	8-09-72	32%	-27%	-8%	10%
	4-22-72	30%	-26%	-16%	24%
	9-25-71	45%	-34%	-15%	21%
	9-26-71	81%	-48%	-16%	21%
	6-28-72	55%	-41%	-15%	21%
Lower Lick	4-25-78	52%	-47%	-9%	11%
	5-22-78	28%	-26%	-8%	11%
Cane Branch	5-07-60	21%	-19%	-12%	15%
	4-05-62	5%	-11%	-5%	4%
Coshocton #10	4-24-70	27%	-24%	-24%	14%
	4-25-61	24%	-24%	-12%	18%
	8-21-60	20%	-19%	-12%	17%
Coshocton #166	4-24-70	28%	-25%	-6%	9%
	7-07-69	64%	-43%	-11%	14%
Coshocton #5	4-24-70	32%	-27%	-11%	15%
	4-25-61	36%	-29%	-11%	18%
	8-21-60	13%	-13%	-8%	13%
Stillwater W-1	9-26-67	23%	-19%	-6%	4%
	4-30-70	27%	-24%	-10%	13%
	6-24-67	21%	-20%	-6%	8%

TABLE 4

Sensitivity of peak discharge to unit hydrograph option and temporal rainfall pattern. Results are given in percent change in predicted peak discharge relative to that predicted with the best estimate of input parameters.

Watershed	Storm date	Unit Hydrograph* Case 1	Unit Hydrograph* Case 2	Rainfall pattern Type 1	Rainfall pattern Type 2
Clays Mill	4-22-72	-20%	-57%	-26%	-10%
	7-24-73	-23%	-62%	-15%	-2%
	6-05-73	-23%	-61%	-23%	-13%
	5-22-73	-23%	-60%	-6%	25%
	9-25-71	-24%	-63%	-27%	-20%
	9-26-71	-24%	-62%	-21%	-15%
	6-28-72	-24%	-62%	-27%	-20%
Lansdowne	8-09-72	-18%	-51%	-26%	-4%
	4-22-72	-23%	-61%	-41%	-21%
	9-25-71	-23%	-60%	-39%	-21%
	9-26-71	-23%	-61%	-31%	-20%
	6-28-72	-22%	-60%	-36%	-28%
Lower Lick	4-25-78	45%	37%	-64%	-64%
	5-22-78	50%	42%	-48%	-7%
Cane Branch	5-07-60	62%	45%	-14%	17%
	4-05-62	32%	19%	NA	NA
Coshocton #10	4-24-70	19%	-39%	-18%	28%
	4-25-61	20%	-37%	-32%	-9%
	8-21-60	19%	-41%	-30%	13%
Coshocton #166	4-24-70	12%	-24%	27%	102%
	7-07-69	23%	-43%	-44%	-25%
Coshocton #5	4-24-70	19%	-40%	-18%	27%
	4-25-61	22%	-40%	-48%	-25%
	8-21-60	12%	-32%	-28%	12%
Stillwater W-1	9-26-67	14%	-25%	8%	61%
	4-30-70	17%	-33%	-23%	13%
	6-24-67	11%	-28%	18%	95%

* Urban Watersheds: Case 1 - Agricultural, Case 2 - Forested
  Agricultural Watersheds: Case 1 - Disturbed, Case 2 - Forested
  Forested Watersheds: Case 1 - Disturbed, Case 2 - Agricultural

sediment control options. In these cases, emphasis is placed on the relative differences on predicted results using different control options rather than on absolute quantities. An example illustrating this type of analysis has been presented by Wilson et al. (1982).

In general, the sedimentology techniques utilized in the model have been tested separately, but remain untested as a unique system of algorithms. The validation of the system is obviously desirable. Unfortunately, an observed data set, which includes a detailed knowledge of sedimentology input parameters for a number of different sediment control structures, is neither readily available nor simple and inexpensive to gather. In addition, since all possible combinations of sediment control options can not be experimentally evaluated, the user still has to rely on the accuracy of individual components for many applications.

In the following discussion, the sedimentology algorithms for predicting the sediment yield, the sediment graph, the particle size distribution, and the method for sediment routing are described along with algorithms used to predict the sedimentation effectiveness of sediment control structures. Algorithms are included for predicting the effectiveness of detention ponds, grass filters, and check dams.

Sediment Yield. SEDIMOT II has two different subroutines called MUSLE and SLOSS which may be used to calculate sediment yield. Sediment yield is calculated for each subwatershed, routed to the specified sediment control structure, and then combined to determine the total sediment yield. The modeling techniques used in each subroutine will be described separately.

In Subroutine MUSLE, sediment yield is estimated using Williams' Modified Universal Soil Loss Equation, MUSLE (Williams, 1975a). Using MUSLE for subwatershed "i", sediment yield is calculated as,

$$Y_i = 95(Q_i q_{p,\,i})^{0.56} K_i LS_i CP_i \tag{9}$$

where Y_i is the sediment yield in tons, Q_i is runoff volume in acre-feet, $q_{p,\,i}$ is peak discharge in cfs, K_i is the soil erodibility parameter, LS_i is the length-slope factor, and CP_i is the control practice parameter. The runoff parameters Q_i and $q_{p,\,i}$ in Equation 1 are predicted from SEDIMOT II's runoff component, whereas K_i and CP_i are input parameters. The length-slope parameter, LS_i, can either be entered by the user or calculated internally by using standard techniques (see Barfield et al., 1981; Wilson et al., 1982).

Sediment yield from each subwatershed is routed to a structure by using Williams' (1975b) model,

$$TY = \sum_{i=1}^{n+1} Y_i \exp(-BT_i \sqrt{D50}_i) \tag{10}$$

where TY is the total sediment yield at any structure, Y_i, T_i, $D50_i$ are the sediment yield, travel time, and median particle diameter for subwatershed " i", respectively; n is the number of subwatersheds, and B is the sediment routing coefficient. The " n+1" subwatershed corresponds to sediment data for the previous structure or junction.

The sediment routing coefficient, B, is calculated as,

$$(Qq_p)_T^{0.56} = \sum_{i=1}^{n+1} \left(Q_i q_{p,i}\right)^{0.56} \exp(-BT_i \, D50_i) \tag{11}$$

where $(Qq_p)_T$ is the product of runoff volume and peak discharge of the total drainage above any structure (including runoff from the previous structure or junction). B is determined in Equation (11) by using the Secant Method, which is an iterative technique.

The accuracy of the sediment yield technique and the sediment delivery method is presented in Williams (1975a) and Williams (1975b), respectively.

In Subroutine SLOSS the user is allowed to divide the flow path into six slope segments. The slope segment can either be an overland flow length or a channel reach. For each segment the sediment load is compared to the transport capacity. If the transport capacity is less than the sediment load, deposition is predicted. In the overland flow segments, detachment is divided into interrill and rill components. The channel segments are assumed to be non-erodible.

Detachment in interrill and rill areas is estimated by the equations given in CREAMS (Knisel, 1980). These equations are based on the Foster et al. (1977a) model with the coefficients evaluated to resemble the USLE. Detachment in interrill areas is predicted by,

$$D_i = 0.21 \, R(S + 0.014) \, K \, CP \tag{12}$$

and in rills as,

$$D_r = 3165.25 M \, Q \, q_p^{1/3} \left(\frac{\ell}{72.6}\right) S^2 \, K \, CP \tag{13}$$

where,

$$M = 2.0 \text{ if } \ell < 150'$$

$$M = 2.0 + 5.011/\ln(\ell) \text{ if } \ell > 150'$$

D_i is soil detachment from interrill area (lb_m/ft^2), D_r is soil detachment from rill areas (lb_m/ft^2), R is the rainfall erosivity factor, S is the sine of the slope angle, Q is runoff volume per unit area (inches), q_p is peak runoff rate per unit area (ft/sec), ℓ is the representative slope length (ft), and K and CP are the erodibility and control practice parameters, respectively. The accuracy of Equations (12) and (13) has been shown by Foster et al. (1977b) to be acceptable.

Sediment transport capacity is calculated from Yang's unit stream power equation (Yang, 1972; Yang, 1973; Yang and Stall, 1976), or,

$$\log C_t = 5.435 - 0.286 \log \frac{\omega d}{\upsilon} - 0.457 \log +$$

(14)

$$(1.779 - 0.409 \log \frac{\omega d}{\upsilon} - 0.314 \log \frac{U*}{\omega}) \log \left(\frac{VS}{\omega} - \frac{V_{cr}S}{\omega} \right)$$

where C_t is the total sediment concentration by weight, ω is the sediment particle terminal fall velocity, d is the median sieve diameter of bed material, υ is the kinematic viscosity, $U*$ is the shear velocity, VS is the unit stream power, and $V_{cr}S$ is the critical unit stream power at incipient motion.

The coefficients in Yang's sediment transport equation were determined from a regression analysis of laboratory data (Yang, 1973). Yang and Stall (1976) compared the observed transport capacity of natural rivers to those predicted by Yang's equation. For a large range of conditions, excellent agreement between predicted and observed sediment discharge was reported. Thus, Yang's equation would be more applicable to the channel flow segments. However, to keep the computational and input format of SLOSS relatively simple, Yang's equation is used on all flow segments. Furthermore, since unit stream power appears to be the dominant factor in sediment transport (Yang, 1972), it is hoped that the results from future research can be easily incorporated into SLOSS.

Load Rate. Subroutines of MUSLE or SLOSS are used in SEDIMOT II to calculate the total sediment yield from each subwatershed. To adequately evaluate the sedimentation effectiveness of a structure, it is necessary to distribute this sediment load with time. In SEDIMOT II the load rate graph is calculated by using an approach given by Ward et al. (1979). In their technique the concentration is assumed proportional to the flow rate, or,

$$C_i = Kq_i^a$$

(15)

where C_i is the concentration at the discharge, q_i; and K and a are constants. Ward et al. (1979) used an "a" of one as a default value. For a constant concentration, such as used by Overton and Crosby (1979), an "a" value of zero is used.

Load rate is estimated as the concentration multiplied by the flow rate, or

$$L_i = C_i q_i = Kq_i^{a+1}$$

(16)

where L_i is the load rate corresponding to a discharge of q_i. Total sediment load is equal to the sum of the load rates, or,

$$Y = \Delta t \sum_{i=1}^{m} L_i = K \Delta t \sum_{i=1}^{m} q_i$$

(17)

where Δt is the time increment between load rates and m is the number of discharge points. From Equation 17, K can be defined as,

$$K = \frac{Y}{\Delta t \sum\limits_{i+1}^{m} q_i^{a+1}}$$

(18)

For a calculated K and an estimated "a", the sediment load rate can be determined from Equation (16).

Particle Size Distribution. An eroded particle size distribution must be specified for each subwatershed. However, instead of entering a distribution for each individual subwatershed, the user initially enters one to fifteen possible distributions. Usually at least two different distributions are entered to represent the disturbed and undisturbed portions of the watershed. Procedures to estimate the eroded particle size distribution are discussed by Barfield et al. (1981) and Warner et al. (1982).

Particle size distribution is allowed to change as sediment is deposited enroute to a specified location. SEDIMOT II uses the Barfield et al. (1979) technique to adjust this distribution. In this procedure it is assumed that the largest particles are trapped. The original fraction finer value is defined as,

$$FF_0(d_i) = \frac{M_{di}}{M_T}$$

(19)

and the delivery ratio of sediment to the downstream point is defined as,

$$DR = \frac{M_D}{M_T}$$

(20)

where M_D is the mass of eroded sediment delivered to the downstream point from upstream, M_T is the total mass of sediment eroded upstream, M_{di} is the mass of sediment finer than particle d_i at the upstream point, DR is the delivery ratio and $FF_0(d_i)$ is the original fraction finer corresponding to particle diameter d_i.

The "new" fraction finer value at the downstream point is defined as,

$$FF_N(d_i) = \frac{M_{di}}{M_D}$$

(21)

where $FF_N(d_i)$ is the fraction finer value at the downstream point. If the largest particles are trapped first, the mass associated with particle diameters that are smaller than a critical diameter (i.e., d_{DR}) is determined by Equation 19. Hence by Equations (19) and (20), $FF_N(d_i)$ can be calculated as,

$$FF_N(d_i) = \frac{FF_o(d_i)}{DR} \quad \text{if } d < d_{DR} \tag{22}$$

and,

$$FF_N(d_i) = 1 \quad \text{if } d \geq d_{DR} \tag{23}$$

where d_{DR} is the particle diameter corresponding to the fraction finer whose value is equal to the delivery ratio.

The particle size distribution is allowed to change as the particles are deposited enroute to the subwatershed exit and from the subwatershed to the structure. Since MUSLE calculates delivered sediment yield, the delivery ratio to the subwatershed outlet is calculated by the procedure recommended by Williams (1975b), or:

$$DR1 = \left(\frac{q_p}{I_{ep}}\right)^{0.56} \tag{24}$$

where q_p is the peak discharge for the subwatershed, I_{ep} is the peak rainfall excess rate in cfs, and DR1 is the delivery ratio for MUSLE to the subwatershed outlet. If SLOSS is used with channel segments, DR1 is output as the delivery ratio from the subwatershed to the structure. Otherwise, the delivery ratio to the structure and between structures is determined from Williams' sediment routing method as,

$$DR2 = \exp(-BT \sqrt{D50}) \tag{25}$$

where DR2 is the delivery ratio from the subwatershed (or previous structure) to the structure. Other terms are as previously defined.

A composite particle size distribution is obtained from particle size data generated for each subwatershed. The composite fraction finer value is a mass weighted average. The new fraction finer value for each subwatershed is calculated by,

$$FF_{NJ}(d_i) = 1 \quad \text{for } d_i \geq d_{DRj} \tag{26}$$

$$FF_{Nj}(d_i) = FF_{oj}(d_i)/(DR1_j \times DR2_j) \quad \text{for } d_i < d_{DRj} \tag{27}$$

where FF_{oj} and FF_{Nj} are the old and new fraction finer, respectively, for subwatershed j, and d_{DRj} is the diameter of the particle corresponding to the fraction finer value that is equal to the product of $DR1_j$ and $DR2_j$. The mass weighted fraction finer can then be calculated as,

$$FF_N(d_i) = \sum_{j=1}^{n} RY_j FF_{Nj}(d_i)/TY \tag{28}$$

where $FF_{Nj}(d_i)$ is the composite fraction finer, RY_j is the routed sediment yield of subwatershed j, TY is the total sediment yield, and n is the number of subwatersheds.

The procedures to adjust particle size distributions are rough approximations that have been verified experimentally on vegetative filters (see Hayes et al., 1982) but not on an entire watershed.

Sediment Control Structures

Detention Ponds. The user of SEDIMOT II has the option of evaluating the performance of a detention pond by using a modified DEPOSITS model (Ward et al., 1977; Ward et al., 1979) or by using a CSTRS (continuous stirred reactors) model (Wilson and Barfield, 1984b). Modifications to DEPOSITS (see Wilson et al., 1982) include different techniques that are used to estimate the sediment load for each inflow plug, determine the effluent particle size distribution, and calculate the outflow hydrograph. DEPOSITS has also been reorganized to improve its computational efficiency. In the CSTRS model the pond is divided into a series of continuous stirred reactors to simulate the effects of mixing. Effluent concentration is predicted by using the sediment settling characteristics and by using reactor theory concepts (see Wilson et al., 1982).

The computational algorithm in the DEPOSITS and CSTRS models can be divided into three steps: (1) pond geometry characteristics, (2) hydrograph routing procedures, and (3) sedimentgraph routing procedures. The basin geometry and hydrograph routing steps are identical for both models. The DEPOSITS and CSTRS models differ in their sedimentgraph routing techniques.

The pond geometry step is used to calculate the characteristics of the pond that are functions only of stage height and are independent of time. Examples of these characteristics include storage volume, average pond depth, and storage routing curves for continuity routing. These values are calculated at each stage point (entered by the user) and are only dependent on the basin geometry and the spillway characteristics.

At each stage height a discharge value can either be entered by the user or it can be calculated by SEDIMOT II for a drop inlet spillway. If calculated by SEDIMOT II, the outflow discharge is predicted by using relationships for weir, orifice, and pipe flow. The smallest predicted discharge is then assigned to the particular stage point. If the user wishes to include an emergency spillway in the design, the entire stage-discharge curve may be entered by the user.

In the DEPOSITS and CSTRS models, the outflow hydrograph is calculated by using continuity routing. This method is based on the hydrologic form of the continuity equation shown below in finite difference form,

$$S_2 + \frac{O_2}{2} \Delta t = S_1 - \frac{O_1}{2} \Delta t + \frac{I_1 + I_2}{2} \Delta t \tag{29}$$

where S is the storage volume, O is the outflow discharge, I is the inflow discharge, and Δt is the routing time increment. The subscripts 1 and 2

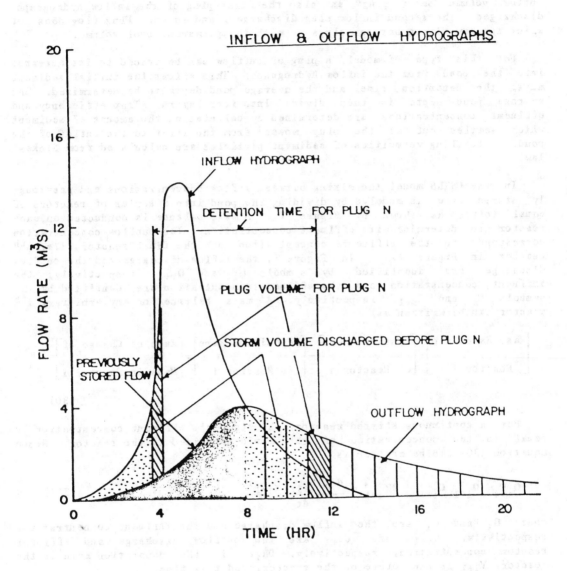

Figure 6. Plug Flow Concept (from Ward et al., 1979)

refer to the beginning and end of the time increment, respectively.

The sedimentgraph routing is done in the DEPOSITS model by assuming plug flow, as shown conceptually in Figure 6. In a plug flow model the first inflow volume or "plug" is also the first plug of the inflow hydrograph discharged, the second inflow plug discharged, and so on. Plug flow does not allow for any mixing between plugs or with the permanent pool volume.

For this type of model, a plug of outflow can be traced to its entrance into the pond from the inflow hydrograph. This allows the initial sediment mass, the detention time, and the average pond depth to be determined. The average pond depth is then divided into four layers. Trap efficiency and effluent concentrations are determined by calculating the amount of sediment which settles out as the plug moves from the inlet to the outlet of the pond. Settling velocities of sediment particles are calculated from Stokes' law.

In the CSTRS model the mixing between inflow concentrations and previously stored flow is modeled by dividing the pond into a series of reactors of equal volume as shown by Figure 7. A mass balance is conducted on each reactor to determine its effluent concentration. The outflow concentration corresponds to the effluent concentration of the final reactor (i.e. n^{th} reactor in Figure 7). In Figure 7, the inflow discharge and the outflow discharge are identified by symbols Q_1 and Q_{n+1}, respectively; the influent concentration and the effluent concentration are identified by the symbols C_1 and C_{n+1}, respectively. A mass balance on any arbitrary i^{th} reactor can be written as,

$$\begin{bmatrix} \text{Mass Rate Into} \\ \text{Reactor i} \end{bmatrix} - \begin{bmatrix} \text{Mass Rate Out} \\ \text{of Reactor i} \end{bmatrix} - \begin{bmatrix} \text{Deposition Rate} \\ \text{in Reactor i} \end{bmatrix} = \begin{bmatrix} \text{Rate of Change of} \\ \text{Mass in Reactor i} \end{bmatrix}$$

$$(30)$$

For a continuous stirred reactor, the reactor's effluent concentration is equal to the concentration of the fluid contained in the reactor. Hence Equation (30) can be written as,

$$Q_i C_i - Q_{i+1} C_{i+1} - DR_{i+1} = \frac{d(V_{i+1} C_{i+1})}{dt} \tag{31}$$

where Q_i and C_i are the inflow discharge and the influent concentration, respectively, Q_{i+1} and C_{i+1} are the outflow discharge and effluent reactor concentration, respectively, DR_{i+1} is the deposition rate in the reactor, V_{i+1} is the volume of the reactor, and t is time.

The trap efficiency of a basin is strongly dependent on the residence time of individual particles, which is dictated by the rate at which a given sediment slug is discharged from the pond. For instance, if a sediment slug is discharged at a rapid rate, the particles have a relatively short residence time; consequently, the trap efficiency of the pond is reduced.

A simple algorithm is used in the CSTRS model to estimate the discharge rate of a given sediment slug. This information indirectly determines the

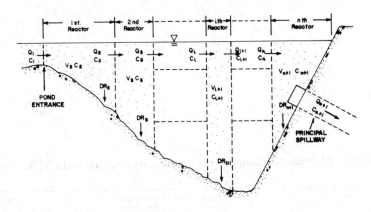

Figure 7.   Pond Divided Into A Series of CSTRS.

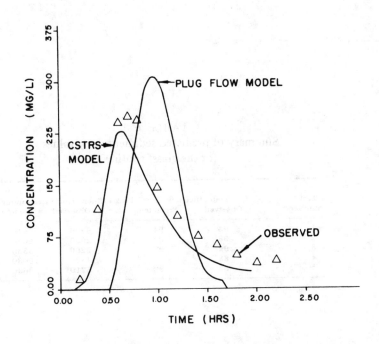

Figure 8.   Comparison of Predicted Effluent TSS
Concentration for CSTRS and Plug Flow
(DEPOSITS) Models.

TABLE 5
Summary of predicted sedimentology values by deposits and CSTRS routines

Test number	Trap efficiency, %			Peak effluent concentration, mg/L			Time to peak, min		
	Observed	DEPOSITS	CSTRS	Observed	DEPOSITS	CSTRS	Observed	DEPOSITS	CSTRS
18	74	77	78	250	310	230	42	57	39
19	85	83	84	1900	3600	2900	30	54	36
22	88	91	88	920	1590	1140	42	66	42
24	85	88	89	2700	2180	2050	18	39	18
26	88	81	80	24000	30500	25900	36	60	36
28	84	74	79	16000	16700	13100	18	45	21

TABLE 6
Summary of predicted sedimentology values for the grassf routine

Test number	Trap efficiency, %		Peak effluent concentration, mg/L	
	Observed	Predicted	Observed	Predicted
1	94	94	6270	7440
2	96	98	2990	2090
3	99	99	2290	3330
4	88	91	14300	10080
5	89	92	5210	3630

residence time of individual particles, and hence is useful in calculating sediment deposition. Although the final result of this algorithm is simple, its mathematical development is quite tedious.

The mass discharged from a pond within a particular time increment can be calculated as a product of the effluent concentration and the flow rate. This mass includes sediment particles from the current slug as well as sediment particles from previous inflow values. To partition the outflow mass an approximate distribution of sediment slugs in the final reactor is obtained by assuming that the series of CSTRS is essentially in equilibrium. For these conditions, the fraction of inflow mass that is discharged into the next reactor is constant with time and is constant between reactors. These conditions, although idealized, provide a simple framework to obtain a rough estimate of sediment distribution. Details are given in Wilson and Barfield (1984b).

The accuracy of the DEPOSITS and CSTRS models was evaluated by using the data presented by Tapp et al. (1981). This data was gathered on a pilot scale pond that was four feet wide and twenty feet long. The accuracy of the DEPOSITS and CSTRS models was evaluated on six data sets which were collected with variable inflow rates and without chemical flocculating agents.

The performance of the CSTRS model was evaluated by using two continuous stirred reactors and by using a dead space volume of fifteen percent. These values were determined by dye studies conducted independently of the verification runs. The DEPOSITS model was evaluated by using a dead space volume of fifteen percent, a uniform withdrawal system, and a uniform initial sediment concentration. Further details about the input parameters for the verification study are given in Wilson et al. (1982).

A summary of predicted and observed sedimentology values is given in Table 5, and a typical result of the predicted and observed sedimentgraph is shown in Figure 8. No attempts were made to calibrate predicted results with observed sedimentology data. In general, the CSTRS model did an excellent job of predicting the shape of the observed sedimentgraph. The DEPOSITS model does a reasonable job of predicting the peak effluent concentration but usually misses the shape and the timing of the observed sedimentgraph. Overall, the observed trap efficiencies of the Tapp et al. runs were adequately predicted by both the DEPOSITS and CSTRS models.

Grass Filter Model. A modified GRASFIL program (Hayes et al., 1982) is used in SEDIMOT II to simulate the sedimentation effectiveness of a vegetative filter. GRASFIL was developed from a series of studies investigating the ability of an erect media to trap sediment. The initial tests on erect media consisted of studies on metal pegs, but were later extended to include real erect grass stems in both laboratory and field studies. Laboratory studies were used to develop predictive equations for bed load transport (Tollner et al., 1982) and for suspended load transport (Tollner et al., 1976). These equations were developed by using a uniformly sized particle. Hayes et al. (1979) and Barfield and Hayes (1980) further extended the model to incude a nonhomogeneous particle size distribution. Modifications to GRASFIL in SEDIMOT II (see Wilson et al., 1982) include: (1) a single algorithm to calculate upslope deposition, (2) a smooth reduction in the trap efficiency of suspended sediment load caused by sediment deposition, and (3) an algorithm to calculate an outflow hydrograph. GRASFIL has also been changed to English units and has been reorganized to improve its computational efficiency and its input-output format.

Based on experimental results from the studies cited above, it was found that the sediment deposition pattern in vegetative filters resembled the profile shown in Figure 9. As sediment laden flow impinges on the filter, a reduction in its velocity causes the transport capacity to be lowered, which allows sediment deposition to occur. In GRASFIL, it is assumed that the bed load material is deposited in the sediment wedge and that the suspended load is trapped in the remaining portion of the filter.

Einstein's bed load dimensionless parameters were calibrated on laboratory data and used to calculate the transport capacity of bed material in the filter (Tollner et al., 1982). Based on these studies, bed load transport in a grass filter can be predicted as,

$$g_b = K \frac{(R_s S)^{3.57}}{d_p^{2.07}} \tag{32}$$

where,

$$K = \frac{6.642 \times 10^7}{(SG - 1)^{3.07}} \tag{33}$$

In the above expression g_b is the bed load transport rate [pounds per second per foot width], R_s is the characteristic hydraulic dimension [feet], and d_p is the diameter of the particle [millimeters]. The characteristic hydraulic dimension in Equation 32 is the spacing hydraulic radius (R_s) and is defined by an analogy between the spacing of grass elements and a rectangular channel of the same width.

As shown by Figure 9, the suspended load zone extends from the base of the deposition wedge to the outlet of the filter. Trapping in the suspended load zone is predicted from the Reynold's number (measure of turbulence) and a dimensionless fall number (measure of settling characteristics). Tollner et al. (1976) predicted the trap efficiency of the suspended load zone as,

$$T_s = \exp\left(-\left(0.00105 \frac{R_e^{0.82}}{N_f^{0.91}}\right)\right) \tag{34}$$

where,

$$N_f = \frac{L \, \omega}{V \, y} \; ; \; R_e = \frac{V \, R_s}{\upsilon} \tag{35}$$

T_s is the trap efficiency of suspended sediment, N_f is the particle fall number, R_e is the Reynold's number, L is the length of the suspended sediment zone, ω is the settling velocity of the particle, V is the flow

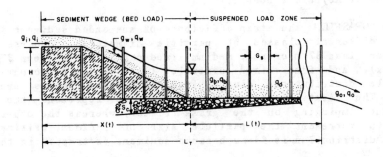

Figure 9. Schematic of Deposition Processes in a Vegetated Media.

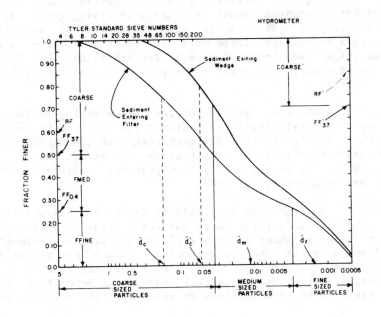

Figure 10. Illustration of Sediment Size Distribution change in a Vegetative Filter.

velocity, y is the flow depth, υ is the kinematic viscosity, and R_s is the spacing hydraulic radius.   In Equation 34, the predicted trap efficiency increases with an increase in particle fall number and decreases with an increase in Reynold's number.

In GRASFIL, the trap efficiency of suspended sediment is calculated for three different classes of particle diameters: Coarse sized particles (greater than 37 microns), medium sized particles (between 37 microns and 4.0 microns), and fine sized particles (less than 4.0 microns). Two different particle distributions, divided into these three classes, are shown in Figure 10.   The first distribution is used to represent the original size of sediment impinging on the grass filter, whereas the other distribution is used to represent the particle size of sediment exiting the wedge. The latter distribution is finer because sediment is trapped in the wedge.

The trap efficiency of sediment in the suspended zone is adjusted for cumulation of sediment on the bed and for infiltration losses. The cumulation of sediment on the filter bed will result in a reduced trap efficiency when the stools and soil indentations are filled. The infiltration losses increase the trap efficiency since suspended sediment particles are carried into the filter's bed with the downward movement of the transporting fluid.

The SEDIMOT II version of GRASFIL was tested on observed data collected by Hayes et al. (1982) on five vegetative strips located on the University of Kentucky experimental farm.   Inflow hydrographs and sedimentgraphs were obtained by spraying an inclined platform containing spoil material. This procedure yielded a nonhomogeneous particle size distribution and an unsteady inflow sedimentgraph.   The inflow flow rate, however, was fairly steady. Four of the five vegetative strips were divided into two slope segments. The outflow hydrograph was known for all for all tests. For further details about the experimental setup and input parameters the reader is referred to Hayes et al. (1982).

A summary of observed and predicted sedimentology values is given in Table 6, and a typical result of predicted and observed effluent concentrations is shown in Figure 11. The performance of the SEDIMOT II version of GRASFIL to predict effluent concentration is quite good.   SEDIMOT II predicted results vary only slightly from previous versions of GRASFIL (Hayes et al. 1982; Barfield and Hayes, 1980).

Check Dam.   An algorithm developed by Warner and Hirschi (1983) is used to predict the trap efficiency of porous check dams. A definition sketch of the geometry used is shown in Figure 12. Deposition of sediment is assumed to occur from a point that is ten percent larger than normal depth to the outlet of the dam.   In Figure 12, these two points are labeled as Point 1 and Point 2, respectively. The horizontal distance of deposition is represented by ΔX. The value of ΔX is determined by using a single step backwater curve.

The detention time of the check dam is defined as the time required for a sediment particle to travel a distance of ΔX. The flow velocity is estimated as an average between normal flow and flow at the outlet of the dam.   In addition, the distance that a particle must fall to be trapped is defined as the average of the normal depth and the depth at the check dam outlet.   For known detention time and fall depth, Hirschi proposed that the trap

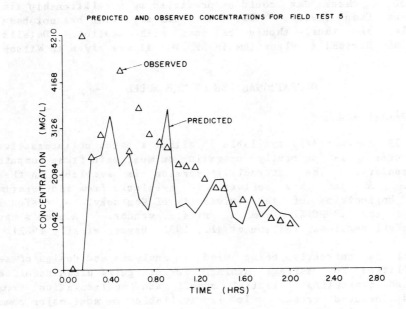

Figure 11. Typical Grass Filter Result

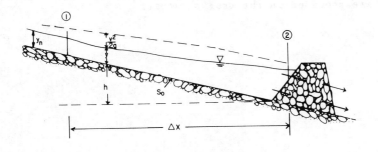

Figure 12. Check Dam Backwater Curve

efficiency of a check dam could be predicted by a relationship similar to that given by Camp (1945). The check dam algorithm has not been tested experimentally, and thus, should be used with caution. Details of the application of Hirschi's algorithm in SEDIMOT II are given in Wilson et al. (1982).

OPERATIONAL USE OF THE MODEL

Model Availability and Use

SEDIMOT II is currently available in either a batch or interactive mode. The batch version is currently operating on most main frame computers and some minicomputers. The interactive version is available on the HP3000 minicomputer. All of this software is available from the Department of Agricultural Engineering of the University of Kentucky. A version is also available for the IBMPC/XT from a private vendor. A user's manual is available for all versions, (Wilson et al., 1982; Warner et al., 1982).

The model is currently being used in analysis and design of sediment control facilities for surface mining, for designing urban runoff control facilities and evaluating potential runoff and sedimentation from agricultural and forested areas. It is available on most major commercial computer systems in the United States of America and in some other countries. A four-day short course is taught by the authors several times each year to familiarize users with the model and to give hands-on experience. At the present, there are in excess of 300 users of the model.

Developing Model Inputs

All model inputs are either available from standard hydrologic tables or are given in the user's manual. In no case is the user required to use optimization procedures to develop calibration constants. Worksheets and coding sheets are provided in the user's manual.

SUMMARY

A discussion is given of the principles to be followed in selecting a nonpoint source model. The principles involve the need for balance between the desires of the modeler for sophistication and the capabilities and requirements of the user. Following these principles, a model known as SEDIMOT II was developed which attempts to balance complexity and ease of use such that a reasonable description of hydrologic responses can be obtained without requiring that the user be an expert hydrologist.

The hydrologic component of SEDIMOT II consists of rainfall, runoff and flow routing modules. The rainfall excess is calculated by the SCS's curve number model and then routed to the watershed outlet by unit hydrograph techniques. Muskingum's routing method is used to route the resulting runoff hydrograph through a channel reach.

The accuracy of the hydrologic component of SEDIMOT II was evaluated on eight different watersheds for a total of twenty-seven storms. Excellent to fair agreement was obtained between predicted and observed hydrographs. The sensitivity of predicted peak discharge was also evaluated for the storms of the verification study. The effect of a variation in curve number, time of concentration, unit hydrograph shape, and temporal rainfall pattern was evaluated.

The sedimentology component of SEDIMOT II was also presented. Sediment yield can be predicted by using Williams' Modified Universal Soil Loss Equation and exponential sediment routing technique, or by using detachment-transport concepts incorporated into an algorithm called SLOSS. The detachment component in SLOSS is predicted by the relationships given in CREAMS and its transport capacity is predicted by a unit stream power relationship.

To evaluate the effectiveness of different sediment control structures, the sedimentgraph and particle size distributions associated with the sediment yield must be calculated. In SEDIMOT II, the sedimentgraph is determined by assuming that the load rate is proportional to the flow rate. The observed range of proportionality ratio is between one and two. The particle size distribution of eroded sediment at the source is entered by the user. This distribution is adjusted for selective deposition by assuming that the largest particles are removed.

In SEDIMOT II the sedimentation effectiveness of three different types of sediment control structures can be evaluated: detention ponds, grass filters, and porous check dams. Two different algorithms are available to evaluate the effectiveness of detention ponds; via, the DEPOSITS model and the CSTRS model. Both models have been validated on a pilot scale pond whereas, the DEPOSITS model has also been tested on larger scale ponds. The sedimentation effectiveness of grass filters is determined by using a modified GRASFIL model. The GRASFIL model has been validated in laboratory studies and in two different field studies. The trap efficiency of porous check dams is calculated by an algorithm proposed by Warner and Hirschi. This algorithm has not been tested experimentally and, hence, should be used with caution.

REFERENCES

Albrecht, S. C. and B. J. Barfield. 1981. Use of a vegetative filter zone to control fine-grained sediments from surface mines. Report No. EPA 600/7-81-117, US EPA Industrial Environmental Research Lab, Cincinnati, Ohio.

Barfield, B. J., I. D. Moore, and R. G. Williams, 1979. Prediction of sediment yield from surface mined watershed. Symposium on Surface Mining Hydrology, Sedimentology, and Reclamation, University of Kentucky, Lexington, Kentucky.

Barfield, B. J. and J. C. Hayes. 1980. Modeling sediment filtration by vegetative filters. Proceedings ASCE Watershed Management Symposium, Boise, Idaho.

Barfield, B. J., R. C. Warner, and C. T. Haan. 1981. Applied hydrology and sedimentology for disturbed areas. Oklahoma Technical Press, Stillwater, Oklahoma.

Beasley, D. B., L. F. Huggins, and E. F. Monke. 1980. ANSWERS: A model for watershed planning. Trans. ASAE, 10(3):485-492.

Betson, R. P., J. Bales, and H. E. Pratt. 1980. User's guide to TVA-HYSIM, a hydrologic program for quantifying land-use change effects. Tennessee Valley Authority, Norris, Tennessee.

Brume, G. M. 1953. Trap efficiency of reservoirs. Trans. Am. Geophy. Union 34(3): 407-418.

Camp, T. R. 1945. Sedimentation and the design of settling tanks. Proc. Am. Soc. Civil Engrs., 71, paper 2285:895-959.

Churchill, M. A. 1948. Discussion of analysis and use of reservoir sedimentation data by L. G. Gottschalk. Proc. Inter-Agency Sedimentation Conference, U. S. Department of Interior, Washington, D. C.

Curtis, D. C. and R. H. McCuen. 1977. Design efficiency of stormwater detention basins. J. Water Resour. Planning and Management Div., Proc. ASCE. 103:125-140.

Dooge, J. C. I. 1973. Linear theory of hydrologic systems. USDA Agricultural Research Service, Technical Bulletin No. 1468, Washington, D. C.

Dooge, J. C. I., W. G. Strupczewski, and J. J. Napiorkowski. 1982. Hydrodynamic derivation of storage parameters of the Muskingum model. Jour. of Hydr., 54:371-387.

Foster, G. R., L. D. Meyer, and C. A. Onstad. 1977a. An erosion equation derived from basic erosion principles. Trans. ASAE, 20(4):678-682.

Foster, G. R., L. D. Meyer, and C. A. Onstad. 1977b. A runoff erosivity factor and variable slope length exponent for soil loss estimates. Trans. ASAE, 20(4):683-687.

Foster, G. R. 1982. The erosion process. Chapter VIII in Hydrologic Models of Small Watersheds. ASAE Monograph No. 5. American Society of Agricultural Engineering, St. Joseph, MI.

Hayes, J. C. 1979. Evaluation of design procedures for vegetal filtration of sediment from flowing water. Unpubl. Ph.D., University of Kentucky, Lexington, Kentucky.

Hayes, J. C., B. J. Barfield, and R. I. Barnhisel. 1979. Filtration of sediment by simulated vegetation II: Unsteady flow with non-homogeneous sediment. Trans. ASAE, 22(5):1063-1067.

Hayes, J. C., B. J. Barfield, and R. I. Barnhisel. 1982. The use of grass filters in strip mine drainage. Volume III: Empirical verification of procedures using real vegetation. Technical Report IMMR 82/070, Institute for Mining and Mineral Research, University of Kentucky, Lexington, Kentucky.

Hirschi, M. C. 1981. Efficiency of small scale sediment controls. Ag. Eng. File Report, University of Kentucky, Lexington, Kentucky.

Knisel W. G., Editor. 1980. CREAMS: A field-scale model for chemical, runoff, and erosion from agricultural management systems. USDA, Conservation Research Report No. 26.

Li, R. M., D. B. Simons, and D. R. Carter. 1977. Mathematical Modeling of Soil Erosion by Overland Flow. In Soil Erosion: Predection and Control, Soil Conservation Society of America, Ankeny, Iowa.

Ohlander, C. A. 1976. Defining The Sediment Trapping Characteristics of a Grassed Buffer. Special Case: Road Erosion. Proceedings Third Federal Interagency Sedimentation Conference, US Government Printing Office, Washington, DC.

Overton, D. E. and E. C. Crosby. 1979. Effects of contour coal strip mining on stormwater runoff and quality. Report to U. S. Department of Energy, Department of Civil Engineering, University of Tennessee, Knoxville, Tennessee.

Schamber, D. R. and B. E. Larock. 1981. Numerical analysis of flow in sedimentation basins. Journal Hydr. Division, Proc. ASCE, Vol. 107, pp. 575-591.

Soil Conservation Service 1964. Computer Program for Project Formulation Hydrology.

Soil Conservation Service. 1972. National Engineering Handbook, Section 4, Hydrology.

Soil Conservation Service. 1973. A method for estimating volume and rate of runoff in small watersheds. Technical Paper No. 149.

Tapp, J. S., B. J. Barfield, and M. L. Griffin. 1981. Prediction of suspended solids removal in sediment ponds utilizing chemical flocculation. IMMR Report 81/063, Institute for Mining and Minerals Research, University of Kentucky, Lexington, Kentucky.

Tollner, E. W., B. J. Barfield, C. T. Haan, and T. Y. Kao. 1976. Suspended sediment filtration capacity of simulated vegetation. Trans. ASAE, 19(4):678-682.

Tollner E. W., B. J. Barfield, and J. C. Hayes. 1982. Sedimentology of erect vegetal filters. Proc. ASCE, 108(HY12):1518-1531.

Ward A. D., C. T. Haan, and B. J. Barfield. 1977. Simulation of the sedimentology of sediment basins. Technical Report No. 103, University of Kentucky, Water Resources Institute, Lexington, Kentucky.

Ward, A. D., C. T. Haan, J. S. Tapp. 1979. The DEPOSITS sedimentation pond design manual. University of Kentucky, Lexington, Kentucky.

Ward, A. D., D. L. Rausch, C. T. Haan, and H. G. Heinemann. 1981. A verification study on a reservoir sediment deposition model. Trans. ASAE, 24(2):340-346, 352.

Warner, R. C. and M. C. Hirschi. 1983. Modeling Check Dam Trap Efficiency Paper No. 83-2082. American Society Agricultural Engineers, St. Joseph, MI.

Warner, R. C., B. N. Wilson, B. J. Barfield, D. S. Logsdon, and P. J. Nebgen. 1982. A hydrology and sedimentology watershed model. Part II: Users' manual. Univesity of Kentucky, Department of Agricultural Engineering, Special Publication, Lexington, Kentucky.

Williams, J. R. 1975a. Sediment-yield prediction with universal equation using runoff energy factor. In present and Prospective Technology for Predicting Sediment Yields and Sources, ARS-S 40:244-251.

Williams, J. R. 1975b. Sediment routing for agricultural watersheds. Water Resources Bulletin, 11(5):965-974.

Williams, J. R. and R. W. Haan. 1973. HYMO, A Problem Oriented Computer Language For Building Hydrologic Models: A Users Manual. USDA-ARS 5-9.

Wilson, B. N., B. J. Barfield, and I. D. Moore. 1982. A hydrology and sedimentology watershed model. Part I: Modeling techniques. University of Kentucky, Department of Agricultural Engineering, Special Publication, Lexington, Kentucky.

Wilson, B. N. and B. J. Barfield. 1984a. Modeling The Performance of Sediment Detention Basins, 2, The BASIN Model Using Diffusion Theory. Technical Report, Institute for Mining and Minerals Research, University of Kentucky, Lexington, Kentucky.

Wilson, B. N., and B. J. Barfield. 1984b. Sediment detention pond model using CSTRS mixing theory. Trans. ASAE, 27(5): 1339-1344.

Wilson, B. N., B. J. Barfield, A. D. Ward and I. D. Moore (1984a). A hydrology and sedimentology watershed model. Part I: Operational format and hydrologic component. Trans. ASAE, 27(5): 1370-1377.

Wilson, B. N., B. J. Barfield, I. D. Moore, and R. C. Warner. (1984b). A hydrology and sedimentology model. Part II: Sedimentology component. Trans. ASAE, 27(5): 1378-1384.

Wolfe, M. L., V. O. Shanholtz, M. D. Smolen, J. N. Jones, and B. B. Ross. 1981. Simulating sediment transport to evaluate sediment control practices in surface mined areas. ASAE Paper 81-2041, ASAE Summer meeting, June 21-24, Orlando, Florida.

Yang, C. T. 1972. Unit stream power and sediment transport. Proc. of ASCE, 98(HY10):1805-1826.

Yang, C. T. 1973. Incipient motion and sediment transport. Proc. of ASCE, 99(HY10):1679-1701.

Yang, C. T. and J. B. Stall. 1976. Applicability of unit stream power equation. Proc. of ASCE, 102(HY5):559-568.

INTEGRATION OF RUNOFF AND RECEIVING WATER MODELS FOR COMPREHENSIVE WATERSHED SIMULATION AND ANALYSIS OF AGRICULTURAL MANAGEMENT ALTERNATIVES

Anthony S. Donigian, Jr.

Anderson-Nichols & Co., Inc.
Palo Alto, CA 94303

Introduction

Mathematical models of agricultural runoff processes are being used to analyze and predict the quantity and quality of runoff from agricultural lands. The ultimate goal is to use these models to develop a Best Management Practice (BMP) plan that will maintain agricultural productivity while minimizing adverse water quality impacts. However, runoff models by themselves are not sufficient to predict water quality effects of BMP's since instream transport and transformations are not usually represented. Although agricultural runoff models can be used to evaluate BMP's effects on runoff and pollutant loadings from small, field-sized areas, they must be linked with receiving water quality models to assess the quality in receiving waters resulting from BMP's applied to the watershed. Moreover, the runoff and receiving water quality models must be compatibly designed, so that the runoff and agricultural pollutant loadings (e.g., sediment, pesticides, nutrients) can be accepted, transported, and transformed by the associated aquatic processes in the receiving water model. Thus predicting water quality in agricultural watersheds requires integrated, comprehensive modeling systems so that conventional practices and changes affected by BMP's can be represented by the runoff models with the resulting water quality conditions determined by the receiving models.

The historical development of modeling as a tool for analyzing water quality problems has progressed along these two distinct and separate avenues: runoff and leaching models developed by the scientific agricultural community and receiving water (instream) models developed by the environmental engineering community. Only recently have model development efforts attempted to specifically integrate runoff (or field-scale) and receiving water components into model packages for comprehensive analysis of complex agricultural watersheds and river basins.

As discussed above, water quality conditions resulting from agricultural practices depend on both surface and subsurface pollutant loadings from fields and instream processes affecting contaminant fate and transport. Consequently, the use of models to properly analyze this cause-effect relationship between water quality and agricultural practices must include both field scale runoff and receiving water models, and linkage or integration procedures. This paper serves as a focal point for two workshop sessions that concentrate separately on field-scale/control practice models and large basin receiving water models.

This paper describes the integration procedures used by the EPA Hydrologic Simulation Program - FORTRAN (HSPF) to link surface runoff, subsurface flow (and quality) and receiving water modules to perform comprehensive watershed simulation. The watershed segmentation process is discussed and various applications of HSPF are cited as examples of analyzing the water quality impacts of Agricultural Best Management Practices (BMP's).

Hydrological Simulation Program - FORTRAN (HSPF)

HSPF is a comprehensive program for modeling runoff, sediment, pesticides, nutrients, and other water quality constituents from urban, agricultural and other land uses. It allows detailed simulation of stream hydraulics, water quality processes, pesticide and nutrient behavior in soil, lake reservoir dynamics, and sediment contaminant transport. It also includes extensive data handling and analysis procedures that support and complement the simulation capabilities.

HSPF is the latest product resulting from some 20 years of process research and model development, testing, refinement, and application. The initial 10 years of this research effort involved the development of the Stanford Watershed Model (Crawford and Linsley 1966) and the Hydrocomp Simulation Program (Hydrocomp 1969). These two models provided the basic theory and framework for the continuous simulation of hydrologic and hydraulic processes in HSPF. During the second decade of development (the 1970's), water quality processes were superimposed on the relevant transport components. Under the sponsorship of the EPA Athens Environmental Research Laboratory extensive model development and associated data collection was performed in an effort to develop effective tools for analyzing agricultural runoff problems and control practices. This 10 year program which led to the development of HSPF has been described by Barnwell and Johanson (1981).

Watershed Segmentation and Simulation

Modeling agricultural nonpoint pollution and resulting water quality conditions in complex watersheds, with varying meteorologic, soils, and land use characteristics, requires the joint use and linkage of surface runoff, subsurface (or leaching), and instream models. Individual models must be run and their output interlinked in an appropriate manner to represent the entire watershed system and its varying characteristics. Figure 1 is a schematic of how land segments and stream reaches can be modeled separately and linked, either by manual data file transfers or automated data base procedures as in HSPF, to simulate complex watershed conditions. The following discussion of watershed segmentation and simulation is derived from experience with HSPF (Johanson et al, 1981) which is one of the few currently available comprehensive watershed models with surface runoff, soil, and instream components.

FIGURE 1: SEGMENTATION OF COMPLEX WATERSHEDS FOR MODELING

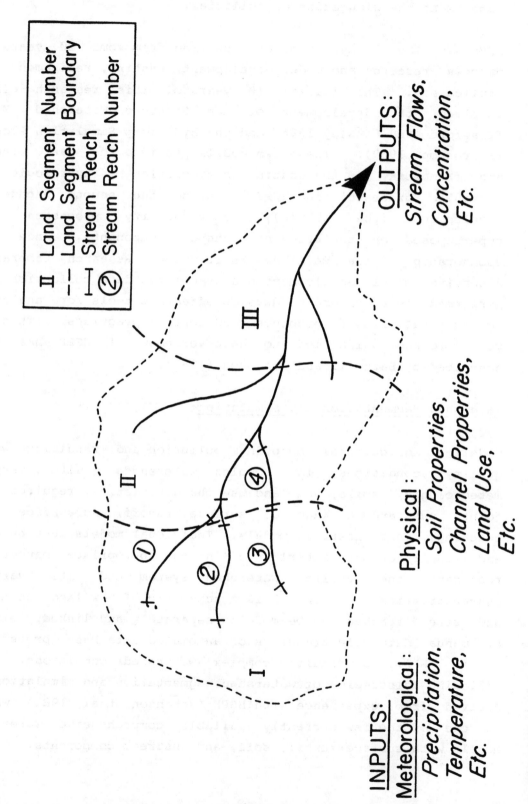

One of the basic components of watershed modeling using HSPF and predecessor models, is the division of the watershed into land segments, each with relatively uniform meteorologic, soils, and land-use characteristics. Similarly the channel system is segmented into 'reaches' as shown in Figure 1, with each reach demonstrating uniform hydraulic properties. The entire watershed is then represented by specifying the reach network, i.e., the connectivity of the individual reaches, and the area of each land segment that drains into each reach. Each land segment is then modeled to generate runoff and pollutant loads (surface and subsurface) per unit area tributary to the stream channel. Multiplying the unit area runoff and pollutant loads by the area of each land segment tributary to each channel reach determines the runoff and pollutant loads to each reach; performing these calculations for each reach in conjunction with modeling the instream hydraulic and water quality processes results in the simulation of the entire watershed.

Table 1 lists the major considerations for watershed segmentation based on inputs, physical characteristics, and outputs. For the purposes of HSPF, a segment is defined as a parcel of land which exhibits a homogeneous hydrologic and water quality response. Hence, one set of hydrologic and water quality parameters can be used to characterize all of the land considered as one segment. For modeling purposes, it is not necessary that all of the land in a segment be contiguous. The only requirements are that the segment parameters reasonably represent the hydrologic and water quality characteristics of all land considered as part of the segment, and that the total area of each segment contributing runoff and pollutants to each hydraulic reach is known.

The hydrologic response of a parcel of land is a function of meteorologic patterns, soils characteristics, land uses, and management practices. In most cases, meteorologic patterns and soils characteristics allow for preliminary division of a basin into segment groups. These are designated as I, II, and III in Figure 1. A segment group is a parcel of land which is exposed

TABLE 1: WATERSHED SEGMENTATION CONSIDERATIONS

	LAND SEGMENTS	CHANNEL SEGMENTS
INPUTS	Meteorologic (Precip., E.T., Other)	Point Discharges
	Man-Made (Irrigation, Chemicals)	Diversions
PHYSICAL CHARAC.	Soils	Slope
	Topography	Roughness
	Land Use	Morphology
		Obstructions (e.g. Dams)
OUTPUTS	Land Use	Gage/Data Locations
	Cropping	Management Alternatives

to meteorologic conditions (rainfall, evaporation, etc.) which for modeling purposes are designated by one set of meteorologic time series. In addition, it is assumed that all of the land in the segment group would exhibit a homogenous hydrologic response if there were uniform land use and management practices. In order to make this assumption, soils characteristics must be reasonably consistent throughout the segment group area; otherwise, further segment divisions would be required. Segment groups are subsequently divided into segments, with each segment representing a different land use or management practice condition.

If land use and management practices vary within a segment group, and the categories are sufficiently different to cause different runoff and pollutant loads, then each segment group must be further divided into separate land use and management categories. For example, if Land Segment II in Figure 1 includes both corn and soybean acreage which demonstrate different runoff and chemical loading conditions, it will be divided into these two land use categories. As described above, the runoff (and subsurface) portions of HSPF are then executed separately for corn conditions and soybean conditions, the resulting contributions from each land use are input to the appropriate channel reach, and instream chemical fate and transport processes are simulated. Figure 2 schematically demonstrates the type of segment linkages required for HSPF simulation of a complex watershed. In Figure 2, PLS represents pervious land segment conditions, ILS represents impervious land segment conditions, and RCHRES represents stream channel (or completely-mixed reservoir) segments. Individual PLSs and ILSs can generate flow and pollutant loads to any number of RCHRES (i.e., channel) segments in order to represent the multiple land use/management practice conditions in large, complex watersheds.

Evaluating Agricultural Management Alternatives

With these procedures, watersheds and river basins are simulated by the linkage of runoff models, providing contributions from

271

FIGURE 2: SCHEMATIC OF EXAMPLE PLS/ILS AND RCHRES CONNECTIONS

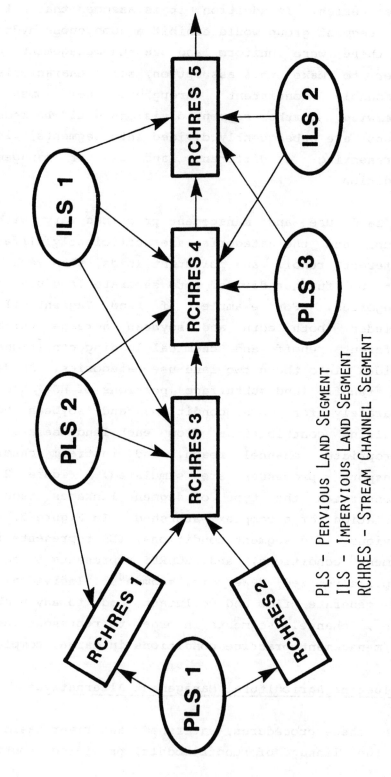

PLS – Pervious Land Segment
ILS – Impervious Land Segment
RCHRES – Stream Channel Segment

Unit area flow/loads converted by drainage area for input to each RCHRES

all land uses in the watersheds, to instream models representing the channel system and associated hydraulic and water quality processes.

Using HSPF and predecessor models, these procedures have been applied extensively throughout the U.S., Canada, South America, and other parts of the world for a wide range of hydrologic, water resource, and water quality management studies (Barnwell and Johanson, 1984; Barnwell and Kittle, 1984). For analysis of agricultural nonpoint pollution and evaluation of management alternatives, HSPF has been applied to both small watersheds and large river basins in Iowa (Donigian et al., 1983b; Imhoff et al., 1983) and Nebraska (Gilbert et al., 1982). The Iowa studies were part of a comprehensive field evaluation program sponsored by the U.S. EPA to test and demonstrate the ability of agricultural BMPs to achieve water quality goals and improvement. HSPF was applied to both a 52 sq. km., intensively monitored, agricultural watershed and an adjacent 7200 sq. km. river basin in Central Iowa to demonstrate its application at both levels. A separate study analyzed the impact and procedures for adjusting HSPF parameters to represent a wide range of agricultural BMPs (Donigian et al., 1983c). Based on these procedures, parameter adjustments were estimated for a "conservation tillage" BMP, the adjustments were applied to PLSs for corn and soybean croplands, and model runs were performed to quantify the impact of the BMPs on runoff, sediment, pesticide, and nutrient loadings, and resulting instream concentrations. In this way, the linkage of runoff and instream modeling components of HSPF were used to analyze water quality conditions in both small and large watersheds resulting from existing and proposed agricultural practices.

Detailed guidelines for watershed segmentation and evaluation of management alternatives, using the Iowa River Study as an example, are available in the HSPF Application Guide (Donigian et al, 1983). Although these linkage procedures have been presented for HSPF and similar models, they are indicative of the representation assumptions needed for comprehensive

273

watershed simulation and analysis of agricultural management
alternatives.

REFERENCES

Barnwell, T.O., and R. Johanson. 1981. "HSPF: A
    Comprehensive Package for Simulation of Watershed Hydrology
    and Water Quality," in Nonpoint Pollution Control - Tool
    and Techniques for the Future, Tech. Pub. 81-1, Interstat
    Comm. on the Potomac River Basin, 1055 First St.
    Rockville, MD 20850, pp. 135-153.

Barnwell T.O., and J.L. Kittle, Jr. 1984. Hydrologic
    Simulation Program - FORTRAN: Development, Maintenance
    and Applications, In: Proceedings of the Third
    International Conference on Urban Storm Drainage, in press
    Chalmers Institute of Technology, Goteborg, Sweden.

Crawford, N.H., and R.K. Linsley. 1966. Digital Simulation in
    Hydrology: Stanford Watershed Model IV. Stanford
    University, Dep. of Civil Eng. Stanford, CA. Tech Rep. 39.

Donigian, A.S., Jr., J.C. Imhoff, B.R. Bicknell, and J.L.
    Kittle, Jr. 1983a. Guide to the Application of the
    Hydrological Simulation Program - FORTRAN (HSPF). U.S.
    Environmental Protection Agency, Athens, GA.

Donigian, A.S., Jr., J.C. Imhoff, and B.R. Bicknell. 1983b.
    Modeling Water Quality and the Effects of Agricultural Best
    Management Practices in Four Mile Creek, Iowa. U.S.
    Environmental Protection Agency. Athens, GA.

Donigian, A.S., Jr., J.L. Baker, D.A. Haith, and M.F. Walter.
    1983c. HSPF Parameter Adjustments to Evaluate the Effects
    of Agricultural Best Management Practices. U.S.
    Environmental Protection Agency. Athens, GA.

Gilbert, D.P., D. Soboshinski, G. Bartelt, D. Razavian. 1982.
    Development of State Water Quality Management Plan for the
    State of Nebraska, Nebraska Water Resources Center,
    University of Nebraska, 310 Agricultural Hall, Lincoln, NE
    68583

Hydrocomp Incorporated. 1969. Hydrocomp Simulation
    Programming: Operations Manual. 2nd ed. Palo Alto, CA.

Imhoff, J.C., B.R. Bicknell, and A.S. Donigian, Jr. 1983.
    Preliminary Application of HSPF to the Iowa River Basin to
    Model Water Quality and the Effects of Agricultural Best
    Management Practices. U.S. Environmental Protection
    Agency. Athens, GA.

Johanson, R.C., J.C. Imhoff, H.H. Davis, Jr., J.L. Kittle, Jr.,
    and A.S. Donigian, Jr. 1981. User's Manual for the
    Hydrological Simulation Program - FORTRAN (HSPF): Release
    7.0. U.S. Environmental Protection Agency. Athens, GA.

Johanson, R.C., and J.L. Kittle, Jr. 1983. Design, Programming
    and Maintenance of HSPF. ASCE J. Tech. Topics in Civil E.,
    109, 41.

FIELD SCALE SIMULATION OF NUTRIENT LOSSES

G.Bendoricchio, Universita` di Padova, Facolta` di
Ingegneria, Istituto di Chimica Industriale, Via F.Marzolo
n.9, 35131 Padova, Italy.

A.Rinaldo, Universita` di Padova, Facolta` di Ingegneria,
Istituto di Idraulica, Via Loredan 20, 35131 Padova, Italy.

ABSTRACT

The rationale for a conceptual model of a unit-mass response function (UMRF) for simulation of nutrient losses from agricultural watersheds is recalled. The model consists of an extension of the well-known hydrological model due to Nash, with agricultural basins represented by a cascade of continuously stirred tank reactors (CSTR). Hysteretical phenomena in pollutant concentration/water discharge relations, appearing both in the mathematics of the model as well as in experimental observations, are discussed. Features of the mathematical model include: optimal transformation of the lumped parameter structure into a large-scale distributed parameter structure (in analogy to IUH hydrological analyses); parsimonious hydrological and chemical input; computational reliability; capacity for continuous simulation and forecasting. Furthermore, the model structure allows for separate estimation of hydrological parameters and pollutant fixed and mobile phase parameters. The results of numerical experiments aimed at emphasizing the sensitivity of the model to chemical parameter variations and the ramifications for field-scale simulation are discussed.

INTRODUCTION

Nutrient loadings to surface and subsurface water bodies have increased dramatically in the last decade. This marked trend calls for identification of critical environmental conditions for aquatic life and, in particular, for exploitation of recreational, agricultural and civil potential of water resources [1], in particular for Northern Italy [2].

Non point source (NPS) contributions play a distinct role in such scenarios, being most difficult to control and relatively newborn.

In fact, attention has been focused on NPS pollution (NPS) phenomena since the beginning of the seventies, when early works in either American or European literature introduced the problem and reported the first environmental effects. Analogous trends have been shown by National Legislation, through which the first recommendations and specifications concerning NPSP have appeared, dating also in the early seventies.

Italian researchers are relative late concerns to this field, with initial efforts made in 1975; nevertheless, the rapidly growing body of Italian work in the area illustrates the recent inclusion of NPSP among industrial and urban point source pollution as major concerns of the scientific, political, and legal communities.

Historically whereas the debate over lake, sea and lagoon eutrophication attracted much attention among the early contributions, the late seventies have brought quantitative results on the magnitude of the phenomena.

In this regards, the Italian research group coorganizing the present workshop has begun to produce working procedures suited to the particular needs of Northern Italy's agricultural watershed.

It appears, and it did so also at the very beginning, that the studies on such complex phenomena as water NPSP need to be supported by tools capable to suggest strategies and control measures aiming, in particular, at large-scale cost-effective response. In this perspectives, mathematical models seem to be perfectly tailored to efficient and cost-effective screening analyses.

NPS contributions to water pollution have been studied experimentally and theoretically, with a view toward overall evaluations of source magnitude. The early results consisted of annual data of vaste loadings of NPS origin, mainly designed for comparison with point source loads.

the environmental conditions of the Venice Lagoon, which itself represents a most controversial site, being subject to national and international attention because of the intrinsic economic and cultural interest, have long been studied and several estimations of annual loads have been proposed [3,4,5,6].

Nevertheless it appears that although annual balances yield indications on potential problems and source magnitudes (together with source areas and delivery locations), they are certainly unsuitable to simulation of time evolution of the phenomena. In fact neither do such approaches portray the mechanisms of NPSP, nor do they suggest control strategies, which are now badly needed.

For these reasons a second-order mathematical model of the UMRF type has been devised, whose rationale is based on the need for integrating or planning models for field-scale detailed simulations of nutrient losses from agricultural watersheds.

The need for modifying the time scale of the models has also arisen in view of the final goal of the simulations, which deals with water quality simulation of the receiving tidal water body (the Lagoon of Venice). In fact [23], the time step required for continuous simulation of the time evolution of receiving water quality is of the order of hour (e.g. intertidal time scale) for which event-based responses are almost useless.

Besides the conceptual interest for the novel approach, its practicality for long-term simulations of the mainland-lagoon system may outweigh that of refined deterministic or distributed parameter models (e.g. [8,9,10], because of the complexity of the mainland ecosystem (2000 km wide, approximately 1 million inhabitants; 27 freshwater delivery locations; 30% of mainland subject to mechanical drainage; a number of industrial sites; flat surface slopes; wide range of variability of hydrologic, geologic and agricultural characteristics).

Integration of runoff and receiving water models, or the "downstream" condition on NPS models, has therefore constituted a dominant issue for the case study at hand.
The theoretical structure of the model is recalled, and two numerical applications are performed: the former emphasizes the effects of single parameters for long-term forecasting and overall verification on real case studies, the latter establishes the capability of the model to portray event-bases rainfall-runoff transformations and the related hysteretical phenomena.

THE MATHEMATICAL MODEL

Nutrient losses from watersheds have been simulated by a number of models. A recent review [7], has classified the models according to level of sophistication, potential field of application and time discretisation. Each class of models is capable of a range of feasible simulations which, in principle, cover any level of detail the user might deem necessary.

Among the foremost contributions is a modeling approach now common for large watersheds, consisting of a suitable subdivision of the catchment into elementary (or homogeneous) generating areas for which the task of event-based or continuous simulations is drastically simplified. Existing literature on the issue has reported some very complex and detailed codes [8, 9, 10, 11] already calibrated on U.S. watersheds. Nevertheless, the transfer of the codes for reliable simulations of other geomorphologic features and agricultural practices (as for European case studies) may apply only after detailed and thorough calibration, in particular for soil erosion patterns.

The lack of field-scale experimental data on water quality (in particular for solid-phase pollutants) and the high density of nonhomogeneities in the watersheds at hand, have suggested the necessity to improve existing first-order [7], approaches of UMRF type for the sake of easy calibration and management.

The further requirements of subdivision into elementary areas have been met by a least-parameter principle, in search for both clear physical meaning and acceptable computational loads. The model makes use of nutrient responses to unit net precipitations, herein assumed as uniformly distributed over the watershed (hence calling for spatial dimension not exceeding the order of 5 Km^2).

Moreover, the validity of UMRF approaches for NPSP simulation has been recently recognized from both theoretical and practical standpoints [7].

The fundamental model structure may be described as the combination of a water quality component, based on simple mass balances, with Nash conceptual hydrologic model of rainfall-runoff transformations [12].

The advantages of the procedure seems to be twofold. In fact:
-limits and validity of the actual conceptual schemes, proposed for water quality features of watershed simulations, depend upon the bounds imposed by Nash' hydrologic model, which have extensively proved both its theoretical and practical value;

-the analytical form of the unit-mass response function derived allows for important theoretical considerations. For instance, hysteretical effects found in experimental curves of pollutant concentration/discharge relations are explained, and a theoretical basis is given for certain absences of correlation.

The model is based upon the following set of conceptual assumptions:

- the system soil-groundwater-surface water can be represented by two phase reactor;

- the watershed can be modelled with a cascade of continuously stirred tank reactors (CSTR), which embody Nash' schemes for water discharge;

- the discharge outflowing from the i-th CSTR obeys a linear dependence on the volume of mobile phase via a runoff coefficient [12];

- an interface equilibrium exists between the concentrations of fixed and mobile phase;

- the driving force for interphase mass transfer depends on the difference between actual and equilibrium concentrations and on a time dependent parameter h(t) related to mass transfer phenomena;

- equilibrium concentrations can be regarded as time dependent parameters $C_E(t)$.

The resulting conceptual schemes of hydraulic and chemical components for watershed simulation are illustrated in Figure 1, and Figure 2 outlines the mathematical models for mobile and fixed phases. Figure 2 shows the i-the CSTR, which would differ from the sketch of the first reactor of the cascade only by presence of the inflowing discharge.

The chemical mass transfer [1, 7] is ruled by:

$$\frac{\partial C}{\partial t} = h (C_E - C) \qquad (1)$$

where C and C_E are current and equilibrium concentrations in mobile phase, respectively. The equilibrium concentration C_E is a time dependent parameter related to the amount of chemical species supplied to the soil, taking into account current reaction processes and complex removal mechanisms due to earlier rainfall events. The equilibrium relationship between the two phases governs the conceptual schemes of figure 2, thus allowing a separate modelling of the phenomena in fixed and mobile phases. Furthermore, according to (1) the mass transfer parameter h accounts also for effects of interphase specific surfaces.

The transfer function of the watershed to rainfall events distributed in time according to $f(\tau)$ is built by convolution as:

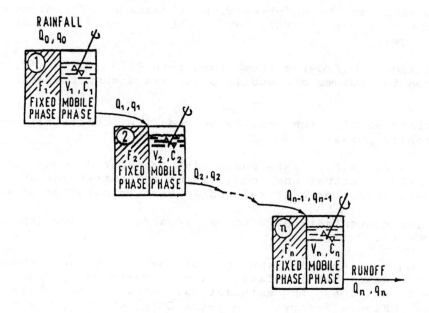

Figure 1. The conceptual model of hydraulic and chemical behavior of the watershed.

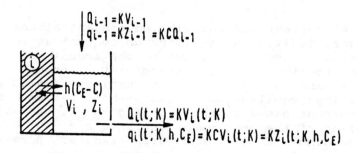

Figure 2. Mathematical model of the i-th CSTR.

$$u(t;n,k) = \frac{A\,k}{\Gamma_{(n)}} (kt)^{n-1} e^{-kt} \qquad\qquad (2)$$

$$Q(t;n,k) = \int_0^t u(t-\tau;n,k)\,f(\tau)\,d\tau \qquad (3)$$

where:

- $u(t;n,K)$ = unit hydrograph, as a function of the parameters n and K;
- $\Gamma(n)$ = Gamma function of argument n;
- A = area of the watershed;
- $f(\tau)$ = impulse hyetograph;
- $Q(t;n,K)$ = water discharge as a response to net precipitations distributed according to $f(\tau)$.

The general case in which C_E and h are time dependent can be integrated as:

$$C = e^{\int_0^t h(t')dt'} \left\{ \int_0^t h(t')\,C_E(t')\,e^{\int_0^t h(t'')dt''}\,dt + const \right\} \qquad (4)$$

For t = 0, C_E = C(0) = const.

The particular case of time-independent h yields:

$$C = e^{-ht} \left\{ \int_0^t C_E(t')\,h\,e^{ht'}\,dt' + C_0 \right\} =$$

$$= e^{-ht} \left\{ \int_0^t C_E(t')\,de^{ht'} + C_0 \right\} =$$

$$= C_E(t) - \left[C_E(0) - C_0 \right] e^{-ht} - e^{-ht} \int_0^t e^{ht'}\,C_E'\,dt' \qquad (5)$$

If C_E is also considered time-independent the C_E' = 0 and $C_E(t)$ = $C_E(0)$ = C_0; under the condition C(0) = 0 = C_0:

$$C = C_E \left(1 - e^{-ht}\right) \qquad (6)$$

A further particular case of (6) in which both C_E and h are constant differs from (5) only by the function

$$f(t; C_E, h) = e^{-ht} \int_0^t e^{ht'} C_E' \, dt'$$

The UMRF yielded by the above assumption is obtained via integration of mass balance equation for the CSTR cascade. The results, whose integration details whenever C_E and h are regarded as constant, are reported elsewhere [13, 14].

The flow rate q and total quantity Z, are:

$$q(t; n, k, h, C_E) = C_E \left\{ Q(t; n, k) - \left(\frac{k}{k+h}\right)^{n-1} Q(t; n, k+h) \right\} \qquad (7)$$

$$C(t; n, k, h, C_E) = C_E \left\{ 1 - \left(\frac{k}{k+h}\right)^n \frac{Q(t; n, k+h)}{Q(t; n, k)} \right\} \qquad (8)$$

$$Z(t; n, k, h, C_E) = \int_0^t q(t; n, k, h, C_E) \, dt =$$

$$= C_E \left\{ V(t; n, k) - \left(\frac{k}{k+h}\right)^n V(t; n, k+h) \right\} \qquad (9)$$

where:
- $q(t; n, K, h, C_E)$ = pollutant flow rate due to water discharge Q;
- $C(t; n, K, h, C_E)$ = pollutant concentration;
- $Z(t; n, K, h, C_E)$ = pollutant total load at time t;
- $V(t; n, K)$ = water volume discharge at time t;
- h = mass transfer coefficient for the pollutant;
- C_E = equilibrium concentration for the pollutant.

quations (2) , (3) and (9) constitute a set of easy-to-use tools
or simulation and forecasting of flood events of water and
ollutants.

wing to the simple analytical formulation of the model, optimal
alibration of the parameters via iterative procedures is
elatively cheap and, in general, the approach is deemed fit to
icrocomputer software application towards which a large portion
f modern ecological modelling is oriented.

HYSICAL ASPECTS OF THE MATHEMATICAL DEVELOPMENTS

ome mathematical aspects of the relationships found seem
oteworthy. Hysteretical effects are phenomena observed in
ertain properties which reflect a memory which that property
olds of the state system under which that property changes. In
articular, flow rates of pollutant concentrations depend upon
tate variables of soil layers and water bodies together with
ypical residence times of the process. Therefore concentrations
f pollutants are nonuniquely related to water discharge.

'rom the physical standpoint hysteretical effects may be due to
nteractions with channel bottom material or to runoff state
ariables.

'he former consists of a phenomena described by Smith and Stewart
15]. The latter can be related to chromatographic effects
leveloped by the soil into the mobile phase.
'he q and C vs. Q curves (calculated for sample values of
arameters) induced by isolated rainfall events show clearly the
henomenon (Figure 3) while the response to two impulses of
recipitation at 12-hour intervals substantiates that the latter
vent modifies the cycle of the former response (Figure 4).

direct consequence of (7) and (8) (Figures 3 and 4) on field
lata acquisition systems is due to the need of gathering
lischarge-weighted measurements of concentrations. Concentration
lifferences between rising and falling stages and the actual lags
n the assessment of equilibrium concentrations with respect to
lischarge peaks are, in fact, two aspects of the same phenomenon,
s mass transfer kinetics enforced into water and pollutant
nass/momentum balances fully justify hysteretical effects.

uniquely determined linear relationship can be identified,
ccording to the model proposed, between total quantities of a
pollutant Z and the total runoff volume V. In fact, according to
(9):

$$Z = \int_0^{\infty} q(t; n, k, h, C_E) dt = C_E \left[1 - \left(\frac{k}{k+h} \right)^n \right] V \qquad (10)$$

285

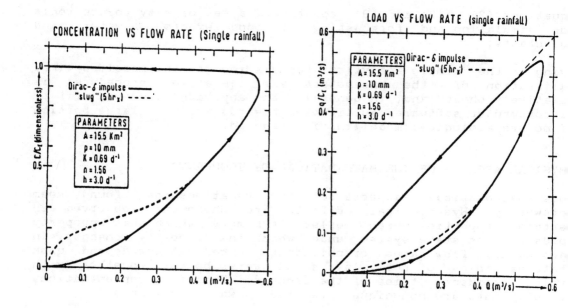

Figure 3. Hysteretical effects shown by the present model response
to Dirac-delta and 5-hour "slug" impulses (values of the
parameters are relative to a case study, [12].

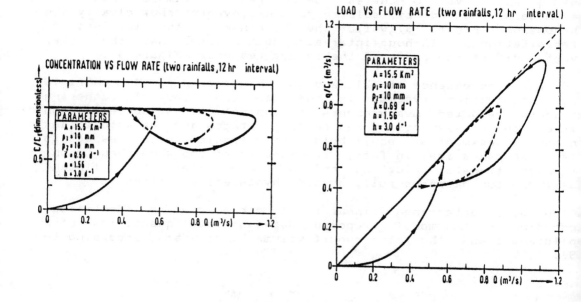

Figure 4. Hysteretical effects shown by the present model response
to two impulses at 12 hour intervals.

or a sequence of j rainfall events, j =1,2,...m, the total
quantity transported Ztot is given by:

$$Z_{tot} = \sum_{j=1}^{m} Z_j = \left[1-\left(\frac{k}{k+h}\right)^n\right] \sum_{j=1}^{m} C_E V_j \qquad (11)$$

ence for C_E independent of j, a strong linear relationship
between Ztot and $\sum_{j=1}^{m}(V_j)$ is found. Possible deviations from
the relationship may be due to truncation of the integral (6)
that are introduced whenever time-averaged (or smoothed)
pollutant and water discharges are considered.

The concept of reference weighted concentration of pollutants in
the runoff (16) is therefore amenable to interesting confirmation
by mass exchange and balance relations provided that long-term,
or event-based, time scales are assumed. According to (8), such a
reference concentration is

$$\bar{C} = \frac{\int_0^\infty CQ\,dt}{\int_0^\infty Q\,dt} = C_E\left[1-\left(\frac{k}{k+h}\right)^n\right] \qquad (12)$$

which is a conceptual form of Haith' average concentration
[16,17,18].

Among the spinoffs of the model, it seems worthwhile to point out
the dependence of pollutant runoff q upon water discharge Q,
which represents a crucial point for the validity of the scheme
and, of course, needs further experimental confirmation. Such a
relationship, which is built in the model assumptions, may
successfully describe quantitatively the NPSP processes provided
that sufficient information and noncontroversial experimental
date are available.

DISCUSSION AND CONCLUSIONS

Two packages of personal computer codes have been tailored to the
study of parameter sensitivity. The former, designed for
long-term forecasting of runoff and pollutant concentration,
transforms real rainfall data 'into net precipitation via
different procedures. A modification of well-known algorithms of
the Soil Conservation Service (S.C.S.) Curve Number [19,20] has
been implemented, which requires, besides total rainfall data,
two characteristics parameters of the watershed.

A further empirical transformation based on regional trends of time evolution of runoff coefficient is implemented [21], aiming at relevant comparison.

Net precipitations are enforced as input of the UMRF, whereas hydrological simulations require further specification of the number n of reservoirs and of the characteristic time (1/k). Nutrient losses are simulated upon suitable choise of mass-transfer coefficient h and equilibrium concentration C_E

A sequence of numerical experiments at daily time steps has been run, designed primarily to test of effects of chemical parameter variance on long-term forecasting, whereas a gauged set of rainfall events has been chosen as input. The case study displays hydrologic behaviour analogues to that used for previous field-scale calibration, [14]. The tracer simulated is dissolved nitrogen, for which field observations are available.

The results imply that the equilibrium concentration C_E is the dominant parameter (Figure 5). In fact, average annual values of C_E would suffice for total annual evaluations although a loss accuracy in event-based simulation may occur.

Simple "eyeball" corrections on average monthly values yield acceptable fit of chemical losses.

The variance of h on over an annual scale (at average C_E values) does not appreciably modify total annual loadings so that a reasonably good fit with the experimental results is maintained. At a daily time scale, the variation of h-values does not yield noteworthy alterations of peak times.

These a posteriori considerations, in addition to certain more rigorous physico-chemical accounts, allow the assumption of constant and time-invariant values of h for long-term forecasting.

The second package focuses on single studies of rainfall-runoff transformation. It implements an hourly time scale which, for the cases considered, was sufficient to portray the phenomena without heavy computational loads.
Input data consists of net values of precipitation while the output is runoff quantity and pollutant concentration.

In analogy to the previous simulation, tests on h-variance effects have been carried out in order to check for the overall effect on nutrient loss from a single event. The results are illustrated in Figure 6, where water and pollutant runoff for three values of h are plotted.

The irrelevance of h values to peak times and the noteworthy effects on total quantities discharged seem to constitute the main drawbacks.

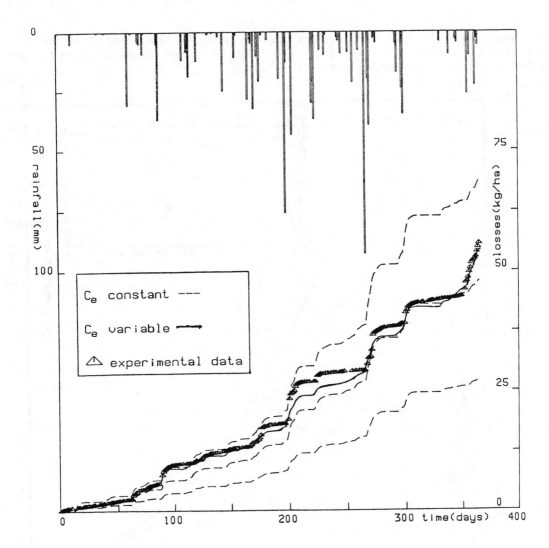

Figure 5. Nutrient loss forecasting (Kg/ha) as by the actual
model. The dotted line represents constant C_E
forecasting (8.5,7,5.5 mg/l): the solid line portrays
the run of monthly C_E variations; the points
portray gauged values. Net precipitations are reported
in the top diagram (mm/m^2).

Loss rates vs. h are also drawn. The wide variability of these
values suggests a word of caution in the assumption of the
parameter h. In fact, at low h-values the variability of load is
marked nevertheless this condition is met at low pollutant
concentrations and large water runoff, and hence errors of
measure may play a major role.

The analyses of parameter sensitivity and of the features of (7) are particularly interesting for the planned extension of the model to more complex forms.

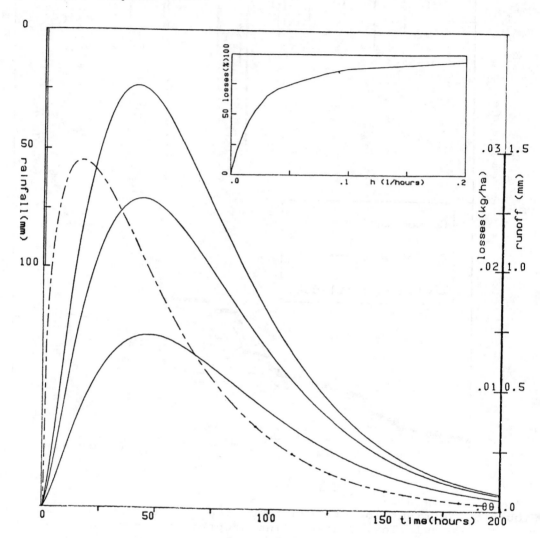

Figure 6. Loss forecasting (solid line) due to a single rainfall event (h=0.005, h=0.01, h=0.015 and C₌ =7); runoff (dotted line) (n=1.5, K=0.03). Total nutrient loss rates vs, h are also shown.

In this respect, it has been shown [22], that a statistical analysis may yield hydrologic responses distributed according to a generalized Gamma function in substitution of the Standard

Gamma (2).

Furthermore, chemical aspects of (7) account for, washout effects upon imposition of $C(0)=0$ on (1) as initial conditions [22]. These extensions imply possible model improvement by the use of two additional parameters (the former hydrological, the latter chemical) which may detail the description of the physical setting of the watershed, although such an expansion may incur further calibration problems.

Since a model extension is unquestionable, as it is desireable to maintain the physical interpretation of the model parameters, inspection of the mentioned runs indicates the possibility of reducing the effective number of parameters. The basis for such simplification lies not only in best fitting of experimental data, but also, and primarly, in the preparation of the model as a forecasting tool suitable to control strategy determination.

The resources management proven structure of the model is tied to the fact that UMRF approaches portray acceptable the hydrology of the catchments. Furthermore, although the mass-transfer law is general, further field of laboratory-scale experimentation is required for linking geomorphologic and agronomic aspects to the values of the parameters.

The relation between C_m and solid-phase concentration F_m allows for deeper insight into the chemical nature of the soils, through which the establishement of more reliable correlations between C_m and the local availability of fertilizers may be obtained.

As mentioned above, a further boundary constraint on NPSP models refers to the so-called "downstream" condition linking the detail of NPS runoff models to that of the receiving water quality model.

In fact in the case of tidal exchange of water (e.g. the Lagoon of Venice) NPS runoff should be specified at subtidal time intervals (as short as half an hour) and should run through large time spans without infeasible use of computational resources [23].

All these requirements are met by the proposed model for field-scale simulations, and should be met also by further applications (as widely experimented for hydrological analyses) of generalized applications to large inhomogeneous watersheds.

AKCNOWLEDGEMENTS

The present work has been supported by funds provided by the Consorzio di Bonifica Dese-Sile (Mestre-Venezia), by the Consiglio Nazionale delle Ricerche (CNR, Roma) and by the Italian Ministero della Pubblica Istruzione (Roma).

REFERENCES

1 - NOVOTNY, V., CHESTERS, G., "Handbook of Nonpoint Pollution", Van Nostrand Reinhold Company, New York, N.Y., 1981.
2 - BENDORICCHIO, G., COMIS, C., ALESSANDRINI, S., "Evoluzione temporale dell'inquinamento da sorgenti diffuse di origine agricola", Inquinamento, (in press).
3 - PERIN, G., "L'inquinamento chimico della Laguna di Venezia: problemi dell'inquinamento lagunare". Tavola Rotonda del Consorzio Depurazione Acque Zona Industriale Porto Marghera, Venezia, ottobre 1975.
4 - RINALDO, A., RINALDO, A., "Determinazione per singoli recapiti delle quantita' di fosforo e azoto che pervengono annualmente in Laguna di Venezia dal proprio bacino scolante relativamente ai territori agricoli". Convegno di Studi Laguna, fiumi, lidı. Cinque secoli di gestione delle acque nelle Venezie. Magistrato delle Acque, Venezia, 10-12 giugno 1983.
5 - ZINGALES, F., et Al., "Inquinamento delle acque da sorgenti diffuse: Analisi Statistiche", Inquinamento, 1, 1981.
6 - COSSU, R., et Al., 'Elementi per il bilancio di nutrienti nella laguna di Venezia", Ingegneria Ambientale, (in press).
7 - JOLANKAI, G., "Modeling of Nonpoint Source Pollution", Application of Ecological Modeling in Environmental Management, Part A". S.E.Jorgensen Ed., Elsevier Scientific Publishing Co., Amsterdam, NL, 283-385, 1983.
8 - KNISEL, W.G., NICKS, A.D., "CREAMS: a Field Scale Model for Chemicals Runoff and Erosion from Agricultural Management Systems". Conservation Research Report No.26. W.G.Kniesel Ed. USDA, 1980.
9 - DONIGIAN, A.S., BEYERLIN, D.C., DAVIS, Jr., H.H., CRAWFORD, N.H., "ARM Model", Report 300/8-77-098, 1977.
10- BEASLEY, D.B., HUGGINS, L.F., MONKE, E.J., "ANSWERS: a Model for Watershed Planing", Transactions of ASAE, 23, 938-944, 1980.
11- HAITH, D.A., "Environmental System Optimization",McGraw-Hill, New York, N.Y., 1981.
12- NASH, J.E., "The Form of the Instantaneous Unit Hydrograph", Bull. Int. Assoc. Scientific Hydrology, 3, 101-126, 1957.
13- BENDORICCHIO, G., RINALDO, A., "Un modello matematico del dilavamento di sostanze chimiche da terreni agricoli", Acc. Patavina SS.LL.AA., Padua (Italy), 93, 137-154, 1981.
14- ZINGALES, F., ALESSANDRINI, S., BENDORICCHIO, G., COMIS, C., MARANI, A., PIANETTI, F., RINALDO, A., SARTORI-BOROTTO, C., "A Model of Nonpoint Source Pollution in Agricultural Runoff", Proceedings of the 13th Pittsbugh Conference on Modelling and Simulation. Pittsburgh, PA (U.S.A.), 13, 4, 1575-1581, 1982.
15- SMITH, R.V., STEWART, D.A., "Statistical Models of River Loadings of Nitrogen and Phosphorus in the Lough Neagh System", Water Research, 11, 631-636, 1977.
16- HAITH, D.A., and DOUGHERTY, J.V., "Non-Point Source Pollution

from Agriculture Runoff", Journal of the Environmental Engineering Division, (102)EE5: 1055-1069, 1976.

7- HAITH, D.A., "Environmental System Optimization".McGraw-Hill, New York, N.Y., 1976.

8- HAITH, D.A., "Developments and Testing of Watershed Loading Functions for Nonpoint Sources". In: Proceedings of the 13th International Conference on Modeling and Simulation at Pittsburgh, PA (U.S.A.), 13, 4, 1463-1467, 1982.

9- MOCKUS, V., National Engineering Handbook, Soil Conservation Service U.S.D.A., 1969.

20- HAWKINS, R.H., "Runoff Curve Numbers Varying Site Moisture", Journal of the Irrigation and Drainage Division, 389-398, 1978.

21- SUPINO, G. "Le reti idrauliche", Patron, Bologna, 1967.

22- MARANI, A., BENDORICCHIO, G., "Models of Statistical Distributions for NPSP Concentrations", this Workshop.

23- RINALDO, A., MARANI, A., "Runoff and Receiving Water Models for NPSP Discharge into the Venice Lagoon", this Workshop.

MODELING SUBSURFACE DRAINAGE AND WATER MANAGEMENT SYSTEMS TO ALLEVIATE POTENTIAL WATER QUALITY PROBLEMS

Dr. R. W. Skaggs, Professor
Department of Biological and Agricultural Engineering
and
Dr. J. W. Gilliam, Professor
Department of Soil Science
North Carolina State University
Raleigh, N.C. 27695

INTRODUCTION

A large percentage of agricultural lands in the humid regions of the world require improved drainage for efficient agricultural production. Pavelis (1) estimates that over 42 million ha of farm land in the USA benefited from artificial drainage as of 1978. This makes up about 22 percent of the nation's cropland. In states like North Carolina, Michigan and Ohio over 40 percent of the cropland requires drainage. Thus any effort to evaluate or affect pollution from nonpoint sources should place some emphasis on these soils.

There are basically two methods of improving drainage on soils that do not have adequate natural drainage. Surface drainage can be provided by estab- lishing drainage outlets and forming, shaping, and smoothing the fields so that excessive water will run off rather than collecting on the surface. Subsurface drainage can be provided by installing drain tubes or open ditches to remove excessive water from the soil profile. The design of a drainage system and the emphasis placed on its surface and subsurface components have a pronounced effect on the rate and quality of water leaving the field. Systems that depend primarily on surface drainage tend to have higher rates of runoff and more sediment than do systems with good subsurface drainage. However, good subsurface drainage increases the outflow of nitrates with the drainage water. Associated water management practices for soils with shallow water tables, such as controlled drainage and subirrigation will also have an effect on both the soil water regime and the quality of drainage water leaving the site.

The purpose of this paper is to describe a water management simulation model, DRAINMOD, and apply a modified version of that model to evaluate the effect of drainage system design on nitrate outflow from agricultural fields. DRAINMOD was originally developed to simulate the performance of drainage and related water management systems over a long period (e.g. 20 to 30 years) of climatological record. It is mainly used for design and evaluation of drain- age and water table control systems, but has also been used to analyze land application systems for wastewater treatment. The model has been accepted by the USDA's Soil Conservation Service and is being implemented on their compu- ters for humid regions of the USA.

The first part of the paper is devoted to a brief description of each of the components of the model. This is followed by a short section on experimental testing of the model. The last section presents an example of the use of the model to analyze the effects of different combinations of surface drainage, subsurface drainage and outlet control on nitrate losses from a poorly drained soil.

MODEL DEVELOPMENT

A schematic of the type of water management system considered is given in Figure 1. The soil is nearly flat and has an impermeable layer at a relatively shallow depth. Subsurface drainage is provided by drain tubes or parallel ditches at a distance, d, above the impermeable layer and spaced a distance, L, apart. When the water level is raised in the drainage ditches, for purposes of supplying water to the root zone of the crop, the drainage rate will be reduced and water may move from the drains into the soil profile giving the shape shown by the broken curve in Figure 1.

Two important criteria were adopted at the outset of the model development process. First, the model should be capable of describing water movement and storage in the profile so as to characterize the soil water regime and drainage rates with time. And second, the model should be developed such that the computer time necessary to simulate long term processes is not prohibitive. The movement of water in soil is a complex process and it would be an easy matter to become so involved with getting exact solutions to every possible situation that the final answer would never be obtained. The guiding principle in the model development was therefore to assemble the linkage between various components of the system, allowing the specifics to be incorporated as subroutines, so that they can readily be modified when better methods are developed.

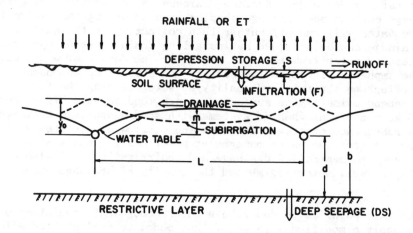

Figure 1.--Schematic of water management system with subsurface drains that may be used for drainage or subirrigation.

The rates of infiltration, drainage, evapotranspiration, and the distribution of soil water in the profile can be computed by obtaining numerical solutions to nonlinear differential equations (2). However, these methods require prohibitive amounts of computer time for long term simulations and thus can

not be used in the model. Instead, approximate methods were used to characterize the water movement processes.

The basic relationship in the model is a water balance for a thin section of
soil of unit surface area which extends from the impermeable layer to the
surface and is located midway between adjacent drains (Figure 1). The water
balance for a time increment of Δt may be expressed as,

$$\Delta V_a = D + ET + SP - F \qquad (1)$$

where ΔV_a is the change in the air volume or water free pore space (cm) in
the section, D is drainage (cm) from (or subirrigation into) the section, ET
is evapotranspiration (cm), SP is seepage to deep or remote lateral sinks
(cm) and F is infiltration (cm) entering the section in Δt.

The terms on the right-hand side of equation 1 are computed in terms of the
water table elevation, soil water content, soil properties, site and drainage
system parameters, crop and stage of growth, and atmospheric conditions. The
amount of runoff and storage on the surface is computed from a water balance
at the soil surface for each time increment which may be written as,

$$P = F + \Delta S + RO \qquad (2)$$

where P is the precipitation (cm), F is infiltration (cm), ΔS is the change
in volume of water stored on the surface (cm), and RO is runoff (cm) during
time Δt. The basic time increment used in equations 1 and 2 is 1 hour.
However when rainfall does not occur and drainage and ET rates are slow such
that the water table position moves slowly with time, equation 1 is based on
Δt of 1 day. Conversely, time increments of 0.1 hr. or less are used to
compute F when rainfall rates exceed the infiltration capacity.

MODEL COMPONENTS

Precipitation

Inputs to the model for precipitation are hourly data which are stored and
automatically accessed from the HISARS data base (3). Although any time
distribution could be used, the rainfall rate is assumed to be uniformly
distributed within each hour.

Infiltration

Infiltration of water at the soil surface is a complex process which has
been studied extensively during the past three decades (4, 5, 6). Infiltration is affected by soil factors such as hydraulic conductivity, initial
water content, surface compaction, depth of profile, and water table depth;
plant factors such as extent of cover and depth of root zone; and rainfall
factors such as intensity, duration, and time distribution.

Approximate methods for predicting infiltration have been proposed by Green
and Ampt (8), Horton (9) Philip (10), Smith (11) and Smith and Parlange (12),
among others. The Green-Ampt equation was selected for use in this model.
It was originally derived for deep homogeneous profiles with a uniform
initial water content. The equation may be written as,

$$f = K_s + K_s M_d S_f / F \qquad (3)$$

where f is the infiltration rate, F is cumulative infiltration, K_s is the hydraulic conductivity of the transmission zone, M_d is the fillable porosity ($M_d = \theta_s - \theta_i$), θ_s and θ_i are respectively the final and initial volumetric water contents, and S_f is the effective suction at the wetting front. For a given soil with a given initial water content equation 3 may be written as,

$$f = A/F + B \qquad (4)$$

where A and B are parameters that depend on the soil properties, initial water content and distribution, etc.

In addition to uniform profiles for which it was originally derived, the Green-Ampt equation has been used with good results for profiles that become denser with depth (13) and for soils with partially sealed surfaces (14). It may also be used for nonuniform initial water contents (15); Morel-Seytoux and Khanji (16) have shown that it retains its original form when the effects of air movement are considered for deep soils.

Mein and Larson (17) used the Green-Ampt equation to predict infiltration from steady rainfall. Their results were in good agreement with rates obtained from solutions to the Richards equation for a wide variety of soil types and application rates. Mein and Larson's results imply that, for uniform deep soils with constant initial water contents, the infiltration rate may be expressed in terms of cumulative infiltration, F, alone regardless of the application rate. This is implicitly assumed in the Green-Ampt equation and in the parametric model proposed by Smith (11). Smith showed that this assumption could be extended to the case of erratic rainfall where the unsteady application rate dropped below infiltration capacity for a period of time followed by a high intensity application. Similar investigations by Reeves and Miller (18) showed that the infiltration capacity could be approximated as a simple function of F regardless of the application rate versus time history. These results are extremely important for modeling efforts of the type discussed herein. If the infiltration relationship is independent of application rate, the only input parameters required are those applying to the necessary range of initial conditions.

The model requires input for infiltration in the form of a table of A and B versus initial water table depth. Methods for estimating S_f based on soil texture were presented by Rawls et al. (19). When rainfall occurs, A and B values are interpolated from the table for the approximate water table depth at the beginning of the rainfall event. The same A and B values are used as long as the rainfall event continues. An exception is when the water table rises to the surface, at which point A is set to A = 0 and B is set equal to the sum of the drainage, ET and seepage rates.

Surface Drainage

Surface drainage is characterized by the average depth of depression storage that must be satisfied before runoff can begin. In most cases it is assumed that depression storage is evenly distributed over the field. Depression storage may be further broken down into a micro component representing storage in small depressions due to surface structure and cover, and a macro component which is due to large surface depressions and which may be altered by land forming, grading, etc. A field study (20) showed that the micro-storage component varies from about 0.1 cm for soil surfaces that have been

smoothed by weathering to several centimeters for rough plowed land. Macro-
storage values for eastern North Carolina fields varied from nearly 0 for
fields that have been land formed and smoothed or that are on a natural
grade to >3 cm for fields with numerous pot holes and depressions or which
have inadequate surface outlets.

Subsurface Drainage

The rate of subsurface water movement into drain tubes or ditches depends on
the hydraulic conductivity of the soil, drain spacing and depth, profile
depth and water table elevation. Water moves toward drains in both the
saturated and unsaturated zones and can best be quantified by solving the
Richards equation for two-dimensional flow. Solutions have been obtained for
drainage in layered soils (21), and for drain tubes of various sizes (22).
Input and computational requirements prohibit the use of these numerical
methods in DRAINMOD, as was the case for infiltration discussed previously.
However, numerical solutions provide a very useful means of evaluating approximate methods for computing drainage flux.

The method used in DRAINMOD to calculate drainage rates is based on the
assumption that lateral water movement occurs mainly in the saturated region.
The effective horizontal saturated hydraulic conductivity is used and the
flux is evaluated in terms of the water table elevation midway between drains
and the water level or hydraulic head in the drains. Several methods are
available for estimating the drain flux including the use of numerical solutions to the Boussinesq equation. However, Hooghoudt's steady state equation, as used by Bouwer and van Schilfgaarde (23) was selected for use in the
present version of DRAINMOD. This equation may be written as,

$$q = \frac{8 \ K_2 \ d_e m + 4 \ K_1 \ m^2}{C \ L^2} \tag{5}$$

where q is the flux in cm/hr, m is the midpoint water table height above the
drain (Figure 1), K_1 and K_2 are the equivalent lateral hydraulic conductivities above and below the drain, respectively, d_e is the equivalent depth
from the drain to the impermeable layer, L is the distance between drains and
C is the ratio of the average flux to the flux at a point midway between
drains. Solutions based on a water balance at the midpoint and a drainage
rate given by equation 5 are compared to numerical solutions of the
Boussinesq equation in Figure 2. These solutions are for an initially horizontal profile so the time required for the elliptical shape to develop and
drawdown begin at the midpoint was computed (24) before equation 5 was
applied.   While good agreement was obtained when a constant C = 1 was used,
almost exact agreement with the Boussinesq solutions was found when C was
allowed to vary with the water table elevation, y. The present version of
the model, however, uses C = 1.0.

Hooghoudt (25) characterized flow to cylindrical drains by considering radial
flow in the region near the drains and applying the D-F assumptions to the
region away from the drains. The Hooghoudt analysis has been widely used to
determine an equivalent depth, d_e, which, when substituted for d in Figure 1
will tend to correct drainage fluxes predicted by equation 5 for convergence
near the drains. Moody (26) examined Hooghoudt's solutions and presented the
following equations from which d_e can be calculated.

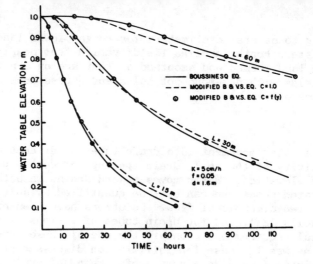

Figure 2.--Midpoint water table drawdown as predicted by solutions to the
Boussinesq equation and by a simplified model which uses a
modified form of the Bouwer and van Schilfgaarde (23) equation.

For $0 < d/L < 0.3$

$$d_e = \frac{d}{1 + \frac{d}{L} \{\frac{8}{\pi}\ln (\frac{d}{r}) - \alpha\}} \tag{6}$$

in which

$$\alpha = 3.55 - \frac{1.6d}{L} + 2 (\frac{d}{L})^2 \tag{7}$$

and for $d/L > 0.3$

$$d_e = \frac{L}{8 \{\ln (\frac{L}{r}) - 1.15\}} \tag{8}$$

in which r = drain tube radius. Usually α = 3.4 can be approximated
with negligible error for design purposes (25).

For real, rather than completely open drain tubes, there is an additional
loss of hydraulic head due to convergence as water approaches the finite
number of openings in the tube. The effect of various opening sizes and
configurations can be approximated by defining an effective drain tube
radius, r_e, such that a completely open drain tube with radius r will offer
the same resistance to inflow as a real tube with radius r (27).[e]

Other methods for calculating the drain flux which consider convergence to
the drains and layered profiles have been summarized by van Beers (28). The
most general is the Hooghoudt-Ernst equation which does not require a separate calculation for d_e. However, it is necessary to determine a geometric
factor from a graphical solution for some layered systems. The modified

Hooghoudt-Ernst equation is also discussed by van Beers (28) and could be easily employed in DRAINMOD.

The methods discussed above for predicting drainage flux assume a curved (elliptical) water table completely below the soil surface except at the midpoint where it may be coincident with the surface. However, in some cases, the water table may rise to completely inundate the surface with ponded water remaining there for relatively long periods of time. Then the D-F assumptions will not hold as the streamlines will be concentrated near the drains with most of the water entering the soil surface in that vicinity. Kirkham (29) showed that in one case, more than 95% of the flow entered the surface in a region bounded by $\pm$ one-quarter of the drain spacing. Drainage flux for a ponded surface can be quantified using an equation derived by Kirkham (29):

$$q = \frac{4 \ K \ (t + b - r)}{gL} \tag{9}$$

where t is the depth of water on the surface, b is the depth from the surface to the impermeable layer, r is the drain tube radius and g is a geometric factor expressed in series form by Kirkham (29).

Use of equations 5 or 9 assumes that drainage is limited by the rate of soil water movement to the lateral drains and not by the hydraulic capacity of the drain tubes or of the outlet. Usually, the sizes of the drain tubes are chosen to provide a design flow capacity, which is called the drainage coefficient, D.C. Typically, the D.C. may be 1 to 2 cm per day depending on the location and crops to be grown. When the flux given by equations 5 or 6 exceeds the D.C., q is set equal to the D.C. in DRAINMOD as suggested by Chieng et al. (30).

Subirrigation

When subirrigation is used, water is raised in the drainage outlet so as to maintain a pressure head above the center of the drain of y_o (refer to the broken curve in Figure 1). Then the equation corresponding to equation 5 for flux is,

$$q = \frac{4K}{L^2}(2 \ h_o \ m + m^2) \tag{10}$$

where $h_o = y_o + d_e$ is the equivalent water table elevation at the drain and m is defined as $m = h_m - h_o$ with h_m being the equivalent water table elevation midway between the drains. For subirrigation $h_o > h_m$ and both m and q are negative. Convergence losses at the drain are treated in the same manner as in drainage by using the equivalent depth to the impermeable layer, d_e, rather than the actual depth, d, to define h_o in equation 10. Equation 10 was derived by making the D-F assumptions and solving the resulting flow equation for steady state evaporation from the field surface at rate q. Since water is flowing from the drains to the soil, q is negative in equation 10. The magnitude of q increases as m becomes more negative, i.e., as h_m becomes smaller, until the water table at the midpoint reaches the equivalent depth of the impermeable layer, $h_m = 0$. For deeper midpoint water table depths, which can occur because the actual depth to the impermeable layer is deeper than the equivalent depth, equation 10 predicts a decrease in the magnitude of q. This is inconsistent with the physics of flow since the maximum subirrigation rate should occur when the midpoint water table reaches the impermeable layer (31). Ernst (31) derived the following equation to correct these deficiencies:

$$q = \frac{4K\ m\ (2h_o + \dfrac{h_o}{D_o}\ m)}{L^2} \tag{11}$$

where $D_o = y_o + d$, d is the distance from the drain to the impermeable layer, and h_o is the same as defined previously, $h_o = y_o + d_e$. Equation 11 is used in DRAINMOD to predict subirrigation flux.

Seepage

The rate at which water seeps via subsurface pathways to or from the field depends on the site, its positions in the landscape and the field boundary conditions, both in the lateral and vertical directions. In many cases, such as in the design of a subsurfce drainage system, seepage may be neglected, since water seeping from the site in addition to drainage would cause the design to be on the conservative or "safe" side. On the other hand, seepage into the field, from an irrigation canal or adjacent lands at higher elevations, should be considered in the drainage system design.

Because seepage may occur under many different circumstances, the methods used for quantifying seepage rates must be modified to fit the given situation. General methods are included as subroutines in one version of DRAINMOD, but these methods would require modifications for different site and boundary conditions. The general site conditions considered are shown in Figure 3, where vertical seepage may occur through the restrictive layer of thickness T and hydraulic conductivity K_v to a groundwater aquifer with constant hydraulic head, H_a. Lateral seepage to a stream located a distance L from the drained field may also take place. Then the total seepage, SP in equation 1, is evaluated as

$$SP = S_v + S_\ell \tag{12}$$

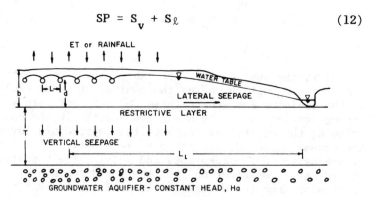

Figure 3.—Schematic of a drainage system with vertical seepage to a ground-water acquifer and lateral seepage to a remote sink.

where S_v and S_ℓ are the vertical and lateral seepages, respectively.

The vertical seepage rate is calculated by a straightforward application of Darcy's law as,

$$q_v = - K_v \frac{H_a - H_t}{T} \qquad (13)$$

where H_t is the average hydraulic head at the top of the restrictive layer which may be approximated as, $H_t = T + d + \bar{y}$, where $\bar{y}$ is the mean water table elevation above the drains. Since the simulation predicts the water table depth at the midpoint on a continuous basis, $\bar{y}$ is approximated by integration of an assumed elliptical water table shape to determine the mean.

The lateral seepage rate to the stream can be approximated by using the D-F assumptions as,

$$q_\ell = K_\ell \frac{(d + \bar{y})^2 - h_\ell^2}{2L} \qquad (14)$$

where K_ℓ is the effective lateral hydraulic conductivity and h_ℓ is the elevation of the water level in the stream. Vertical and horizontal seepage volumes for a given time increment are simply calculated as $S_v = q_v \Delta t$ and $S_\ell = q_\ell \Delta t$. Where Δt is the basic time increment used in applying equation 1 and is usually one hour.

Evapotranspiration

The determination of evapotranspiration (ET) is a two-step process in the model. First the daily potential evapotranspiration (PET) is calculated in terms of atmospheric data and is distributed on an hourly basis. The method developed by Thornthwaite (32) is usually applied to calculate PET, although other methods (33, 34) may be used if input data are available. PET is distributed at a uniform rate for the 12 hours between 6:00 AM and 6:00 PM. Hourly PET is set equal to zero for any hour in which rainfall occurs. After PET is calculated, checks are made to determine if ET is limited by soil water conditions. If soil water conditions are not limiting, ET is set equal to PET. When PET is higher than the amount of water that can be supplied from the soil system, ET is set equal to the smaller amount.

When the water table is near the surface or when the upper layers of the soil profile have a high water content, ET will be equal to PET. However, for deep water tables and drier conditions, ET may be limited by the rate that water can be taken up by plant roots. Gardner, (35) analyzed the factors controlling steady evaporation from soils with shallow water tables by solving equations for unsaturated upward water movement. He presented simplified expressions for the maximum evaporation rate in terms of water table depth and unsaturated hydraulic conductivity function parameters. For steady unsaturated flow, upward flux is constant and may be written as,

$$q = -K(h) \frac{dh}{dz} + K(h) \qquad (15)$$

where h is the soil water pressure head K(h) is the unsaturated conductivity function, and z is measured downward from the surface. For any given water table depth, the rate of upward water movement will increase with soil water suction (-h) at the surface. Therefore the maximum evaporation rate for a given water table depth can be approximated by solving equation 15 subject to a large negative h value, say h = -1000 cm, at the surface (z = 0) and h = 0 at z = d, the water table depth. Simple, rapidly converging numerical methods were developed by Alexander (36) to solve equation 15 for layered soils and for functional or tabulated K(h) relationships. By obtaining solutions for a range of water table depths, the relationship between maximum rate of upward water movement and water table depth can be estimated. Relationships for seven North Carolina soils are shown in Figure 4.

Relationships such as these shown in Figure 4 are read as inputs to the model in tabular form. Then, for example, if the PET is 5 mm/day for a Wagram l.s., the ET demand could be satisfied directly from the water table for water table depths less than about 0.64 m. For deeper water tables, ET for that day would be less than 5 mm or the difference would have to be extracted from root zone storage. The root zone is allowed to "dry out" if ET proceeds at a faster rate than can be supplied by upward movement from the water table. The root depth will be discussed in a later section. However, it should be pointed out here that the roots are assumed to be concentrated within an effective root depth, and that the surface boundary condition may be shifted to the bottom of the root zone as indicated by the abscissa label in Figure 4.

Soil Water Distribution

The basic water balance equation for the soil profile (equation 1) does not require knowledge of the distribution of the water within the profile. However, the methods used to evaluate the individual components such as drainage and ET depend on the position of the water table and the soil water distribution in the unsaturated zone. One of the key variables that is determined at the end of every water balance calculation in DRAINMOD is the water table depth. The soil water content below the water table is assumed to be essentially saturated; actually it is slightly less than the saturated value due to residual entrapped air in soils with fluctuating water tables. In some earlier models, the water content in the unsaturated zone was assumed to be constant and equal to the saturated value less the drainable porosity. However, numerical solutions to the Richards equation (22, 37) have shown that, except for the region close to drains, the pressure head distribution above the water table during drainage may be assumed nearly hydrostatic for many field scale drainage systems. The soil water distribution under these conditions is the same as in a column of soil drained to equilibrium with a static water table. This is due to the fact that, in most cases in fields with artificial drains, the water table drawdown is slow and the unsaturated zone in a sense "keeps up" with the saturated zone. This implies that vertical hydraulic gradients are small. The assumption of a hydrostatic condition above the water table will generally hold for conditions in which the D-F assumptions are valid. This will be true for situations where the ratio of the drain spacing to profile depth is large but may cause errors for deep profiles with narrow drain spacings.

Water is also removed from the profile by ET which results in water table drawdown and changes in the water content of the unsaturated zone. In this case the vertical hydraulic gradient in the unsaturated zone is in the upward direction. However when the water table is near the surface, the vertical

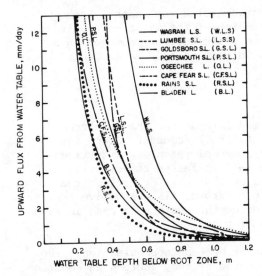

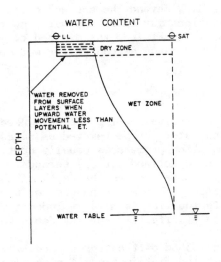

Figure 4.—Effect of water table depth on steady upward flux from the water table for seven North Carolina soils.

Figure 5.—Schematic of soil water distribution when a dry zone is created near the surface

gradient will be small and the water content distribution will still approximate the equilibrium distribution.

The maximum water table depth for which the approximation of a drained to equilibrium water content distribution will hold depends on the hydraulic conductivity functions of the profile layers and the ET rate. The maximum depth will increase with the hydraulic conductivity of the soil and decrease with the ET rate. Because the unsaturated hydraulic conductivity decreases rapidly with water content, large upward gradients may develop near the surface, or near the bottom of the root zone, when the soil water distribution departs from the equilibrium profile. At this point, the upward flux cannot be sustained for much deeper water table depths, and additional water necessary to satisfy the ET demand would be extracted from storage in the root zone creating a dry zone as discussed in the ET section. This is shown schematically in Figure 5.

For purposes of calculation in DRAINMOD, the soil water is assumed to be distributed in two zones – a wet zone extending from the water table up to the root zone and possibly through the root zone to the surface, and a dry zone. The water content distribution in the wet zone is assumed to be that of a drained to equilibrium profile. When the maximum rate of upward water movement, determined as a function of the water table depth, is not sufficient to supply the ET demand, water is removed from the root zone storage creating a dry zone as discussed in the ET section.

The assumptions made concerning soil water distribution may cause errors during periods of relatively dry conditions in soils with deep water tables and low K in the subsurface layers. Deep water tables may result from vertical seepage into an underlying aquifer or because of deep subsurface drains. For such conditions, the soil water at the top of the wet zone just beneath the root zone may be depleted by slow upward movement and by roots extending beyond the assumed depth of the concentrated root mass. Such conditions may cause the water content at the top of the wet zone to significantly depart from the drained to equilibrium distribution. However this will not cause a problem for wet conditions in most shallow water table soils.

Rooting Depth

The effective rooting depth is used in the model to define the zone from which water can be removed as necessary to supply ET demands. Since the simulation process is usually continuous for several years, an effective depth is defined for all periods. When the soil is fallow the effective depth is defined as the depth of the thin layer that will dry out at the surface. When a second crop or a cover crop is grown, its respective rooting depth function is also included. The rooting depth function is read in as a table of effective rooting depth versus day of year.

This method of treating the rooting depth is at best an approximation. The depth and distribution of plant roots is affected by many factors in addition to crop species and time after planting. These factors include physical barriers such as hardpans and plow pans, chemical barriers, fertilizer distribution, tillage treatments and others. One of the most important factors influencing root growth and distribution is soil water. This includes both depth and fluctuation of the water table as well as the distribution of soil water during dry periods. Since the purpose of the model is to predict the water table position and soil water content, a model which includes the complex root growth processes is needed to accurately characterize the change of the root zone with time.

FIELD EVALUATION OF THE MODEL

Field experiments were conducted over a five-year period at three locations in the N.C. Coastal Plains to test the reliability of the model. Three soil types and five different drainage system designs were included in the experiment from which 21 site years of data were obtained. Subirrigation was practiced on all of the sites during some years. Rainfall intensity and water table elevations were measured continuously at each site and the observed day end water table elevations were compared to predicted values. Effective lateral hydraulic conductivity values were measured in the field using both auger hole and water table drawdown methods. Numerous other field and laboratory measurements were made for each soil to determine input soil property and site parameter data.

In general, comparison of predicted and measured water table elevations were in good agreement. The average absolute deviation between predicted and observed water table depths for 21 site-years of data (approximately 7400 pairs of daily predicted and measured values) was 10.3 cm. This agreement was not the result of data fitting. Input parameters were determined independently as noted above. Results are described in detail by Skaggs (38).

DRAINMOD has also been tested using eight years of field data from Ohio (39). In that study predicted and measured outflow volumes were compared for field plots with surface drainage alone, subsurface drainage alone and for plots with both surface and subsurface drainage. Further testing was conducted for a tight silt loam soil in Louisiana (40) and for irrigated California soils (41). Agreement between predicted and observed values was good in all field studies. It was concluded, in each case, that the model was reliable for use in design and evaluation of agricultural water management systems.

APPLICATION: EVALUATING THE EFFECT OF DRAINAGE DESIGN ON NITRATE LOSSES

Movement of nitrates from agricultural fields via drainage waters is an important factor in nonpoint source pollution of surface waters. The amount of nitrate present in the soil water below the rooting zone and lost from the fields via subsurface drainage water is dependent on many factors including rate of fertilization and type of crops grown (42, 43). However, a more important factor is the natural drainage of the soil profile. Data from North Carolina show that poorly drained soils with relatively high water tables (0.3 to 1.5 m below the surface) lose less nitrate to drainage waters than do naturally well drained soils (44) because of denitrification in the subsoil of the poorly drained soils (45).

Artificial drainage increases nitrate movement from agricultural fields even when no additional fertilizer is applied (46). When combined with the increased farming intensity which drainage allows, nitrogen losses in drainage waters may be compounded. However, artificial drainage is necessary to farm some of the world's most productive soils. To attempt to farm many poorly drained soils without artificial drainage would result in total crop loss in some years and severely reduced yields in others.

There are many alternative designs of surface and subsurface drainage systems for most agricultural lands that will satisfy the drainage objectives. Some of these alternatives are better than others in reducing outflow of nitrates and other nutrients to drainage waters. In addition, other water management techniques may be used to reduce drainage outflows during offseason periods and, in some cases, during the growing season. Controlled drainage has been used for years (47) to raise water tables during the growing season and increase water supplies to growing crops. Water management procedures have been suggested (48, 49) to promote denitrification and reduce the amount of nitrate entering ground and surface waters. Field studies in North Carolina on poorly drained soils (50) showed that controlled drainage decreased nitrate outflow by about 50 percent.

The use of controlled drainage and other water management practices to reduce amounts of nitrate leaving agricultural lands is not necessarily inconsistent with conventional drainage objectives. However such systems must be carefully designed to satisfy the multiple objectives.

In this example, simulations are conducted for various drainage system designs and operational criteria for a poorly drained soil. The results are analyzed to show the effects of soil, site and drainage system parameters on nitrate loss and on the ability of the system to satisfy drainage requirements for crop production.

Nitrate Concentration

The simulation model predicts water table position, water content distribution, surface runoff, subsurface drainage volume, and seepage volumes at the end of each time increment. If nutrient concentrations of each of the outflow sources is known or can be predicted, total outflows can be calculated on a day to day basis for a given drainage system design and mode of operation. Although nitrate concentrations in drainage waters will generally depend on the drainage system design, there is evidence that, for some soils, concentrations will remain relatively constant. Gilliam et al. (50) found that for poorly drained soils with high natural water tables, denitrification occurred in the subsoil and the nitrate concentration was always zero in horizons at depths greater than 0.9 to 1.2 m. Water leaving the field by deep or lateral seepage must pass through this reduced zone and thus can be assumed to have a nitrate concentration of zero. The depth of the reduced zone may depend on the drainage design, particularly the depth of the drain tubes. However seepage is slow for these soils and a permanent water table will exist at about the drain depth; so denitrification below that depth would be expected with the consequent negligible nitrate in seepage waters.

Nitrate concentrations used in this example for surface and subsurface drainage waters were based on field experiments (50, 51) on soils which had normal fertilizer applications for a corn-soybean rotation in North Carolina. The weighted average concentration in the surface runoff water for all runoff events for three years was approximately 1.0 mg/L, so this was the value assumed for surface runoff concentration in the simulations conducted in this paper.

Although nitrate concentrations in subsurface drainage waters generally depend on several factors including fertilizer history and soil profile characteristics, Gilliam et al. (50) found that concentrations are relatively constant for a given North Carolina soil in a corn-soybean rotation. Concentrations in outlet drainage ditches (which included surface as well as subsurface drainage waters) of poorly drained soils were essentially constant during all seasons of the year. Furthermore, these concentrations were not affected by raising the water level in the outlet to reduce drainage rates. Ditch water concentrations were nearly the same for a system in which drain tubes were spaced 30 m apart as for a similar soil with a drain spacing of 80 m. Measurements on subsurface drainage waters of poorly drained soils in the NC Coastal Plains indicate NO_3^--N concentrations of approximately 10 mg/L. For moderately well drained soils which do not have a permanent water table within 3 m of the surface, concentrations as high as 30 mg/L were measured. For the poorly drained soil considered in this study we will assume a constant NO_3^--N concentration in the subsurface drainage wter of 11 mg/L. Concentrations may be somewhat higher than this for closely spaced drains or lower for very low drainage intensities. However, these differences are neglected in this example.

Soil, Crop and Drainage System Input Data

Simulations were conducted for a uniform Typic Umbraqualt soil which is similar to the poorly drained soils studied by Gilliam et al. (50). The soil is relatively flat with poor surface drainage in its natural state. Downward water movement is restricted by a tight layer at a depth of 3 m. Vertical seepage through the 30 m thick restrictive layer to an aquifer with a constant hydraulic head can occur but is slow. Lateral seepage to a stream channel located 2000 m from the site also occurs. Soil properties and site parameters used as inputs in the simulation model are listed in Table I. Multiple entries in Table I indicate the different parameter values for which simulations were conducted.

All simulations were conducted for corn production in a corn-soybean rotation, at a location near Wilmington, NC. Simulations were not conducted for the soybeans because drainage requirements for corn exceed those for soybeans

in North Carolina. The growing season was assumed to be 120 days in duration with planting on April 15. The depth and distribution of roots were
assumed to be time dependent only and the effective root depth-time relationship was obtained from the data of Mengel and Barber (52).

In addition to conventional drainage systems, simulations were also conducted
for controlled drainage during the winter months and for controlled drainage
during the entire growing season. Controlled drainage was simulated by
raising the drainage outlet elevation to within 15 cm of the soil surface
during the winter months and to within 50 cm of the surface during the
growing season (Table II). In both cases the outlet elevation was lowered to
the drain depth on March 15 for rapid drainage prior to seedbed preparation,
and on Aug. 15 for drainage during harvest.

Climatological input data consisted of hourly precipitation records and
maximum and minimum daily temperatures. These data were obtained from
storage in HISARS (3) for Wilmington, NC for the 26 year period, 1950-1975.

TABLE I. Soil property and site parameter inputs for the simulations.

Depth to restricting layer, b	3.0 m
Saturated hydraulic conductivity, K	1.0 cm/h
Saturated water content	0.41 cm^3/cm^3
Water content at lower limit available to plants	0.15 cm^3/cm^3
Minimum water-free pore space for spring tillage	30 mm
Minimum daily rainfall to stop tillage operations	10 mm
Minimum time after rainfall before can restart tillage	2 days
Thickness of restrictive layer, T	30 m
Hydraulic head in the ground water aquifer, H_a	29 m
Effective vertical hydraulic conductivity of restrictive layer, K_v	1.2 m/day
Distance to remote lateral sink	2000 m
Effective lateral hydraulic conductivity to lateral sink, K_h	2.4 m/day
NO_3-N concentration in drainage water	11 mg/L
NO_3-N concentration in surface runoff	1 mg/L
NO_3-N concentration in seepage water (both lateral and vertical)	0.0 mg/L
Drain diameter	102 mm (4 in.)
Surface depression storage, S	2.5 mm, 12.5 mm, 25 mm
Drain spacing, L	7.5 m, 15 m, 22.5 m, 30 m, 60 m
Drain depth	1 m

Water Management Objectives

The performance of a water management system design was evaluated in terms of
total annual nitrate outflow and its ability to satisfy trafficability and
crop protection requirements. Three parameters were calculated and tabulated
for each year of simulation: (a) total NO_3-N outflow (kg/ha); (b) number of
working days during the one month period prior to planting; and (c) SEW_{30}

which provides a measure of excessive soil water conditions during the growing season.

A working day is defined as a day in which (a) rainfall is less than a given maximum value which depends on the soil, (b) a minimum number of days have elasped since that given amount of rainfall occurred, and (c) the air volume (drained volume) in the profile exceeds a limiting value which is also dependent on the soil. The starting and stopping working hours are also inputs to the simulation model and are used to compute partial working days.

The concept of SEW_{30}. It was originally defined by Sieben (53) to evaluate the influence of high fluctuating water tables during the winter on cereal crops. It is used herein to quantify excessive soil water conditions during the growing season and may be expressed as,

$$SEW_{30} = \sum_{i=1}^{n} (30 - x_i)$$

where x_i is the water table depth on day i, with i = 1 being the first day and n the number of days in the growing season. Negative terms inside the summation are neglected. Although this method has rather severe limitations (54) it does provide an approximate means of quantifying the crop protection aspects of drainage. For our purposes it is assumed that drainage is sufficient for crop protection if SEW_{30} is less than 100 cm-days.

TABLE II. Weir depths in the drainage outlet for conventional and controlled drainage conditions.

Date	Stage	Conventional drainage	Controlled drainage winter only	Controlled drainage all year
March 15 – April 15	seedbed preparation	1.0 m	1.0 m	1.0 m
April 15 – August 15	growing season	1.0	1.0	0.50
August 15 – October 15	harvest	1.0	1.0	1.0
October 15– March 15	fallow	1.0	0.15	0.15

RESULTS AND DISCUSSION

Drain spacing and the quality of surface drainage both have large effects on total annual outflows via surface runoff, subsurface drainage and seepage. This is illustrated by outflow predictions plotted in Figure 6 for a relatively wet year (1969).

The effect that changing the relative amounts of runoff, drainage and seepage waters has on NO_3^--N outflows is shown in Figure 7 for a 5 year recurrence interval (5 YRI). Since the nitrate concentration is less in surface runoff water than in subsurface drainge water, total NO_3^--N outflows is less for good than for poor surface drainage because the amount of subsurface drainage is decreased. Larger drain spacings also decrease subsurface drainage (Figure 6) and nitrate outflow (Figure 7) regardless of surface drainage quality.

The results in Figure 7 show clearly that nitrate outflows for this soil are heavily dependent on the drainage system design, and that nitrate movement from the field can be minimized by using good surface drainage and wide subsurface drain spacings. However, it is also necessary to design the drainage system to satisfy trafficability and crop protection requirements. The number of working days (5 YRI basis) during the month prior to seedbed preparation is plotted versus drain spacing in Figure 8. The number of working days needed would depend on the specific farming operation. Arbitrarily selecting 10 working days as a design value would require drain spacings of 22 m for poor surface drainage and 28 m for good surface drainage. Drainage for crop protection would not be a problem for good surface drainage as SEW_{30} values of less than 20 cm days were obtained for all drain spacings. However, in order to hold the 5 YRI SEW_{30} value to 100 cm days for poor surface drainage, a spacing of 17 m would be necessary (Figure 9). Therefore, the drainage requirements can be satisfied with poor surface drainage and a 17 m drain spacing or with good surface drainage and a 28 m spacing. The two alternatives would have 5 YRI annual NO_3^--N outflows of 39 kg/ha and 20 kg/ha, respectively. Thus, by using good surface drainage and wider drain spacings, nitrate outflow can be reduced by about 50 percent for this soil.

Fields with good surface drainage will have less nitrate outflow than those with poor surface drainage because the amount of subsurface drainage with its high concentration of nitrate, is reduced while surface drainage, with low nitrate concentration, is increased (Figure 6). Another way to reduce subsurface drainage and increase seepage and surface runoff is to raise water levels in the outlet during periods when intensive drainage is not necessary. Predicted results for NO_3^--N outflow are plotted in Figure 10 for conventional drainage, controlled drainage during the winter only and controlled drainage during the entire year. These results for good surface drainage show that nitrate outflows can be substantially reduced by using controlled drainage during the winter and further reduced, although by a smaller amount, by raising the outlet water level during the growing season. Controlled drainage has a somewhat negative effect on trafficability because of higher water tables at the beginning of the seedbed preparation period. For good surface drainage, a drain spacing of 22.5 m would be required to give 10 working days on a 5 YRI basis if controlled drainage is used. This would give a 5 YRI annual NO_3^--N outflow of 14.5 kg/ha for controlled drainage during the winter and 12.2 kg/ha for controlled drainage during the entire year. The 22.5 m spacing gave (5 YRI) SEW_{30} values less than 40 cm days for both controlled drainage situations. More details on the results including for controlled drainage on soils with poor surface drainage are given by Skaggs and Gilliam (55).

Results for various drainage system designs and operational procedures that will satisfy trafficability and crop protection requirements are summarized in Table III. Note that for good surface drainage the drain spacing selection was controlled by the trafficability requirement, while, for poor surface drainage, the limiting factor for spacing was the SEW_{30} value of 100 cm—

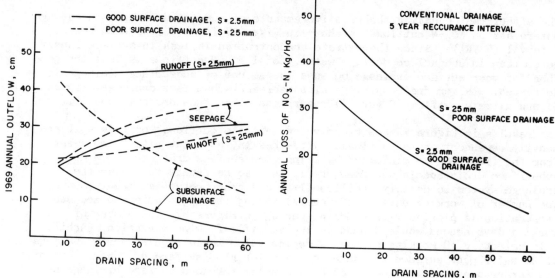

Figure 6.--Effect of drain spacing and surface drainage on runoff, seepage and sub-surface drainage for 1969, a relatively wet year.

Figure 7.--Effect of drain spacing on annual loss of NO_3-N for both good and poor surface drainage.

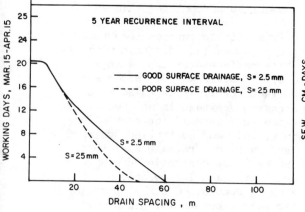

Figure 8.--Working days as affected by drain spacing for two levels of surface drainage.

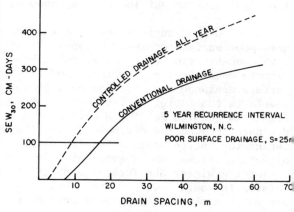

Figure 9.--Effect of drain spacing on SEW_{30} for conventional and controlled drainage when the surface drainage is poor.

days. The results show clearly that the design and operation of a drainage system may have a tremendous effect on the movement of nitrates from agricultural lands. Of the six alternatives given in Table III, NO_3^--N outflows vary from a maximum of 39 kg/ha for poor surface drainage without controlled outlets to 12 kg/ha for good surface drainage and water table control during the entire year. The more than 3-fold reduction in NO_3^--N outflows may be accomplished without sacrificing trafficability or crop protection objectives of the drainage system. The reader should note that water management practices identified herein to diminish the amount of nitrate leaving fields are those that tend to increase surface runoff and seepage and decrease subsurface drainage. Increasing surface runoff also increases the movement of any pollution carried in the runoff. Thus the use of these practices to reduce nitrate outflow may increase sediment and phosphorus in the drainage waters.

It should also be emphasized that the success of the drainage system, so far as reducing nitrate outflows is concerned, depends on it being operated as it was designed. For example, we found that the 5 YRI NO_3-N outflow for conventional drainage with drain spacings of 28 m was 20 kg/ha. By placing the drains closer together (22.5 m) and using controlled drainage during the entire year, the outflow could be reduced by 39 percent to 12.2 kg/ha. However, if the system is designed and installed for this method of operation and the operator does not raise the weir in the outlet as planned, NO_3-N outflow will be 23 kg/ha (Figure 7) on a 5 YRI basis. The system used in this way would be over designed and have higher nitrate outflows than the conventional drainage system.

TABLE III. Summary of 5 YRI NO_3-N outflow for various drainage system designs and operational procedures.

	Conventional drainage	Controlled drainage, winter	Controlled drainage, all year
Good surface drainage			
Drain spacing, m	28	22.5	22.5
NO_3-N outflow, kg/ha	20	14.5	12.2
Working days*	10	10	10
SEW_{30}, cm days	<100	>100	<100
Poor surface drainage			
Drain spacing, m	17	17	9
NO_3^- outflow, kg/ha	39	33	39
Working days	>10	>10	>10
SEW_{30}*	100	100	100

*Factor which controlled the drain spacing.

SUMMARY AND CONCLUSIONS

A computer model for simulating the operation of drainage and water table control systems was described and applied to evaluate the effects of drainage system design and operation on nitrate outflow. The model, DRAINMOD, was developed for design and evaluation of multicomponent systems which may include facilities for subsurface drainage, surface drainage, subirrigation or controlled drainage and irrigation of wastewater onto land. The model is based on a water balance in the soil profile and is composed of a number of components, incorporated as subroutines, to evaluate the various mechanisms of water movement and storage in the profile. In order to simplify the required inputs and make them consistent with available data, approximate methods are used for most components.

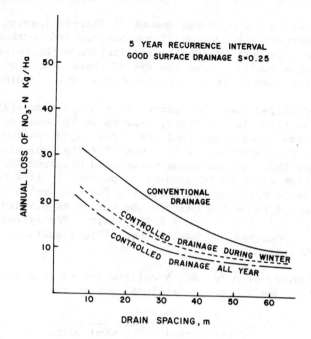

Figure 10.--Annual loss of NO_3^--N as influenced by drain spacing for good
surface drainage for entire year except during planting and
harvest seasons.

Several field studies to evaluate the reliability of the model were conducted. Experiments were conducted to test the model in North Carolina, Ohio,
Louisiana, Florida and California with good results in each location.

As an example of the application of the model to nonpoint source pollution
problems, it was modified to predict nitrate movement from artificially
drained soils with high water tables. Nitrate concentrations in surface
runoff, subsurface drainage and seepage waters were assumed to be constant
and independent of the drainage system design for soils with poor natural
drainage. Simulations were conducted for various designs of both conventional drainage and controlled drainage systems. The results were analyzed
to determine the effects of various alternative drainage systems on total
nitrate outflows to drainage waters.

The study resulted in the following conclusions for a Typic Umbraqualt soil
under North Carolina climatic conditions.

1. Trafficability and crop protection requirements of a drainage system
system can usually be satisfied by several drainage system designs and operational procedures. Some of these alternatives will cause less nitrate outflow than others. For the poorly drained soil considered in this study,
there was a 3-fold difference in NO_3^--N outflow among drainge systems that
satisfied drainage objectives.

2. Nitrate outflows for lands with good surface drainage will be considerably less than for poor surface drainage. Outflows for good surface drainage
in the example considered here were about 50 percent of those predicted for
poor surface drainage.

3. The amount of NO_3^--N that leaves the field through drainage waters can be reduced by using controlled drainage during the winter months and during the growing season. Use of controlled drainage requires somewhat closer drain spacings than conventional drainage systems in order to meet trafficability requirements. Therefore nitrate outflows will be increased over that of conventional drainage if outlet water levels are not controlled as planned in the system design.

REFERENCES

1. Pavelis, G. A. "Drainage Development: Area, Investment and Capital Values, Ch. 8 in USDA Bulletin: Farm Drainage in the United States: History, Status and Prospects. In Review.

2. Freeze, R. A. "Three Dimensional Transient Saturated – Unsaturated Flow in a Groundwater Basin", Water Resources Research, Vol. 7, 1971, pp. 347-366.

3. Wiser, E. H. "HISARS – Hydrologic Information Storage and Retrieval System, Reference Manual", North Carolina Agricultural Experiment Station Tech. Bull. No. 215, 1975, 218 pp.

4. Phillip, J. R. "Theory of Infiltration", Advances in Hydroscience, Vol. 5, 1969, 215-296 pp.

5. Hadas, A., D. Swartzendruber, P. E. Rijtema, M. Fuchs and B. Yaron. "Physical Aspects of Soil Water and Salts in Ecosystems", Springer-Verlag, New York.

6. Skaggs, R. W. and R. Khaleel. "Infiltration", Chapter 4 in ASAE Monograph Hydrologic Modeling of Small Watersheds, C. T. Haan, ed, American Society of Agricultural Engineers, St. Joseph, MI, 1972, pp. 121-166.

7. Richards, L. A. "Capillary Conductivity of Liquid Through Porous Media", Physics, Vol. 1, 1931, pp. 318-333.

8. Green, W. H. and G. Ampt. "Studies of Soil Physics, Part I – The Flow of Air and Water Through Soils", J. Agricultural Science, Vol. 4, 1911, pp. 1-24.

9. Horton, H. N., C. B. England and V. O. Shanholtz. "Concepts in Hydrologic Soil Grouping", Transactions of the American Society of Agricultural Engineers, Vol. 10(3), 1967, pp. 407-410.

10. Philip, J. R. "The Theory of Infiltration: 4. Sorptivity and Algebraic Infiltration Equations", Soil Science, Vol. 84, 1957, pp. 257-264.

11. Smith, R. E. "The Infiltration Envelope: Results from a Theoretical Infiltrometer", Journal of Hydrology, Vol. 17, 1972, pp. 1-21.

12. Smith, R. E. and J.-Y. Parlange. "A Parameter-Efficient Hydrologic Infiltration Model", Water Resources Research, Vol. 14(3), 1978, pp. 533-538.

13. Childs, E. C. and M. Bybordi. "The Vertical Movement of Water in Strati-

fied Porous Material - 1". Infiltration, Water Resour. Res. Vol. 5, 1969, pp. 446-459.

14. Hillel, D. and W. R. Gardner. "Steady Infiltration into Crust Topped Profiles", Soil Science, Vol. 108, 1969, pp. 137-142.

15. Bouwer, H. "Infiltration of Water into Nonuniform Soil", J. Irrigation and Drainage Division, American Society of Agricultural Engineers, Vol. 95, 1969, pp. 451-462.

16. Morel-Seytoux, H. J. "Derivation of an Equation of Infiltration", Water Resources Research, Vol. 10(4), 1974, pp. 795-800.

17. Mein, R. G. and C. L. Larson. "Modeling Infiltration During a Steady Rain", Water Resources Research, Vol. 9(2), 1973, pp. 384-394.

18. Reeves, M. and E. E. Miller. "Estimating Infiltration for Erratic Rainfall", Water Resources Res., Vol. 11(1), 1975, pp. 102-110.

19. Rawls, W. J., D. L. Brakensiek and N. Miller. "Green-Ampt Infiltration Parameters from Soils Data", J. Hydraulic Engineering, Vol. 109(1), 1983, pp. 62-69.

20. Gayle, G. A. and R. W. Skaggs. "Surface Storage on Bedded Cultivated Lands", Transactions of the American Society of Agricultural Engineers, Vol. 21(1), 1978, pp. 102-109.

21. Tang, Y. K. and R. W. Skaggs. "Subsurface Drainage in Soils with High Hydraulic Conductivity Layers", Transactions of the American Society of Agricultural Engineers, Vol. 21(3), 1978, pp. 515-521.

22. Skaggs, R. W. and Y. K. Tang. "Effect of Drain Diameter, Openings and Envelopes on Water Table Drawdown", Transactions of the American Society of Agricultural Engineers, Vol. 22(2), 1978, pp. 326-333.

23. Bouwer, H. and J. van Schilfgaarde. "Simplified Method of Predicting Fall of Water Table in Drained Land", Transactions of the American Society of Agricultural Engineers, Vol. 6(4), 1963, pp. 2188-291.

24. Skaggs, R. W., "Factors Affecting Hydraulic Conductivity Determinations from Drawdown Measurements", American Society of Agricultural Engineers Paper No. 79-2075, 1979, 22 pp.

25. van Schilfgaarde, J., "Nonsteady Flow to Drains", In Drainage for Agriculture, J. van Schilfgaarde, ed. American Society of Agronomy, Madison, WI. pp. 245-270.

26. Moody, W. T., "Nonlinear Differential Equation of Drain Spacing", Journal of the Irrigation and Drainage Division, American Society of Civil Engineers, Vol. 92, 1966, pp. 1-9.

27. Mohammad, F. S. and R. W. Skaggs, "Drain Tube Opening Effects on Drain Inflow", J. of Irrigation and Drainage Engineering, Vol. 109(4), 1983, pp. 393-404.

28. van Beers, W. F. J., "Computing Drain Spacings", Bulletin 15, Interna-

tional Institute for Land Reclamation and Improvement/ILRI, P. O. Box 45, Wageningen, The Netherlands. 1966.

29. Kirkham, Don. "Potential Flow into Circumferential Openings in Drain Tubes", Journal of Applied Physics, 1950, pp. 665-660.

30. Chieng, Sie-Tan, R. S. Broughton and N. Foroud, "Drainage Rates and Water Table Depths", Journal of the Irrigated and Drainage Division, ASCE, Vol. 104(IR4), 1969, pp. 413-433.

31. Ernst, L. F., "Formulae for Groundwater Flow in Areas with Subirrigation by Means of Open Conduits with a Raised Water Level", Misc. Reprint 178, Institute for Land and Water Management Research, Wageningen, The Nether-lands, 1975, 32 pp.

32. Thornthwaite, C. W., "An Approach Toward a Rational Classification of Climate", Geog. Rev., Vol. 38, 1948, pp. 55-94.

33. Penman, H. L., "Evaporation - an Introductory Survey", Netherlands Journal of Agricultural Science, Vol. 4, 1956, pp. 9-29.

34. Jensen, M. E., H. R. Haise and R. Howard, "Estimating Evapotranspiration from Solar Radiation", Journal of the Irrigation and Drainage Division, ASCE, Vol. 89(IR4), 1963, pp. 15-41.

35. Gardner, W. R., "Some Steady Solutions of the Unsaturated Moisture Flow Equation with Application to Evaporation from a Water Table", Soil Sci., Vol. 85, 1957, pp. 228-232.

36. Alexander, L., "Predicting Steady Upward Flux from the Water Table", M.S. Thesis, North Carolina State University, Raleigh, 1983, 207 pp.

37. Skaggs, R. W. and Y. K. Tang, "Saturated and Unsaturated Flow to Parallel Drains", Journal of Irrigation and Drainage Division of American Society of Civil Engineers, Vol. 102(IR2), 1976, pp. 21-238.

38. Skaggs, R. W., "Field Evaluation of a Water Management Simulation Model", Transactions of the American Society of Agricultural Engineers, Vol. 25(3), 1982, pp. 666-674.

39. Skaggs, R. W., N. R. Fausey and B. H. Nolte, "Water Management Model Evaluation for North Central Ohio", American Society of Agricultural Engineers Paper No. 79-2070 presented at the 1979 Summer Meeting of the ASAE, Winnipeg, Canada, 1981.

40. Gayle, G. A., R. W. Skaggs and C. E. Carter, "Predicting Effects of Excessive Soil Water Conditions on Sugarcane Yield", American Society of Agricultural Engineers, Paper No. 83-2564, 14 pp.

41. Chang, A. C., R. W. Skaggs, L. F. Hermsmeier and W. R. Johnston, "Evalua-tion of a Water Management Model for Irrigated Agriculture", Transactions of the American Society of Agricultural Engineers, Vol. 26(2), 1978, pp. 413-433.

42. Muir, John, J. S. Boyce, E. C. Seim, P N. Mosher, E. J. Deibert and R. A. Olson, "Influence of Crop Management Practices on Nutrient Movement Below

the Root Zone in Nebraska Soils, J. Environ. Qual. Vol. 5, 1976, pp. 255-259.

43. Pratt, P. F. and D. C. Adriano, "Nitrate Concentrations in the Unsaturated Zone Beneath Irrigated Fields in Southern California", Soil Sci. Soc. Am. Proc., Vol. 37, 1973, pp. 321-322.

44. Gambrell, R. P., J. W. Gilliam and S. B. Weed, "Nitrogen Losses from Soils of the North Carolina Coastal Plains", J. Environ. Qual., Vol. 4, 1975, pp. 317-323.

45. Gambrell, R. P., J. W. Gilliam and S. B. Weed, "Denitrification in Subsoils of the North Carolina Coastal Plan as Affected by Soil Drainage", J. Environ. Qual. Vol. 4, 1975, pp. 311-316.

46. Baker, J. L. and H. P. Johnson, "Impact of Subsurface Drainage on Water Quality", Proceedings of the 3rd National Drainage Symposium, American Society of Agricultural Engineers Publication No. 77-1, 1977.

47. Renfro, G., Jr., "Applying Water Under the Surface of the Ground", Yearbook of Agriculture, USDA, 1955, pp. 273-278.

48. Willardson, L.S., B. D. Meek, L. B. Grass, G. L. Dickey and J. W. Bailey, "Nitrate Reduction with Submerged Drains", Transactions of the American Society of Agricultural Engineers, Vol. 15(1), 1972, pp. 84-90.

49. Raveh, A. and Y. Avnimelech, "Minimizing Nitrate Seepage from the Hula Valley into Lake Kinneret (Sea of Galilee): I. Enhancement of Nitrate Reduction by Sprinkling and Flooding", J. Environ. Qual., Vol. 2, 1973, pp. 455-458.

50. Gilliam, J. W., R. W. Skaggs and S. B. Weed, "Drainage Control to Diminish Nitrate Loss from Agricultural Fields", J. Environ. Qual., Vol. 8, 1979, pp. 137-142.

51. Gambrell, R. P., J. W. Gilliam and S. B. Weed, "The Fate of Fertilizer Nutrients as Related to Water Quality in the North Carolina Coastal Plains", Report No. 93, Water Resources Research Institute of the University of North Carolina, N. C. State University, Raleigh. 1974.

52. Mengel, D. B. and S. A. Barber, "Development and Distribution of the Corn Root System Under Field Conditions", Agronomy Journal Vol. 66, 1974, pp. 342-344.

53. Sieben, W. H., "Het Verban Tussen Ontwatering en Opbrengst Bij de Jonge Zavelgronden in de Noordoostpolder", Van Zee tot Land, 40, Tjeenk Willink V. Zwolle, The Netherlands. (as cited by Wesseling, 1974) 1964.

54. Bouwer, H., "Developing Drainage Design Criteria", Ch. 5 in Drainage for Agriculture, J. van Schilfgaarde, ed., American Society of Agronomy, Madison, WI, 1974.

55. Skaggs, R. W. and J. W. Gilliam, "Effect of Drainage System Design and Operation on Nitrate Transport", Transactions of the American Society of Agricultural Engineers, Vol. 24(4), 1981, pp. 929-934.

RUNOFF AND RECEIVING WATER MODELS FOR NPS DISCHARGE INTO THE VENICE LAGOON

Andrea Rinaldo^ and Alessandro Marani^^

^ Istituto di Idraulica "G. Poleni", Università di Padova, via Loredan, 20 - 35131 Padova, Italy
^^ Dipartimento di Spettroscopia Elettrochimica e Chimica Fisica, Università di Venezia, D.D. 2137 30123 Venezia, Italy

ABSTRACT   -   The   paper deals with selection and application of runoff and receiving   water   models   for nonpoint source pollution (NPSP) of the Venice Lagoon   (Italy).   Annual   average nutrient loadings (either in dissolved or solid-phase   form)   are   recognized together with source areas and spatial location   of   delivery. Novel theoretical grounds for mathematical modeling aimed   at   simulation of water quality within the Lagoon are presented. The approach,   which   focuses   on   long-term NPS impacts, yields a critical discussion concerning the features of NPSP models suitable to comprehensive simulations.   The models consist of multiple-box mass balances of water and pollutants   at   nonconstant   flow   rates   portraying the tidal hydraulic behavior   of   the   water body. Spectral analyses of calculated and measured tidal   elevation   data   have been performed and comparison of the hydraulic regime   of   the   proposed   model   with   transfer functions of static models (successfully   used   to   evaluate   water   volumes   exchanged by the sea with   lagoon) is carried out. The nonnegligible effects of propagation are also discussed and modeled. The chemical behavior of the storages in Lagoon is assimilated to that of a continuous stirred tank reactor. Sample runs of the   model   for pollutant propagation through the multiple-box net are also presented. Limits   and   validity   of   the modeling approaches, a review of noteworthy   references   and   comparison   with   features of analogous models available in the literature are discussed.

1. INTRODUCTION.

In the review and analysis to follow, attention is focused on modeling nonpoint source pollution (NPSP) for large contributing areas and complex receiving water bodies. In particular, special attention is directed to prediction of nutrient losses from agricultural watersheds and of their impacts, even though, under certain restrictions, the mass conservation principles implemented apply to a wider range of simulations.

NPSP model selection is affected by the characteristics and the quality standards of the receiving water body. In fact, the output of NPS model simulations serves usually as input for water quality models, whose levels of sophistication must be tailored to foreseeable predictive ability and scaling-timing of NPS models. Quality and extent of the data base either for the source areas or for the hydrodynamic regimes of the water body also play a major role in model selection.

The case study refers to long-term pollution of the Venice Lagoon (Italy), a result of agricultural activities in the large (2000 km2) contributing mainland (Figure 1).

Some novel aspects of mathematical modeling of water quality variations of the receiving Lagoon are introduced; nevertheless, the strongly varying spatial and temporal characteristics of the hydrodynamic regime studied indicate the need for further research.

The paper deals also with a discussion of feasible integrations of runoff and receiving water models especially for long-term simulations, to which the interest of most control measures is confined.

2. PREDICTION OF MAINLAND DISCHARGE INTO THE VENICE LAGOON.

The task of predicting overall nutrient and sediment fluxes into the receiving Lagoon may seem overwhelming, given the geomorphologic and hydraulic complexity of the watersheds constituting the mainland.

In fact, up to 27 freshwater delivery locations have been identified, whose size and type hold a wide variability. The mainland itself is roughly 2000 km^2 in area, whereas approximately 30% of the land requires mechanical drainage via a number of pumping stations.

The complex nature of the watershed and the wide range of land uses have called for a thorough model selection phase, in view also of the difficulties of calibration. The screening of NPSP models, crucial to this phase of the studies, has been recently aided by updated state-of-the-art contributions (Jolankai, 1983; Jorgensen, 1983).

The possibility of use of zero-order approximation (like the empirical runoff-concentration curves (e.g. Hock, 1974; Manchak, 1974; Porter, 1973; Yu et al., 1973, reported in Jolankai, 1983) has been discarded, in view of

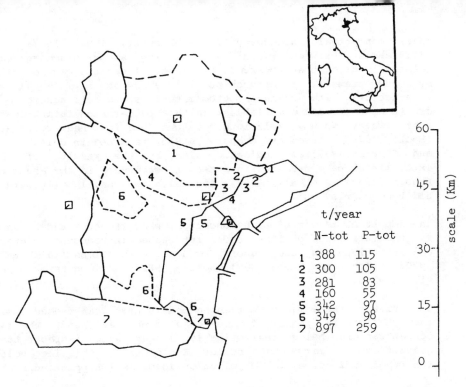

	N-tot	P-tot
1	388	115
2	300	105
3	281	83
4	160	55
5	342	97
6	349	98
7	897	259

t/year

scale (Km)

Figure 1. Schematic layout of the 2000 Km^2 mainland discharging into the Venice Lagoon, with indication of delivery location and average annual intensity of agricultural NPSP loadings.

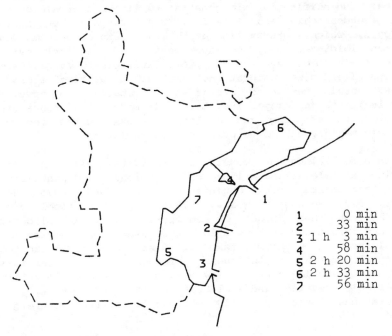

1		0 min
2		33 min
3	1 h	3 min
4		58 min
5	2 h	20 min
6	2 h	33 min
7		56 min

Figure 2. Phase delay of tidal elevations in 7 gage stations (the reference is Diga Sud Lido, 1) for the semidiurnal frequency band (after Goldmann et al.,1975).

their conceptual inadequacy (Zingales et al., 1984). In fact, significant differences between rising and falling stage discharge/concentration relations have been observed. Hysteretical effects can be fully explained by mass transfer relationships (Zingales et al., 1984) referring to interactions with channel bottom materials, to soil adsorption mechanisms or to runoff state variables. In this pattern, pollutant load/discharged total water volume relations seem amenable to linear correlation schemes (Hock, 1970; Betson and McMasker, 1975, reported in Jolankai, 1983; Smith and Stewart, 1977), as indicated also by the results of smoothing of experimental data gathered in a testing watershed in the environment of the Lagoon (Zingales et al., 1981), for which two annual daily data collections had been carried out.

An important result of such conceptual mass transfer and balance relations is the validation of first-order approaches introducing reference weighted concentrations in the runoff, or loading functions (Haith and Dougherty, 1976; Haith and Tubbs, 1981; Haith, 1982) for long-term, or event-based, time scales.

In view of these facts, screening models have been deemed appropriate for the preliminary phase of the case study at hand, the flow-averaged concentrations being realistically linear with respect to total water quantities at large time scales. Incidentally, this fact holds true also for other indices, e.g. BOD pollution loadings (Whipple and Hunter, 1977).

A second-order class of modeling approaches, suitable to large scale simulations of runoff and dissolved nutrient losses, consists of the development of unit-mass response functions (UMRF) for the source area. Such development is either of experimental nature (based, for instance, on parameter indentification via constrained linear search), or of conceptual nature (Bendoricchio and Rinaldo, 1981; Zingales et al., 1984). Such approaches, which portray in depth most features of time development of pollution loadings, rely on hydrologic and chemical parameters which require calibration and hence prove suitable to large scale prediction based upon field-size calibrations. Although predictive errors of screening or UMRF models may seem excessive if compared to observed values, they should be placed in perspective. Errors in model predictions are frequently comparable to sampling and analysis errors, given the difficulty in accurately measurement of nonpoint source pollution.

A third-order class of models refers to basically deterministic models involving detailed simulations of hydrology, sedimentology and water quality components (e.g. Donigian and Crawford, 1975; Beasley, 1976; Kniesel, 1979, Novotny et al., 1978). Since these models are designed to show the watershed impact or sensitivity to spatial placement of soils, crops and practices, they allow the user to describe thoroughly the physical setting of the watershed via distributed parameter techniques. At the actual spatial scale such approaches do not seem tailored to the quality of available data, in view also of the need of long-term monitoring for reliable testing and calibration (Hartigan et al., 1983). Same considerations apply to fully stochastic approaches (e.g. Bogardi and Duckstein, 1978).

The rapid changes in crop practice and the progressive urbanization of part of the territory constitute a further obstacle for the predictive ability via deterministic approaches (Gburek, 1983).

The model adopted, as a preliminary screening phase of source areas, is a modification of the loading function approach (Haith and Dougherty, 1976; Haith and Tubbs, 1981; Haith, 1982) suited to the available data. The waste loading $L_{N,P}$, either of solid-phase or dissolved (or total) Nitrogen and Phosphorus, is defined as a nutrient export from unit source area k, due to runoff event on day t, eventually integrated over a time span T. Computational sequences for dissolved and solid-phase chemicals are separated at small time scales, reflecting the different mechanisms of leaching from croplands (runoff-erosion); nevertheless both phases can be treated jointly for long-term total nutrient loss evaluations. Hence:

$$L_{N,P} = \sum_{k=1}^{MS} \sum_{j=1}^{MC} C_j \, Q_{jkT} \, A_{kj} \, S_{kT} \qquad (1)$$

where: the function L is based on a standardized model for the annual runoff Q_{ikT} (the Curve Number Equation of the U.S. SCS, e.g. Chow, 1964); C_j is the weighted average nutrient concentration either in dissolved or solid-phase form as from the above stipulations; A_{kj} is the area of homogeneous cells, or source areas (as far as average land use, geomorphology and hydrology are concerned); S_{kT} is a transport or attenuation factor (Haith, 1982). MS, MC and T are, respectively, the number of soil types, crop practices and precipitation events over the typical year.

An example of typical average concentrations, taken from the U.S. literature, is presented in Table 1. Nevertheless such data need be used with caution because the wide variability of these parameters has also been pointed out (Jolankai, 1983).

Table 1 - Typical dissolved Nitrogen and Phosphorus Concentration in Runoff from Selected Crops (after Haith, 1982).

Crop	Warm Weather Concentrations		Cold Weather Concentrations	
	Nitrogen	Phosphorus	Nitrogen	Phosphorus
		(mg/l)		
Pasture	2.0	0.3	3.0	0.3
Hay	1.0	0.2	3.0	0.2
Small Grains	3.0	0.3	1.0	0.3
Corn	2.0	0.2	3.0	0.3
Inactive	1.1	0.2	1.6	0.2

Nevertheless experimental evidence gathered in the 15 km^2 wide test catchment of Zuccarello (Zingales et al., 1980; Bendoricchio et al., 1984) have substantiated the validity of the values of Table 1 for the case study at hand. Snowmelt has not been included and dissolved phase has been explicitly considered for gross verifications with experimental data (after Zingales et al., 1981; Haith, 1982). As it turned out, observed and predicted average loading factors are consistenly of the same order of magnitude. The results, whose details are reported elsewhere (Rinaldo, 1982), are outlined in Figure 1.

Figure 1 illustrates also source areas, delivery location and intensity of the seven major agricultural nutrient contributors (out of 27 delivery locations) to the Lagoon.

The approach allows to point out the main sources and to simplify the structure of NPSP within the Lagoon. In fact, the Venice Lagoon is essentially made up by three subbasins relatively independent: Northern, or Bacino di Lido (765 Km^2 wide); Central, or Bacino di Malamocco (536 Km^2 wide); Southern, or Bacino di Chioggia (708 Km^2 wide). Total nutrient loads and loads per unit area of Lagoon within the subbasins are illustrated in Table 2.

Upon comparison of source magnitudes for the 7 major delivery locations (shown in Figure 1), some considerations can be drawn:
- the Northern basin (Bacino di Lido) has intermediate total load but low specific load. NPSs are concentrated (85% of total) at four delivery locations;
- the Central basin (Bacino di Malamocco) holds the lowest total load and intermediate loads per unit area of Lagoon. NPSs are distributed along several delivery locations among which the top-ranked source carries about 30% of the total load;
- the Southern basin (Bacino di Chioggia) is the most endangered with respect to either total or specific loads. Delivery locations are concentrated (up to 80% of total) at two sites.

Table 2 - Summary of results obtained by the loading function approach (after Rinaldo and Rinaldo, 1982).

	Source area (km^2)	Total-N (t/year)	Total-P (t/year)	N input load per ha of Lagoon	P input load per ha of Lagoon
Bacino di Lido	765 (38%)	1331 (34%)	422 (36%)	60	19
Bacino di Malamocco	536 (27%)	1072 (27%)	309 (27%)	84	24
Bacino di Chioggia	708 (35%)	1493 (39%)	427 (37%)	149	43

The marked discretization of input load delivery locations seems relevant to the receiving water model. In fact, the limited number of NPSP sources yields a lowered priority for spatial transport modelling, at least in the scale needed for evaluation of residence time within the Lagoon. On the other hand, integration of runoff and receiving water models needs a time discretization which enables the user to consider tidal effects on overall dispersion of the water body.

Therefore, once long-term nutrient discharges are pointed out via screening models, the logical further step is related to the need of shrinking the time scale of event-based responses (which, for the mainland of the Lagoon, range from one day to the dozen of days) to, at least, intertidal instantaneous values. Such a time scale is deemed necessary for water quality modelling of the receiving water body, given the tidal hydrodynamic regime at hand. Experimental results (Zingales et al., 1980; 1981 a,b; Bendoricchio, et al., 1984) have allowed calibration of a conceptual unit-mass response function approach which would better fit the level of detail required for prediction of lagoonal conditions (on the order of the hour). The transfer function of the watershed to rainfall events distributed in time according to $i(\tau)$ is built by convolution as:

$$u(t; n,K) = (AK/\Gamma(n))(Kt)^{n-1} \exp(-Kt), \tag{2}$$

$$Q(t; n,K) = \int_0^t u(t-\tau; n,K)i(\tau)d\tau, \tag{3}$$

where: u = unit hydrograph (after Nash, 1957), as a function of the parameters n,K; $\Gamma(n)$ = Gamma function of the argument n; A = area of the watershed; $i(\tau)$ = impulse hyetograph. The conceptual UMRFs (Zingales et al., 1981; Bendoricchio and Rinaldo, 1982) obtained for dissolved phase, in terms of X-th pollutant flow rate qX and concentration C, are:

$$qX(t; n,K,hX,C_EX) = C_EX(Q(t; n,K) - (\frac{K}{K+hX})^n Q(t; n,k+hX)) \tag{4}$$

where: X = X-th pollutant index; qX = X-th pollutant discharge due to water discharge Q; C = X-th pollutant instantaneous concentration; hX, C_EX = respectively mass transfer coefficient and equilibrium concentration for the X-th pollutant.

In this pattern, the concept of reference weighted concentration of pollutants in the runoff has been recently linked to the determination of parameters of conceptual UMRF (Zingales, Marani, Rinaldo and Bendoricchio, 1984) via the relation:

$$C_j = C_{Ej} \left(1 - \left(\frac{K}{K+h_j}\right)^n\right)$$

(5)

where C_j is Haith's average concentration, C_{Ej}, h_j, K, n are
the parameters of the conceptual UMRF portraying mass transfer kinetics
and flow rate generation. Even though any assumptions concerning
extrapolations of calibrated parameters seem in principle somewhat
arbitrary, it is certainly possible to tailor UMRFs to total quantities
predicted via screening models for given events, hence yielding acceptable
time specializations for modeling of the receiving water body.

3. NPSP IMPACTS ON TIDAL RECEIVING WATER BODIES

Remarkable efforts have been put, in the recent past, in establishing
the theoretical grounds for simulation of the hydrodynamic and
environmental regimes of the Venice Lagoon. These approaches (summarized in
thorough state-of-the-art contributions, Ghetti, 1980; Di Silvio and
Fiorillo, 1980) range from zero-order lumped parameter mathematical models
to complex formulations of the water quality model, each being fit to
responses of differing level of sophistication. For instance (Ghetti, 1980;
Ghetti et al., 1972), the simpler models of tidal behavior of the Lagoon
work satisfactorily with respect to prediction of water level and global
water balances, while detailed prediction of local features (velocities,
concentrations), especially in certain zones, needs further specializations
of the mathematical model (Ghetti, 1980; Di Silvio and Fiorillo, 1980).
Detailed modelling of intensive quantity movements and transformations
(Davis and Donigian, 1977; Dejak et al., 1981; Jørgensen, 1983) like
temperature, salinity or nutrient concentrations, induced by the decay or
reaction properties of given tracers of water quality, casts further
difficulties for the prediction ability of the theoretical tools. In view
of these facts, it seems that the representation of mass transport in a
model is inseparably related to the spatial structure of the physical
system (Shanahan and Harleman, 1984), which therefore needs be resolved.

Furthermore:
- while the mechanism of leakage of urban and industrial point sources is
  not affected by the hydrodynamic conditions and is characterized by
  relatively high local concentrations, non-point source tracers are
  characteristically discharge-dependent as far as either concentration or
  total quantities are concerned. The higher dissolved or solid-phase loads
  are carried, in fact, by the floods, which at the same time, yield
  important dilution of the loadings. Low concentrations, difficult
  simulations of movements and transformations of its components, time
  variability (dependent upon hydrologic and hydraulic parameters) and
  large total quantities involved in mass balances are therefore the main
  features of NPSP;
- to simulate long-term concentration variations, a lagoon water quality
  model must consider the influence of mass transport. Long-term lagoon
  water quality modeling should implement fast and efficient algorithms,
  and subtidal time discretizations. Certain discontinuities at the
  sea-Lagoon interface need be accounted for, because generally outflowing

water holds higher nutrient concentrations than the inflowing one. It seems that dispersion models, as the ones already implemented for environmental studies on the Lagoon (Di Silvio and Fiorillo, 1981; Dejak et al., 1982) are tailored to simulations of current and tracer patterns in a specified tidal or wind condition as only a relatively short simulation pattern is required (Di Silvio and Fiorillo, 1981). Week-long simulations of build-up of concentration all over the lagoon, produced by continuous output of different wastes, has been achieved up to quasi-steady state conditions (Di Silvio and Fiorillo, 1981).

At the present stage, it is clear the difficulty in meeting all requirements of predictions/simulations characterized by short time step and long time span. Zero-dimensional (Ghetti et al., 1972; Lamberti et al., 1980), one-dimensional (Di Silvio and D'Alpaos, 1972) and two-dimensional models (Leendertse, 1975; Di Silvio and Fiorillo, 1981; Sguazzero et al., 1978) of the propagation of tidal waves inside the lagoon have already been discussed. While a simple zero-dimensional model serves well in reproducing overall water balances, the effects of propagation inside the water body cannot be neglected (as shown in Figure 2 as from spectral analyses of recorded level data, after Goldmann et al., 1975).

Interestingly enough, although the governing differential equations hold a non-linear resistance effect, the spectral analysis of level data has shown that the spectra of the gage meters data, spread in 15 locations all over the lagoon, mantain their energy approximately in the same bands. Thus, the bands differences in the values of the spectra from station to station have been related to damping and rotation of the frequency amplitudes, or to damping and translation of the physical waves. In other words, little energy has been shown to spread onto modes not affected by the tidal forcing term, and hence the mechanism of propagation is essentially linear (Goldmann et al., 1975).

The characters of the zero-dimensional hydrodynamic formulation are essentially non-linear, and that, of course, casts serious obstacles to analytic solutions which, in turn, are those of interest for long term simulations. In search for reliable simplifications, a spectral screening of static models has been performed. A sample of the results is outlined in Figure 3, whereas a three-day recorded series of tidal elevations in the outer lagoon has been analyzed through static model runs with the insertion of differing openings of the three mouths (after Ghetti et al., 1972).

The spectral analysis has been carried out via replacement of the water elevation functions $y(t)$ by its sample chain $y_j(t) = y((j \pi 2\Delta t/N),$ $j = 1, N$. The complex discrete Fourier transform (DFT) coefficients f_k are therefore:

$$y_j(t) = \sum_{0K}^{N-1} f_K \left(e^{i2\pi\Delta t} \right)^{jK/N} = \sum_{0K}^{N-1} f_K \, W_N^{jK} \tag{6}$$

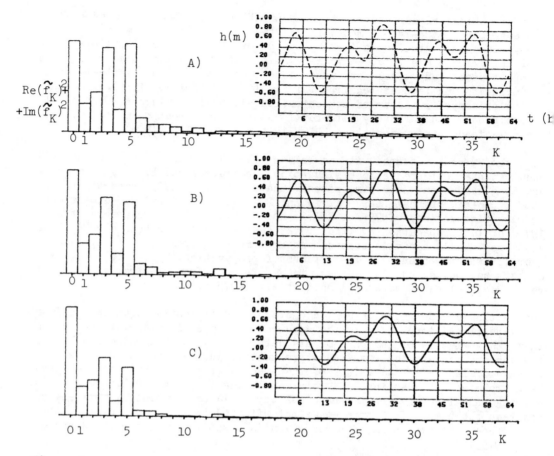

Figure 3. DFT analysis of measured and computed data of elevation :
A) measured elevations at Diga Sud Lido, May 11-12-13,1968;
B) tidal elevations within the Lagoon as by a static model
run with restricted openings at the mouths (C$_q$A$_o$=10124,after
Ghetti et al.,1972); C) tidal elevations within the Lagoon
as by a static model run with further restriction of the mouths
(C$_q$A$_o$=4034, after Ghetti et al.,1972).

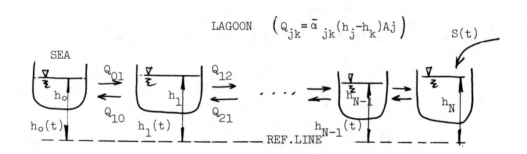

Figure 4. Conceptual scheme of multiple-box Lagoons.

where W_N is the first N-th root of one. Interestingly enough, the modification induced on the forcing spectrum by a zero-dimensional differential operator can be essentially related to damping and translation of the physical waves. Since negligible energy is transferred to the spectrum at different wave numbers, the possibility of linearizing zero-dimensional models is' precisely confirmed. Analogous results have been achieved upon analyses of the spectra of more refined zero-dimensional models (Lamberti et al., 1980).

The relative importance of the results of Sguazzero et al., 1975 and of the spectral analyses of zero-dimensional model results lies in the fact that the overall process of tidal surge propagation is reduced to linear damping and translation, for which wide analogies with electrical engineering circuits or chemical engineering cells-in-series are available. In view of the fatal recourse to approximations, called by the use of input NPSP models and by the need of cheap and rugged tools for years-long simulations, a novel resistance-impedance model of water balance has been devised, as schematized in Figure 4.

For N reservoirs the water balances are given by:

$$
\begin{cases}
\dfrac{d\,h_1}{d\,t} = \alpha_1\,(\,h_o - h_1\,) - \alpha_2\,(\,h_1 - h_2\,) \\[4mm]
\cdots\cdots\cdots\cdots\cdots\cdots \\[2mm]
\dfrac{d\,h_j}{d\,t} = \alpha_j\,(h_{j-1} - h_j) - \alpha_{j+1}(h_j - h_{j+1}) \\[4mm]
\cdots\cdots\cdots\cdots \\[2mm]
\dfrac{d\,h_N}{d\,t} = \alpha_N\,(\,h_{N-1} - h_N\,) + S(t)
\end{cases}
\qquad (7)
$$

where $h_j(t)$ is the water elevation in the j-th reservoir and α_j is the complex impedance between (j-1)-th and j-th box. Besides the resistance-impedance meaning of the complex constants, the physical rationale of the formulation is to relate a linear net flux per unit surface of reservoir to the level rise. In this pattern, the complex constants could be related to the terms of the governing boundary value problem (inertia, linearized friction), although calibration based on propagation lags seems in any case necessary.

The first reservoir (i = 0) sets the boundary conditions together with the input onto the N-th reservoir.

It is worthwhile mentioning that the simple connection between adjacent reservoirs can be arbitrarily made complex (at cost, as it will be discussed in what follows, of heavier formal solutions) in order to meet the requirements of the model spatial structure. These further connections, for the j-th tank, can be accounted for by the j-th reservoir water balance as:

$$\frac{d\,h_j}{d\,t} = -\sum_{k=1}^{MZ} \alpha'_{jk}\,(\,h_j - h_k\,)$$

(8)

MZ being the total number of cells simply connected and α_{jk} a complex number analogous to α_j.

By means of DFT expansion, (7) transforms as:

$$h_j(t) = \sum_{K=0}^{M-1} h_K^j\,e^{iKt'}, \qquad t' = t\,2\,\pi/T$$

(9)

$$\begin{cases} iK\,h_K^1 = \alpha_1\,(h_K^o - h_K^1) \quad - \quad \alpha_2\,(h_K^1 - h_K^2) \\ \cdots\cdots\cdots\cdots \\ iK\,h_K^j = \alpha_j\,(h_K^{j-1} - h_K^j) \quad - \quad \alpha_{j+1}(h_K^j - h_K^{j+1}) \\ \cdots\cdots\cdots\cdots \\ iK\,h_K^N = \alpha_N\,(h_K^N - h_K^{N-1}) \quad + \quad S_K \end{cases}$$

(10)

for every mode K up to the wave number cutoff M; the period T is arbitrary. Solving for h_K^j the j-th equation yields:

$$h_K^j = \frac{\alpha_j\,h_K^{j-1}}{iK + \alpha_j + \alpha_{j+1}} + \frac{\alpha_{j+1}\,h_K^{j+1}}{iK + \alpha_j + \alpha_{j+1}}, \qquad j=1,N$$

(11)

The mode h_K^j, in general, can be written as:

$$h_K^j = \Phi_{jm} \, h_K^m + \Psi_{jl} \, h_K^l , \qquad (0 < m < j < l < N) \qquad (12)$$

where the transfer functions Φ_{jm} and Ψ_{jl} combine the effects of transport from the m-th to the j-th tank and from the l-th to the j-th respectively. A specialization of (12) for $m = 0$ and $l = N$ yields the transfer function of the outer tidal forcing wave with its characteristic frequencies.

If, for semplicity sake, the mainland forcing term S_K is neglected, a signal flow diagram analogous to (11) is obtained, as outlined in Figure 5, where:

$$a_j = a_{jK} = \alpha_j / (iK + \alpha_j + \alpha_{j+1})$$

and

$$b_j = b_{jK} = \alpha_{j+1} / (iK + \alpha_j + \alpha_{j+1}).$$

The flow diagram of Figure 5 reduces to that of Figure 6 with the usual procedure. The transfer functions Φ_{j0} and Ψ_{jN} are built by product of transfer functions from node to node according to the scheme of Figure 6, where the quantities $\phi_j = \phi_K^j$ and $\psi_j = \psi_K^j$ are defined by recursive formulas

$$\phi_j = \frac{1}{1 - b_j a_{j+1} \phi_{j+1}} , \qquad (j=1,2,\ldots,N-2) \qquad (13)$$

$$\psi_j = \frac{1}{1 - b_{j-1} a_j \psi_{j-1}} , \qquad (j=2,3,\ldots,N-1)$$

where, furthemore, $\phi_1 = \psi_{N-1} = 1$. Hence, the transfer functions Φ_{j0} and Ψ_{jN} are:

$$\Phi_{j0} = a_1 \phi_1 a_2 \phi_2 \cdots \cdots a_j \phi_j \qquad (14)$$

$$\Psi_{j0} = b_j \psi_j b_{j+1} \psi_{j+1} \cdots \cdots b_{N-1} \psi_{N-1}$$

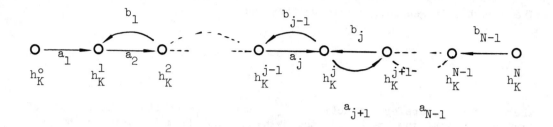

Figure 5. Reduced Signal Flow Diagram of eq.(11).

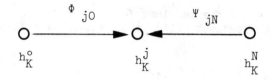

Figure 6. Reduced Signal Flow Diagram of eq.(12)

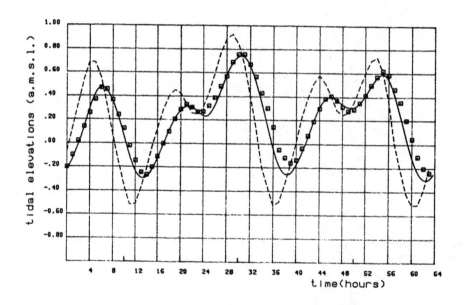

Figure 7. Computed and measured tidal elevations. The dotted line portrays the
forcing tide at Diga Sud Lido (May 11-13, 1968); the experimental
points portray the results of a static model run at maximum contraction
of the mouths (after Ghetti et al., 1972); the solid line is yielded
by eq. (19) with $\alpha_1 = 3.3 + i\,(.25)$.

Table 3 illustrates the explicit form of transfer coefficients ϕ_j and ψ_j, whereas the formalism of continuous fractions has been implemented.

Table 3 - Nodal transfer coefficients ϕ_j and ψ_j.

j	ϕ_j	ψ_j
1	$\cfrac{1}{1-}\ \cfrac{b_1 a_2}{1-}\ ..\ \cfrac{b_{N-3}a_{N-2}}{1-b_{N-2}a_{N-1}}$	1
j	$\cfrac{1}{1-}\ \cfrac{b_j a_{j+1}}{1-}\ .....\ \cfrac{b_{N-3}a_{N-2}}{1-b_{N-2}a_{N-1}}$	$\cfrac{1}{1-}\ \cfrac{b_j a_j}{1-}\ ..\ \cfrac{b_2 a_3}{1-b_1 a_2}$
$N-1$	1	$\cfrac{1}{1-}\ \cfrac{b_{N-2}a_{N-1}}{1-}\ \cfrac{b_2 a_3}{1-b_1 a_2}$

The general solution of (10), (for $S_K = 0$) is:

$$h_K^j = \frac{\Delta_j}{D} = a_1 a_2 \cdots\cdots a_j\ \frac{D_0^j}{D}\ h_K^0 + b_j b_{j+1} \cdots\cdots b_{N-1}\ \frac{D_{N-j}^0}{D}\ h_K^N$$

$$h_j(t) = \sum_{K=0}^{N-1} h_K^j\ e^{iKt'} \tag{17}$$

where: D_K^j, D_{N-j}^0 and D are suitable determinants of the matrix of the coefficients (10); h_K^0, h_K^N, are the forcing spectra, respectively from the sea and from the mainland.

4. PRELIMINARY APPLICATION

The application refers to the easiest case portrayed in Figure 1, or the 1-reservoir Lagoon, whose conceptual scheme (besides the account for the hydraulic behavior of the mouths) had been adopted for the zero-dimensional tidal model (Ghetti et al., 1972). The particular solution for $N = 1$ and $S(t) = 0$ is:

$$h_K^1 = \frac{\alpha_1}{\alpha_1 + iK}\ h_K^0 \tag{18}$$

If the complex constant α_1 is set equal to $\alpha_1 = \alpha + i\beta$, the resulting real and imaginary part of the DFT coefficients are:

$$Re(h_K^1) = \frac{\alpha(\alpha\ Re(h_K^0) - \beta\ Im(h_K^0)) + (\beta+K)\ (\ \beta\ Re(h_K^0) - \alpha\ Im(h_K^0)\ )}{\alpha^2 + (\ \beta + K\ )^2}$$

$$Im(h_K^1) = \frac{\alpha(\beta\ Re(h_K^0) + \alpha\ Im(h_K^0)) - (\ \beta + K\ )\ (\ \alpha\ Re(h_K^0) - \beta\ Im(h_K^0)\ )}{\alpha^2 + (\ \beta + K\ )^2}$$

and, in the physical space, the inverse DFT transformation yields:

$$h(t) = \sum_{K=0}^{M-1} \frac{\alpha_1}{\alpha_1 + iK} h_K^0\ e^{iKt} \tag{19}$$

Calibration has been performed via a suitable filtering of the spectrum of tidal elevation. The reference results (Ghetti et al., 1972) are related to a three-day gaged tide at Diga Sud Lido and to a run of the static model with a restricted mouth configuration. The rationale for this choice was to test the transfer function (18) on extreme cases for which the possible nonlinear effects might have turned out to be crucial. The parameter identification consisted of assuming known the spectrum h_K^j (the spectral analysis of the results of Ghetti et al., 1972) solving (18) for α_1.

The results of the solution for are outlined in Table 5.

TABLE 7 - Identification of the complex constants via the results of the static-model run with restricted mouths (after Ghetti et al., 1972).

Wave number	α	β
0	2,91	0,49
1	3,04	0,25
2	3,05	- 0,49
3	3,29	0,097
4	2,37	- 0,099
5	3,33	0,503
6	- 0,463	2,94
. . . .		

The results of Table 7, together with the results of Figure 4, seem of interest. In fact, notwithstanding the variability of real and imaginary parts of the complex constant , it seemed that the most effective modes k = 0, 3, 5) are damped and rotated in acceptable agreement with the given relation. Higher modes, which nevertheless are rapidly damped by the differential operator, do not respect the given relation. Since in particular the semidiurnal frequency has turned out to be important, an average value of $\alpha_1 = 3.3 + i(.25)$ has been tentatively adopted.

The surprising fit obtained is reported in Figure 7, where the experimental points portray the results of the reference run of the static model (Ghetti et al., 1972).

It seems worthwhile mentioning that eq. (18) is the easiest possible model configuration, and that the calibration has been a most approximate one. Nevertheless the degree of approximation is deemed acceptable, the maximum difference (in the physical space) is of the order of 15%. Furthermore, besides its analytical structure, the solution found has a great advantage over the common numerical procedures because the solution quantities are a periodical extension of period T of the forcing functions.

Figure 8 illustrates the effects induced on the tidal elevations by artificial mainland forcing discharge at same average value and differing frequencies.

As a final comment on water balances, little complication is induced on the solving equations by the possible account for evaporation from the lagoonal surface. In fact, the introduction of the velocity of evaporation as uniform over the whole surface allows the same general solution of (17) upon suitable changes of variable. This holds true provided that the velocity of evaporation, although a function of wheather conditions, does not depend on tidal phases.

5. MATHEMATICAL FORMULATION OF MASS BALANCE

The equation of mass conservation for a dissolved or suspended constituent in the receiving water is a well-known second order elliptic partial differential equation portraying two transport phenomena: that due to organized large-scale motion such as wind-driven or tidal circulation (advective transport) and the residual due to small-scale turbulent fluctuation (diffusive transport).

The conceptual scheme of Figure 4 of the hydrodynamic regime seems perfectly tailored to a multiple-box appoach. Such approaches assume that each tank (or, completely mixed volume element) behaves as a continuous stirred reactor. Concentration is determined by simple mass-balance, or by integration of the advection-diffusion equation over the k-th volume element, so that the multiple-box conceptual scheme results in ordinary differential equations for the volume elements. The balance equation for the k-th element is:

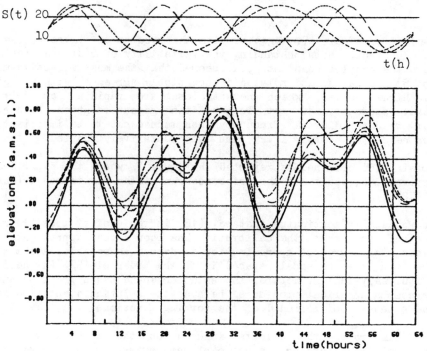

Figure 8. Effects of artificial mainland forcing discharges at same mean value and differing frequency.

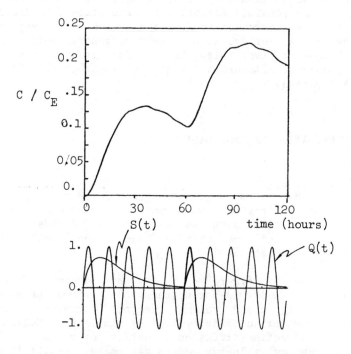

Figure 9. Response of the single box model to an impulse of pollutant at equilibrium concentration distributed as $f(t)=t^2\exp(-t)$, periodic of period $T=60$ hours. The tidal oscillation is $V=V_o+\sin(\pi t/6)$, $V_o= 10$ m^3, t is expressed in (hours).

$$V_k \frac{dC_k}{dt} + C_k \frac{dV_k}{dt} = \sum_j \{Q_{jk} C_j - Q_{kj} C_k + D_{jk}(C_j - C_k)\} + S_k \tag{20}$$

where: $C_k(t)$ = average concentration of the k-th element; $V_k(t)$ = water volume of k-th element; j = summation index extended to all the boxes connected with the k-th element; D_{jk} = dispersion coefficient between elements j and K; S_k = sum of all sources and sinks within element k.

As implicitly stated by eq. (8), multiple box models may have one, two or three dimensional configurations.

The foremost advantage introduced herein consists of the fact that neither discharges Q_{jk} and Q_{kj} need be imposed as constant nor need they be calculated via numerical integration. In fact, the analytical solution, albeit in a spectral form which implies a periodical extension of period T of the solution, allows a thorough description of tidal events.

Some special considerations are needed in order to account properly for the discharges Q_{jk} and Q_{kj}. In fact, according to the solution of (17), such quantities are calculated regardless of the sign, accepting that a possible reversal of sign of Q_{jk} would mean Q_{kj}, or through-flow is from element k to j. Since for the mass balance (20) this fact implies an outgoing or ingoing discharge at differing concentrations, a preprocessing of the solution is necessary. If the discharge exchanged between the adjacent elements j and k is:

$$Q(t) = A\alpha(h_j(t) - h_k(t)) = \sum_k q_k e^{ikt} \tag{21}$$

(where A = the surface of the box under study), a filtering is made as:

$$Q_{kj}(t) = \frac{|Q(t)| - Q(t)}{2} = \sum_m Q_m^- e^{imt} \tag{22}$$

$$Q_{jk}(t) = \frac{Q(t) + |Q(t)|}{2} = \sum_m Q_m^+ e^{imt}$$

Since the absolute value operator is a convolution operator for the DFT expansion, from the spectral standpoint some effort is required for fast and accurate calculation of Q_m^+ and Q_m^- via FFT-related algorithms (Cooley and Tukey, 1965; Giorgini and Travis, 1969).

The resulting general solution of (20) for the same conceptual scheme of (17) needs further specializations aimed at the resolution of the differential operator d/dt.

A further matter of concern is related to the choice of the number of tank-in-series which should properly account for implicit dispersion built in the conceptual scheme (Shanahan and Harleman, 1984).

It is worthwhile mentioning that the source-sink term appearing in (20) must simulate nutrient movements and transformations tied, for instance, to the interactions with the sediments. Albeit a wide literature is available in the field (e.g. Jørgensen, 1983), the task of mathematical modeling is twofold:
- on one side the importance of the source-sink terms is crucial, and needs be resolved;
- on the other reaction, decay or adsorption kinetics do not cause modifications to the solving procedure of the differential equations, at least according to the linearized schemes of chemical reactor engineering.

6. PRELIMINARY APPLICATION

For consistency with the 1-box example, the analytical solution of (20) is presented, under the assumptions that no dispersion is taken into account ($D_{jK} = 0$) and that no reaction or decay kinetics are accounted. Eq. (20) can therefore be written as:

$$V_k \frac{dC_k}{dt} + C(\frac{dV_k}{dt} + Q_{ko}) = S_k(t) \tag{23}$$

Let $1/V_k = \sum_m \gamma_m e^{imt}$ (again, the result of a convolution sum), hence:

$$\frac{dC_k}{dt} + C_k \; NCCONV(\gamma_m | i m V_m + Q_m^-) = NCCONV(\gamma_m | S_m) \tag{24}$$

where: NCCONV $(|)$ is the noncyclic DFT convolution; V_m, S_m = DFT coefficients of $V_k(t)$ and $S(t)$. If, furthermore,

$$\sum_k f_k \, e^{ikt} = \text{NCCONV}(\gamma_m \,|\, im V_m + \bar{Q}_m)$$

(25)

$$\sum_k \phi_k \, e^{ikt} = \text{NCCONV}(\gamma_m \,|\, S_m)$$

the analytical solution is:

$$C_k(t) = e^{-(f_0/2)t} \, e^{-\sum_k (f_k/ik)e^{ikt}} \, \{I(t) + C(0)\}$$

(26)

$$I(t) = \int_0^t (e^{(f_0/2)x} \, e^{\sum_n (f_n/in)e^{inx}} \, \sum_k \phi_k \, e^{ikx}) dx$$

While further mathematical manipulations of (20) are forthcoming, the results of the response of a single-box model to three impulses of mass at equilibrium concentrations are illustrated in Figures 9, 10 and 11 (where $S(t) = (C_E/C_E) \cdot f(t)$).

4. CONCLUSIONS

The research has brought to light some relevant points concerning the integration of runoff and receiving water models. They refer to:

- the use of a combination of screening and UMRF approaches for identification of potential NPSP source areas and resolution of inputs for the receiving water model;

- the impractical use of zero-order approaches or refined distributed parameter modeling for the mainland under study;

- the possibility of consistent linearizations of water mass and momentum balance for the receiving water body, based on spectral analyses of experimental data and of computed results;

- the computational advantages of the proposed multiple-box approach for long-term simulations.

Preliminary results, designed to portray the capabilities of the approach in the easiest possible model configuration, have been presented and discussed.

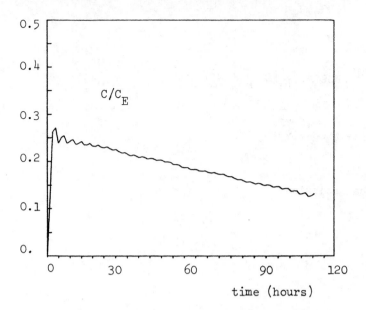

Figure 10. Response of a single-box model to a Dirac-delta impulse of
pollutant at equilibrium concentration (specifications as in
Figure 9).

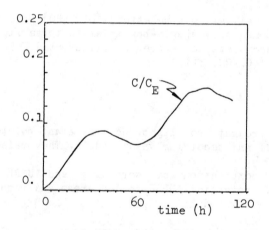

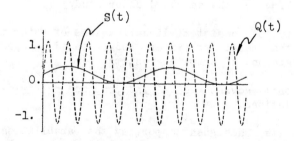

Figure 11. Response of a single-box model to a sinusoidal impulse of
pollutant at equilibrium concentration distributed according
to $f(t)=.1 + .1 \sin(\pi t/30)$.

References

1 Beasley, D.B., 1976. Simulation of the environmental impact of land use on water quality. In: Best management practices for non point source pollution control. EPA-905/9-76-005 Report, U.S.EPA, Washington D.C.:225-240.

2 Beasley, D.B.; Monke, E.J. and Huggins, L.F., 1977. The ANSWERS model: a planning tool for watershed research. ASAE Paper 77-2332, American Society of Agricultural Engineers, St. Joseph, Michigan, 20 pp.

3 Bendoricchio, G. and Rinaldo, A., 1981. Un modello matematico del dilavamento di sostanze chimiche da terreni agricoli. In: Accademia Patavina di Scienze, Lettere ed Arti, Vol. XCIII:137-154.

4 Bendoricchio, G.; Comis, C. and Alessandrini, S., 1984. Evoluzione temporale dell'inquinamento da sorgenti diffuse di origine agricola. Inquinamento, in print.

5 Beyerlin, D.C. and Donigian, A.S., 1979. Modeling soil and water conservation practices. In: R.C. Loher et al. (Editors), Best management practices for agriculture and silviculture. Ann. Arbor Science, Ann. Arbor, Michigan: pp. 18.

6 Bird, R.B.; Stewart, W.E. and Lightfoot, E.N., 1960. Transport phenomena. J. Wiley & Sons, New York, N.Y., 575 pp.

7 Blank, D.; Delleur, D.W. and Giorgini, A., 1971. Oscillatory Kernel functions in the linear hydrologic models. Water Resourc. Res., 7(5):1102-1117.

8 Bogardi, I. and Duckstein, L., 1978. Input for a stochastic model for P loading: Ecological Modeling, 4:15-32.

9 Bragadin, G.L.; Lamberti, A. and Montefusco, L., 1977. Un nuovo modello matematico statico della Laguna di Venezia. L'Energia Elettrica, (LIV)7-8:341-349.

10 Cooley, J.W. and Tukey, J.W., 1965. An algorithm for the machine calculation of complex Fourier series. Mathematics of Computation, (9):114-121.

11 Chow, V.T., 1964. Handbook of Applied Hydrology. McGraw Hill, New York, N.Y., 850 pp.

12 Crawford, N.H. and Linsley, R.K., 1966. Digital simulation in hydrology: the Stanford watershed model IV. Tech. Rept. 39, Dept. of Civil Eng., Stanford, CA, 210 pp.

13 Crawford, N.H. and Donigian, S.A., 1974. Pesticide transport and runoff model for agricultural lands. Environmental Protection Agency, Technology Series. EPA-660/2-74-013, 80 pp.

14 D'Alpaos, L., 1978. Modello unidimensionale per lo studio della propagazione della marea nella Laguna di Venezia (Verifica dell'accuratezza del modello). Proceedings of Congress "Metodologie numeriche per la conservazione e difesa della Laguna di Venezia", (VII):79-87.

15 Dejak, C.; Mazzei Lalatta, I. and Molin, M., 1981. Dispersione di inquinanti nella Laguna di Venezia (modello bidimensionale avvettivo-diffusivo). Proceedings of the Seminar "Processi di mescolamento in ambienti lagunari e loro modellazione", Roma, (1):25-55.

16 Di Silvio, G. and D'Alpaos, L., 1972. Il comportamento della Laguna di Venezia esaminato con il metodo propagatorio unidimensionale. Istituto Veneto di Scienze, Lettere ed Arti - Rapporti e Studi, (V), 107 pp.

17 Di Silvio, G. and Fiorillo, G., 1980. Modelling of Lagoons (the experience of Venice). Proceedings of the Symposium on Predicting Abilities of Surface Water Flow and Transport Models, Berkeley, California, 1:91-109.

18 Donigian, A.S. and Crawford, N.H., 1976. Modeling nonpoint pollution from the land surface. U.S.Environmental Protection Agency, EPA-600/3-76-083, 279 pp.

19 Donigian, A.S.; Beyerlin, D.C.; Davis, H.H. and Crawford, N.H., 1977. Agriculture runoff management (ARM) model version II: refinement and testing. U.S.Environmental Protection Agency, EPA-600/3-77-098, 294 pp.

20 Dooge, C.I., 1977. Problems and methods of rainfall-runoff modelling. In: T.A. Cirioni, U. Maione and J.R. Wallis (Editors), Mathematical Models for Surface Water Hydrology. John Wiley and Sons, New York: 197-252.

21 Dronkers, J.J., 1969. Tidal computations for rivers, coastal areas and seas. Journal of the Hydraulic Division, ASCE, (95)1:29-78.

22 Frere, M.H.; Ross, J.D. and Lane, L.J., 1980. The nutrient submodel. In: W.G. Kniesel, CREAMS: A Field-Scale Model for Chemical Runoff and Erosion from Agricultural Management Systems. U.S. Departement of Agriculture, Conservation Research Report No. 26, 65 pp.

23 Ghetti, A.; D'Alpaos, L. and Dazzi, R., 1971. La regolazione delle bocche della Laguna di venezia per l'attenuazione delle acque alte indagata con metodo statico. Proceedings of the Accademia Patavina di Scienze, Lettere ed Arti, (XVII)7:11-41.

24 Ghetti, A., 1979. Etudes concernant les problemes hydrodynamiques de la Lagune de Venise. General Lecture, Proceedings of the XVII Congress of IAHR, Cagliari, (6):56-94.

25 Giorgini, A. and Travis, J.R., 1969. A short convolution. Tech. Rep. No. 26, School of Civikl Engineering, Purdue University, West Lafayette, IN 47907, 120 pp.

26 Goldmann, A., Rabagliati, R. and Sguazzero, P., 1975. Characteristics

of the tidal wave in the Lagoon of Venice. IBM Tech. Rep. CRV009/513-3539, 47, 52 pp.

27 Gburek, W.J., 1983. Hydrologic delineation of nonpoint source contributing areas. Journal of the Environmental Engineering Division, ASCE, (109)5: 1035-1048.

28 Haith, D.A. and Dougherty, J.V., 1976. Non-point source pollution from agriculture runoff. Journal of the Environmental Engineering Division, ASCE, (102)EE5: 1055-1069.

29 Haith, D.A., 1976. Land use and water quality in New York rivers. Journal of the Environmental Engineering Division, ASCE, (102)EE5: 24-38.

30 Haith, D.A., 1982. Development and testing of watershed loading functions for nonpoint sources. In: Proceedings of the 13th International Conference on Modeling and Simulation, April 10-12, 1982 at Pittsburgh, PA (U.S.A.), 4: 1463-1467.

31 Hartigan, J.P.; Quasebarth, T.F. and Southerland, E.,1983. Calibration of NSP model loading factors. Journal of the Environmental Engineering Division, ASCE, (109), 6: 1259-1272.

32 Hill, C.G., 1977. An introduction to chemical engineering kinetics and reactor design. J. Wiley & Sons, New York, N.Y., 380 pp.

33 Himmelblau, D.M. and Bischoff, K.B., 1968. Process analysis and simulation: deterministic systems. J. Wiley & Sons, New York, N.Y., 210 pp.

34 Jolankai, G., 1977. Field experiments and modeling of non-point source pollution (with special respect to plant nutrients) on a watershed of Lake Balaton. In: Proceedings of the 17-th Congress of IAHR, August 15-19, at Baden Baden (WG), 301-316.

35 Jolankai, G., 1983. Modelling of nonpoint source pollution. In: S.E. Jorgensen (Editor). Application of Ecological Modeling in Environmental Management, Part A. Elsevier Scientific Publishing Co., Amsterdam (N): 283-385.

36 Jorgensen, S.E., 1983. Application of Ecological Modeling in Environmental Management. Part A. Elsevier Scientific Publishing Co., Amsterdam (N), 735 pp.

37 Kniesel, W.G., 1978. A system of models for evaluating nonpoint source pollution - an overview. IJASA Publication, CP-78-11.

38 Kniesel, W.G. and Nicks, A.D., 1980. Introduction. In: W.G. Kniesel (Editor), CREAMS: a Field-Scale Model For Chemicals, Runoff and Erosion from Agricultural Management Systems. U.S. Department of Agriculture, Conservation Research Report, No. 26, 12 pp.

39 Novotny, V.; Tran, H.; Simsiman, G.V. and Chesters, G., 1978. Mathematical modeling of land runoff contaminated by phosphorus.

Journal of WPCF, Jo(1): 101-112.

40 Novotny, V. and Chesters, G., 1981. Handbook of nonpoint source pollution. Van Nostrand Reinhold Co., New York, N.Y., 555 pp.

41 Overcash, M.R. and Davidson, J.M., 1980. Environmental impact of nonpoint source pollution. Ann Arbor Science, Ann Arbor, Michigan, 449 pp.

42 Rinaldo, A. and Rinaldo, A., 1983. Determinazione per singoli recapiti delle quantita' di Azoto e Fosforo che pervengono annualmente in Laguna di Venezia dal proprio Bacino scolante relativamente ai terreni agricoli. Convegno di studi "Laguna, fiumi, lidi: cinque secoli di gestione delle acque a Venezia". Magistrato alle Acque, Venezia, 10-12 giugno, 25 pp.

43 Rinaldo, A., 1982. Nonpoint source pollution from agricultural runoff: a large scale application of mathematical modeling. In: Proceedings of the Thirteenth Pittsburgh conference on Modeling and Simulation, (13) 4: 1570-1574.

44 Selim, H.M. and Iskandar, I.W., 1981. Modeling Nitrogen transport and transformations in soils: 1. Theoretical considerations. Soil Science, (131): 233:241.

45 Shanahan, P. and Harleman, R.F., 1984. Transport in lake water quality modeling. Journal of Environmental Engineering, ASCE, (110)1: 42-57.

46 Smith, H.M. and Stewart, D.A., 1977. Statistical models of river loadings of Nitrogen and Phosphorus in the Lough Neagh system. Water Research, (11): 631-636.

47 Whipple, W.Jr. and Hunter, J.V., 1977. Nonpoint sources and planning for water pollution control. WPCF Journal, (3): 15-23.

48 Zingales, F.; Alessandrini, S. Bendoricchio, G.; Marani, A.; Pianetti, F.; Rinaldo, A., Sartori-Borotto, C. and Zanin, S., 1980. Inquinamento dovuto alle acque di un bacino agricolo sversate nella laguna di Venezia. Inquinamento, (12): 25-31.

49 Zingales, F.; Alessandrini, S.; Bendoricchio, G.; Comis, C.; Marani, A.; Pianetti, F.; Rinaldo, A. and Sartori-Borotto, C., 1981. Simulazione del dilavamento di sostanze chimiche da terreni agricoli. Inquinamento, (10): 21-27.

50 Zingales, F.; Alessandrini, S.; Bendoricchio, G.; Comis, C.; Marani, A.; Pianetti, F.; Rinaldo, A.; Sartori-Borotto, C., 1982. A model of nonpoint source pollution in the agricultural runoff. In: Proceedings of the Thirteenth Pittsburgh Conference on Modeling and Simulation, Pittsburgh, PA (U.S.A.), (13) 4: 1575-1581.

51 Zingales, F.; Marani, A.; Rinaldo, A. and Bendoricchio, G., 1984. A conceptual model of unit-mass response function for nonpoint source pollutant runoff. Submitted for publication on to Ecological Modelling.

DISTRIBUTED PARAMETER HYDROLOGIC AND WATER QUALITY MODELING

DISTRIBUTED PARAMETER HYDROLOGIC AND WATER QUALITY MODELING

David B. Beasley
Associate Professor
Agricultural Engineering Department
Purdue University
West Lafayette, Indiana 47907

ABSTRACT

One of the most effective tools that planners and researchers in the water resources and water quality areas can have is an accurate, comprehensive watershed hydrology and pollutant yield model. ANSWERS (Areal Nonpoint Source Watershed Environment Response Simulation), a distributed parameter model which was developed at Purdue University, is one such model and is detailed in this paper. General concepts behind and strengths of distributed parameter models are presented. A new, more descriptive sediment detachment and transport sub-model, which can predict the particle-size distribution of the sediment load, is described. Applications on an agricultural watershed of approximately 2,000 hectares and on a small agricultural area being converted to home sites demonstrate the the various capabilities of the ANSWERS model.

INTRODUCTION

Rational, effective planning to control nonpoint source pollution is ultimately dependent upon the wisdom and ingenuity of planners, i.e., the human resources employed. However, the efficiency and effectiveness of these key, technically trained individuals is very much influenced by the scientific tools made available to assist them. One of the most useful tools that can be employed in the complex task of planning control measures for nonpoint source pollution is an accurate, comprehensive watershed model capable of simulating all effects of proposed and/or applied control measures.

One of the largest problems a planner or user of one of these tools faces is the adaptation of the model to the specific situation at hand. Most watershed models are "lumped" in nature and describe an overall or average response of the watershed (Woolhiser, 1973). Since nonpoint sources are, by definition, spatially variable, these models often do a less than adequate job of describing the physical situation. The use of calibration does, to some degree, offset the inability of the model to take into account spatially varying processes. However, calibration data is extremely rare and is not very useful when the watershed under study is being extensively modified. In addition, a model calibrated to a particular watershed is generally not transferable to another watershed, even one nearby, unless the two drainage areas are essentially identical in all respects (a highly unlikely eventuality).

In order to be a useful planning or evaluation tool, a watershed model must be able to describe accurately the effects of changing topography, land use, management, soil responses and meteorological inputs. Thus, the model should be able to discern the varying impacts of watershed modifications made in different places. Additionally, the model should be deterministic in nature and use physically-measurable data to simplify or remove any need to calibrate or "adjust" the model to a particular watershed.

While no single model is currently available which can handle the entire range of complicated nonpoint source pollution problems, several have reached a state of development that makes them effective planning tools for specialized problem areas. The fundamental concepts of the ANSWERS model (Beasley, 1977; Beasley et al., 1980; Beasley and Huggins, 1982), particularly those dealing with the expanded sediment detachment/transport portion, are outlined. Additionally, example applications are presented which demonstrate most of the capabilities of the ANSWERS model.

DISTRIBUTED PARAMETER MODELING PHILOSOPHIES

A distributed parameter modeling approach is one which attempts to incorporate, to a practical degree, data concerning the areal distribution of parameter variations. Additionally, the computational algorithms which are included which attempt to evaluate the influence of this distribution on the simulated behavior. Because of the infinite scale on which variations exist in the real world, all distributed modeling approaches utilize approximations to spatial variability. Thus, any distributed parameter modeling approach incorporates some degree of averaging or lumping approximation.

Simply stated, a distributed parameter model attempts to increase the accuracy of the resulting simulation by preserving and utilizing information concerning the areal distribution of all spatially variable, non-uniform processes incorporated into the model. At the opposite end of the spectrum, a lumped parameter model attempts to reduce computational requirements for a simulation, without an unacceptable loss in simulation accuracy, by using "effective" parameter values to approximate spatial influences. From a practical standpoint, the distributed approach has become feasible only in recent years as a result of continuing rapid improvements in the size, speed and general availability of modern computers.

One research advantage of a distributed parameter model is its ability to evaluate the significance of varying degrees of lumping. Figure 1a is a topographic-soils map of a small (1 hectare), experimental, gauged watershed located about 12 kilometers south of Lafayette, IN. Table I lists the infiltration properties of the three soils present in the watershed. Figure 1b corresponds to a hypothetical situation, chosen to illustrate the significance of parameter distribution on the hydrologic response of the

346

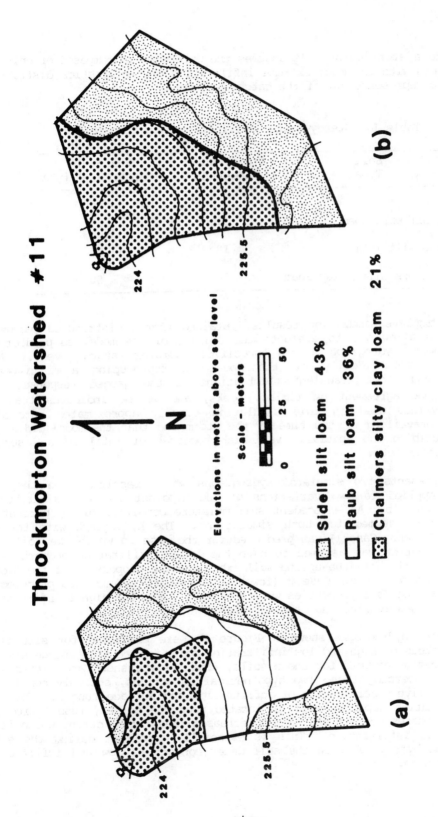

Figure 1 – Throckmorton Experimental Watershed Showing Actual (a) and Hypothetical (b) Soil Data

watershed to a storm event.   It assumes the watershed is composed of only the two soil types with the most extreme infiltration characteristics distributed so that each represents 50% of the catchment area.

Table I.   Steady-State Soil Infiltration Properties

Soil Type	Steady-state Rate-mm/hr	Rate @ 10% Saturation - mm/hr
Sidell silt loam	20	100
Raub silt loam	18	80
Chalmers silty clay loam	10	50

Figure 2a displays simulation results, obtained from a distributed parameter model, which allow one to evaluate the ability of the model to predict the gauged runoff hydrograph of a specific, complex storm event.   Note particularly the fidelity of the model in reproducing a multi-peaked hydrograph and other transient fluctuations of the gauged response.   The generally good agreement on timing, peaks, and volume indicates that the parameter values used in this simulation closely approximated antecedent conditions prevailing at the time of the storm and that the model did a good job of describing the spatially variable processes of infiltration, surface storage, runoff, etc.

Figure 2b presents two simulated hydrographs which resulted when the same storm was applied to two variations of the hypothetical soil distribution shown in Figure 1b.   An antecedent soil moisture condition of 40 percent of saturation was assumed for both simulations.   The hydrograph with the two nearly equal peaks (Case 1) was predicted for the case in which the soil area nearest the outlet was assumed to have the lower infiltration properties of the Chalmers soil while the upland soil exhibited the properties of the Sidell soil.   The hydrograph for Case 2 (lower total runoff volume) was produced by assuming that the soil properties were reversed, i.e., the higher infiltration capacity soil was located near the outlet.

Both simulated hydrographs shown in Figure 2b were predicted for situations with equal areas of high and low infiltration soils -- the only change was the physical location assumed for these soils.   The results shown are, after some reflection concerning these two hypothetical situations, entirely consistent with an intuitive understanding of infiltration and surface runoff.   When the high infiltration capacity soil is located near the outlet, runoff from the upland area passes over it and has an increased travel time during which it is available for infiltration.   This is especially important during the early phase of the hydrograph when the soil water content is low and infiltration capacity is high.

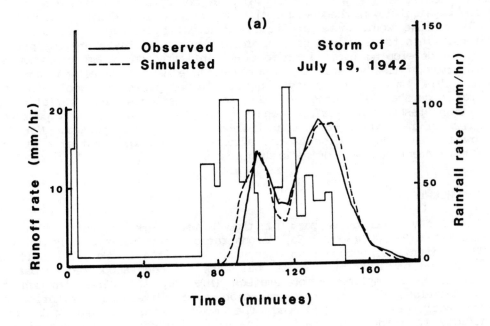

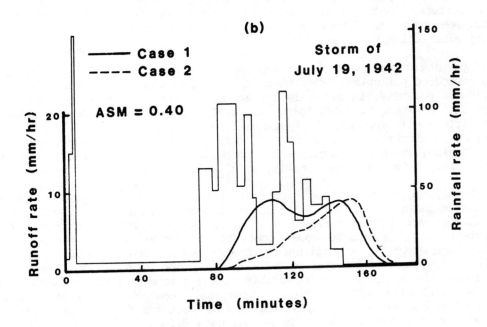

Figure 2 - Hydrographs for the Two Scenarios in Figure 1

The significant point of this example is that, while a distributed model produced a result entirely in keeping with our conceptual understanding of the governing physical laws, the overwhelming majority of lumped hydrologic models currently in use would have predicted identical hydrographs for these two situations. This result would have occurred because of the averaging process involved in developing overall "effective" infiltration parameters when the relative areas of the two soil types were unchanged with hypothetical position changes. This "implied linearity" in most lumped parameter models does not allow for accurate description of the spatial effects that variation of infiltration responses, surface characteristics, land uses, topography, structures, or rainfall patterns can have on hydrologic responses. Nor does it allow for easy description of the effects that changing patterns of land use and management can have on watershed response.

ANSWERS CONCEPTS

ANSWERS was developed as a part of the "Black Creek Project" which was supported by PL 92-500 (Sec. 108a) demonstration grant funds from Region V of the U.S. Environmental Protection Agency to the Allen County (IN) Soil and Water Conservation District. Additional work has been supported as part of the Indiana Model Implementation Project (MIP) by the USEPA through the Indiana Heartland Coordinating Commission. The watershed monitoring and research activities that produced the model were carried out under subcontracts to Purdue University. The resulting model was designed as a planning tool for persons concerned with nonpoint source pollution originating on agricultural lands. Recent work has been aimed at the extension of the application areas to disturbed soil situations, $\underline{e.g.}$, construction sites, surface mine reclamation, etc.

Since hydrologic processes are viewed as the driving force for agricultural nonpoint source pollution, the fundamental framework of a previously developed distributed parameter runoff modeling concept (Huggins and Monke, 1966) was chosen as the basis for developing the ANSWERS model. Hydrologic simulation capabilities were significantly improved with the addition of channel and subsurface drainage components and sediment production/transport relationships, which were incorporated to build a model applicable to national water quality planning needs (Beasley et al., 1980).

Conceptually, a watershed must first be subdivided into a grid of square elements as shown in Figure 3. The physical properties (parameter values) for each element are then described in a watershed data file. In order to be consistent with distributed model concepts, the size of an element must be small enough that all significant parameter values are essentially uniform within the boundaries of each element. Thus, the size of an element will depend upon the degree of non-uniformity of the catchment and the intended use for the simulation output. From a practical standpoint, we are currently recommending element sizes ranging from 1 to 4 hectares for most applications. Smaller elements require larger watershed data files. Thus, the modeler must take computer capabilities and simulation costs into account when specifying element sizes.

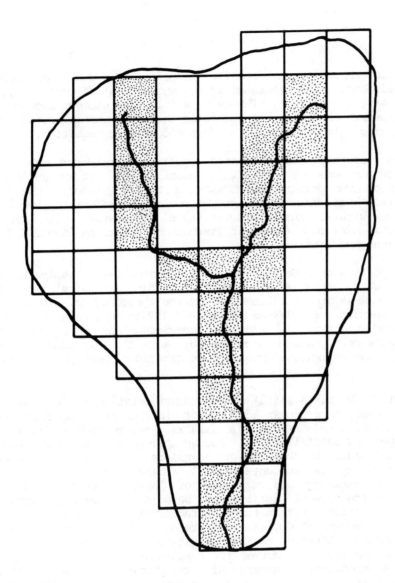

Figure 3 - Watershed Divided into Elements with Channel Elements Shaded

The internal structure of the model is best understood in terms of a single element. Essentially, the model is composed of mathematical relationships intended to simulate, for a single element with uniform physical characteristics, all physical processes (components) considered significant to the dependent (predicted) variables: water, sediment and chemical movements within and from an element. Since element characteristics vary in an unrestricted manner from one element to another, the parameter values (equation coefficients) also change between elements. Therefore, the simulated response of each element varies in a manner directly related to actual watershed characteristics.

The link which allows the composite response of this collection of "independent" elements to be simulated is the equation(s) of continuity, i.e., the law of conservation of mass. The outputs from each element become sources of input to adjacent elements. The computer does the necessary "accounting" by using integration algorithms applied to the continuity equation.

This conceptual approach to watershed modeling represents one of the fundamental contributions of the ANSWERS model to the art of environmental modeling. One of its primary advantages is that it reduces the task of developing process simulation equations to one of characterizing the response of a small area with uniform characteristics. Parameter value variations between elements and integration of continuity equations account for all spatial variations and interactions.

The hydrologic portion of ANSWERS (Huggins and Monke, 1966) was developed as a spatially-responsive, event-oriented watershed hydrology simulator. It was designed to provide detailed information on the hydrology of ungauged areas. The original version of ANSWERS (Beasley, 1977) incorporated subsurface drainage, channel routing, and sediment detachment and transport into the hydrologic framework. Present versions of ANSWERS also contain improved channel routing and structural practice descriptions (Beasley and Huggins, 1982).

A modified form of Holtan's (1961) infiltration relationship coupled with a subsurface drainage term which allows for infiltration capacity recovery (Huggins and Monke, 1966) is used in ANSWERS. Since flow is continuously routed downslope, any rainfall excess from an upslope area has the opportunity to become infiltration at some point downslope, if a soil with greater infiltration capacity is encountered. A surface roughness term, which accounts for the frequency and magnitude of "peaks" and "valleys" on the surface, modifies the rate of infiltration based on the ratio of inundated to total area. In addition, surface shape characteristics are taken into account in describing the amount of depressional storage. The continuity equation is solved using Manning's equation as the definition for flow depth. The surface storage and outflow rate are solved for explicitly using a piecewise linear segmented curve technique, instead of an iterative finite differences approach. Detailed discussions of the component relationships and their application are given by Beasley (1977) and Beasley and Huggins (1982).

Sediment yield is modeled by equations which predict soil detachment rates (erosion) due to rainfall impact and overland flow. The detached soil, when combined with the amount contributed by inflows from adjacent elements, becomes available for movement out of an element. The actual rate of movement is determined by sediment transport relationships which are functions of overland flow depths and rates. Relationships employed were based upon information obtained by several researchers (Yalin, 1963; Meyer and Wischmeier, 1969; Foster and Meyer, 1972; Wischmeier and Smith, 1978) as well as simulated rainfall erosion studies carried out during the Black Creek Project in cooperation with the USDA-ARS (Mannering et al., 1975; Mannering, 1976).

A research version of ANSWERS currently can predict the particle size distribution of the eroded sediment. This ability allows for assessment of the "clay enrichment" of the outflowing sediment (Dillaha, 1981). Essentially, the detachment relationships are the same in the expanded sediment model as they are in the "original" model. Several additional assumptions are made concerning the transport of "washload", i.e., primary clay and silt particles. The most notable differences exist in the transport relationships, which are based on the complete Yalin relationships (Yalin, 1963) and use shear stress, rather than flow rate, as the independent parameter. Since the availability of sediment in any particle-size class is directly influenced by the distribution in the original soil, the model allows for the partitioning of transport across classes based on the ease of transport and settling velocities. The model provides information at the end of the simulation that describes the effective particle-size distribution of the outflowing sediment. The difference between the outflowing particle-size distribution and the "average" of the original soil distributions yields an enrichment ratio for the fine particles.

Chemicals associated with eroded sediment are, in the current version of the model, predicted by correlation relationships between chemical concentrations and sediment yields. Another research version of the model (Amin-Sichani, 1982) uses the "clay enrichment" information mentioned above and a very descriptive phosphorus fate model to predict total, particulate, and soluble phosphorus yields. A detailed discussion of the development of component relationships for the ANSWERS model is given in the ANSWERS User's Manual (Beasley and Huggins, 1982) and Dillaha (1981).

EXAMPLE PLANNING APPLICATIONS

ANSWERS was designed to simulate the hydrology, erosion response, chemical yield, etc. of ungauged agricultural watersheds. Through the use of a comprehensive data file, distributed parameters and physically-based deterministic relationships, the model predicts the consequences or benefits of land use and/or management changes.

In order to best utilize the very limited monetary and personnel resources presently available, a planning methodology has been developed which should simplify treatment selection and evaluation tasks. The planning methodology was divided into four phases:

1. Establishment of a "baseline condition" and definition of "critical areas",

2. Planning of structural, tillage, and management changes necessary for treating "critical areas",

3. Determination of water quality impacts caused by "critical area" treatment,

4. Prioritization of cost sharing monies for practices on a cost effective basis.

Finley Creek Example

The Finley Creek watershed in Boone and Hamilton counties, Indiana is typical of much of the agricultural upper Midwest. The drainage area encompasses approximately 1964 hectares and has a generally rolling topography with an average slope of less than 0.7 percent with extremes of 0.1 and 3.5 percent. The soils are predominately silt loams. Over 77 percent of the area is planted to row crops (corn and soybeans) with both conventional and conservation tillage systems. Grasslands (pasture and hay) account for about 10 percent of the land area, while wooded areas occupy more than 8 percent of the watershed. Small grains, built-up areas and specialty crops account for the rest of the land uses. This multiple land use watershed drains directly into Eagle Creek. Eagle Creek, in turn, empties into Eagle Creek Reservoir, the centerpiece of the nation's largest city park and a future water supply for Indianapolis. Although the sediment and chemical yields from this watershed are not severe by most standards, the fact that they immediately impact a reservoir led to the selection of Eagle Creek as one of the two Indiana MIP study areas.

In order to model Finley Creek, a data file was constructed which described the watershed as it existed in 1979. An element size of approximately 2 hectares was chosen and the topographic, soils, land use and management data for each small area was entered. Channel descriptions, as well as subsurface drainage information were also added to the descriptive data.

Once this information had been gathered, a simulation was run using a hypothetical storm as the basis for later comparisons. The hypothetical storm corresponds to a 1.5 hour event with a return interval of approximately 8 years. This particular event has been shown to produce predicted sediment yields that approximate the average annual yield for several continuously monitored, agricultural watersheds in northeastern Indiana. The intensity distribution is typical of late spring and early summer storms for the eastern 2/3 of the United States. The land uses and cropping patterns approximate those that would be expected to exist during seedbed preparation to about one month after planting. This combination of intense rainfall and erosive surface conditions is probably the most severe test of any type of management practice or system.

Table II presents the "baseline" and several alternative management systems that could be used to reduce sediment and nutrient yields from the Finley Creek watershed. The cost information given is for comparison purposes only and does not take into account many of the important factors to be considered in a total cost-benefit analysis.

Simulation 1 provides the basis for determining the effectiveness of the various management changes to be tried. The sediment and phosphorus yields are consistent with monitored information in the area. As Figure 4 indicates,

Table II. Simulation Results for Alternative Strategies for Finley Creek -- 1980

Strategy*	Area Affected by BMPs PTO (ha)	Cons. Till. (ha)	Sediment (kg/ha)	Total Yield at Watershed Outlet Total P (kg/ha)	Avail. P (kg/ha)	Sed. N (kg/ha)	Sol. N (kg/ha)	Sediment Reduction (%)	Cost** ($/tonne reduced)
1	0	664	3820	5.7	1.4	34	1.5		
2	0	740	3560	5.3	1.2	32	1.5	7	.32
3	200	1383	2300	3.2	.8	21	1.2	40	3.94
4	200	740	2460	3.5	.9	22	1.2	36	3.89
5	64	1383	2940	4.3	1.1	26	1.4	23	2.79
6	0	1383	3250	4.8	1.2	29	1.4	15	1.39
7	0	1383	1320	1.8	.4	12	.8	14	3.61

* The strategies simulated are defined below:

1. Baseline condition: 1979 land use and management.
2. 1980 Actual land use (increased area in conservation tillage and reduced total area in row crops).
3. PTO terraces installed where sediment yield was in excess of 20 tonnes/hectare. Additionally, all row crop areas using conventional management were converted to conservation tillage.
4. PTO terraces installed where sediment yield was in excess of 20 tonnes/hectare.
5. PTO terraces installed in three specific, high yield areas. Additionally, all row crop areas using conventional management were converted to conservation tillage.
6. All row crop areas using conventional management were converted to conservation tillage.
7. Same as Strategy 5 except that a storm with 25% lower intensity and total volume was used. The "baseline condition" for this storm gave a total sediment yield of 1540 kg/ha.

** Cost information was based on 1979 construction costs for PTO terrace systems in Allen County, Indiana. The cost is based on total area benefited (both above and below terraces). The figure used in these calculations was $510.80 per hectare benefited. A 10-year life was assumed, which yielded an annual cost of $51.08 per hectare benefited. The chisel plow was also assumed to have a 10-year life. The average annual cost per hectare, based on the cost of a new plow, was $2.17. Since the "design storm" used in this example produced approximately the annual sediment yield, the cost per tonne of reduced yield at the watershed outlet is, essentially, the annual cost. However, due to simplifying assumptions and unique local conditions, these cost figures should not be considered to be generally applicable to other planning situations. They were included in an effort to give the reader a feeling for the type of analysis which can be performed by ANSWERS.

355

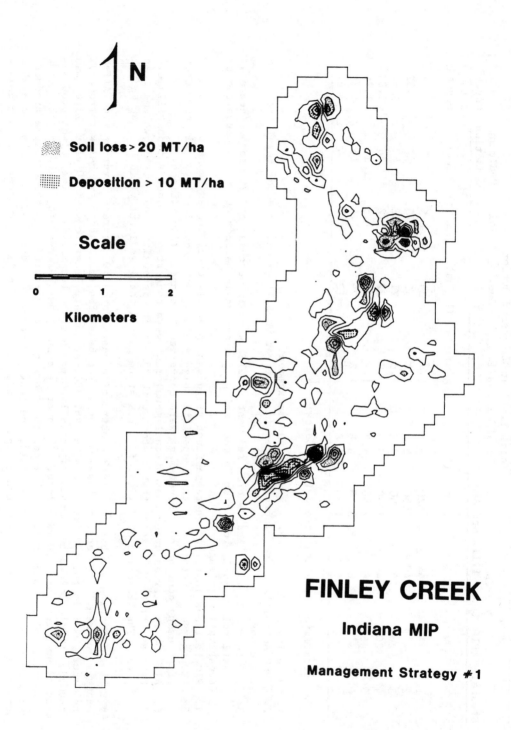

N

Soil loss > 20 MT/ha

Deposition > 10 MT/ha

Scale

0   1   2

Kilometers

FINLEY CREEK

Indiana MIP

Management Strategy #1

Figure 4 - Soil Loss/Deposition Map for Determining "Critical
Areas" Within Finley Creek Watershed

there are several areas of intense erosion and deposition. The areas of high erosion are the logical starting places for application of management strategies. Some of these areas may not be contributing directly to water quality problems though, since the sediment may be depositing prior to reaching the stream. For those areas that are questionable as to impact, specific simulations can quickly determine whether or not they are actually contributing to the outlet.

Simulations 2 through 6 show the impacts of specific combinations of tillage and structural practices. The structural practice used is the Parallel Tile Outlet (PTO) terrace, a detention-type structure. Conservation tillage is considered to include residue management and either no fall tillage or use of a chisel plow as opposed to a turning plow.

Strategy 2 was included to show the effect that the pattern of land use could have on watershed yields. With increased conservation tillage, reduced total row crop area and a spatially modified cropping pattern, a 7% reduction in sediment yield was predicted even with no additional practice application. Strategies 3-6 have been listed in terms of decreasing effectiveness for reduction of sediment yield at the outlet of the watershed. Their locations were fixed by the 1979 baseline simulations. Obviously, some of the areas that had high yields in 1979 were not as bad in 1980 simply due to the changes noted in strategy 2. However, the ranking would be quite different if annual unit cost of achieving a sediment yield reduction was employed. Still different results would be obtained if nutrient yields or concentration levels in the stream are chosen. All of these water quality improvement criteria and others are valid for developing a control program. Generally, several of them would be given consideration.

The ranking of strategies is also influenced by the choice of baseline conditions, as is illustrated in Table II by Strategy 7. The only difference between results for Strategies 6 and 7 is the severity of the hypothetical storm used to drive the simulation. For Strategy 7, a storm with 25 percent lower intensities and total volume was used. This gave a sediment yield of 1540 kg/ha for the same land use in Strategy 1. The smaller storm gave lower total yields and smaller reduction percentages. Because of this result, the unit cost ($/tonne) of sediment yield reduction increased substantially. This result again illustrates the complexity of analyzing nonpoint source pollution and its control.

Construction Site Example

A residential construction site within the Eagle Creek watershed near Zionsville was chosen for detailed model analysis to study the model's ability to evaluate proposed sediment management plans. The watershed consists of approximately 53 hectares and has slopes ranging from zero to 17 percent. The soils of the area are silt loams and silty clay loams. A 9.5 hectare subwatershed of the main watershed was selected for detailed analysis because it was the only portion of the study area with ongoing construction activities. Only intermittent grab samples and no flow data were available.

Thus, some concentration information was available for limited verification checking.

To assess the impact of construction activities on the study area, construction scenarios were simulated and compared with the predicted sediment yields from the watershed before construction. Prior to construction, the subwatershed consisted of 85 percent row crops (normally corn) and 15 percent woods and swampy (cattails) areas. Conventional tillage techniques were used in the row crop area, where slopes ranged from 0.2 to 15.3 percent with an average of 5.3 percent.

A series of model simulations were made for the watershed management scenarios shown in Table III. Runs 1, 2, and 3 represent different pre-construction agricultural scenarios. Runs 4 through 8 represent different phases of the construction process or different sediment management scenarios. The design storm produced average sediment yields which ranged from 5,100 to 16,300 kg/ha. The sediment yield for the control agricultural scenario (Run 1, crop stage 1 corn) was 13,500 kg/ha.

Runs 4, 5 and 6 represent conventional construction site scenarios. In all of these situations, the former row crop areas have been scalped and the roadway areas have been graded down to the subsoil. Run 4 represents the scalped row crop area and bare roadways scenario. The localized high erosion rates in this simulation also cause high rates of deposition just downslope of some of the high erosion areas in areas with slight slopes. Run 5 is the same as Run 4 except gravel sub-base has now been applied to the roadways in preparation for road paving. The soil loss for this scenario was 14,500 kg/ha. Runs 4, 5 and 6 show that significant localized erosion can be controlled and still have little overall impact on the overall soil loss from the watershed. Applying gravel and sod to the most critical areas only brought about a 14 percent decrease in the average sediment yield. This is because the greatest source of sediment was the former row crop area which was left untreated.

Runs 7 and 8 represent scenarios which significantly decrease the average soil loss from the watershed because they also control soil loss in the former row crop area. Run 7 (Figure 5) demonstrates the dramatic effects of treating the scalped row crop area by seeding, mulching and fertilizing while leaving the roadways bare. The roadways now stand out as high erosion areas and the other areas have low erosion rates. The average soil loss from the watershed for this simulation was 12,700 kg/ha or about 21 percent less than the original scalped, bare roadway scenario.

Table III also shows the effects of various management scenarios on the particle-size distribution of the eroded sediment leaving the watershed. The percentage of fines is shown to increase with increasing disturbance of the soil surface. The practices which decreased the erodibility of the surface also decreased the enrichment of fines. This is due to the fact that this steep watershed was, in most places, detachment limited in the case of sediment movement. Thus, more erodible areas produced much higher sediment loads. When these high loads reached elements with smaller slopes, transport capacity became limiting, and deposition occurred. Since the larger particles deposited easier, enrichment of the fine materials was a direct result of deposition, unless everything was deposited. Each successive deposition area simply increased the percentage of fines in the sediment load. If the watershed had few or no deposition areas, the suspended material leaving the watershed would have a particle size distribution quite similar to the original, eroded soil. Thus, those practices which left exposed soil and had large sediment yields also had higher percentages of nutrient and chemical carrying capacity due to the larger percentage of fines. This means that doubling the sediment yield could triple or quadruple the nutrient or chemical yield. It is, therefore, imperative to reduce overall erosion rates!

Table III. Summary of Construction Site Simulations

| Run No. | Simulation Scenario | Total Runoff, mm | Peak Flow mm/h | Sediment Yield, kg/ha | Particle Size Distribution, % | | | | |
| | | | | | Size Classes, mm | | | | |
					0.002	0.010	0.030	0.500	0.200
1	Sl Corn	31	84	13,500	4	11	39	28	18
2	Fallow	30	77	16,300	7	18	59	6	10
3	Chisel Plow	28	62	5,100	3	9	30	45	13
4	SC, BR	31	72	16,000	7	17	59	7	10
5	SC, GR	30	61	14,500	6	17	57	8	12
6	SC, PA	32	81	15,400	5	16	50	12	17
7	SE, BR	30	68	12,700	5	9	41	29	16
8	SE, GR	30	58	9,800	3	9	30	45	13

where: Sl Corn   = conventional corn, crop stage 1
       Fallow   = conventional tillage, no surface residue
       Chisel Plow = chisel plowed corn, surface residue
       SC   = scalped topsoil (some cover remaining)
       BR   = bare roadways, compacted soil
       GR   = roadways covered with gravel subbase
       PA   = roadways paved
       SE   = scalped areas seeded, mulched and fertilized

and the particle size classes are described as follows:

Size (mm)	Type	Specific Gravity
.002	primary clay	2.65
.010	primary silt	2.65
.030	small aggregate	1.80
.500	large aggregate	1.60
.200	primary sand	2.65

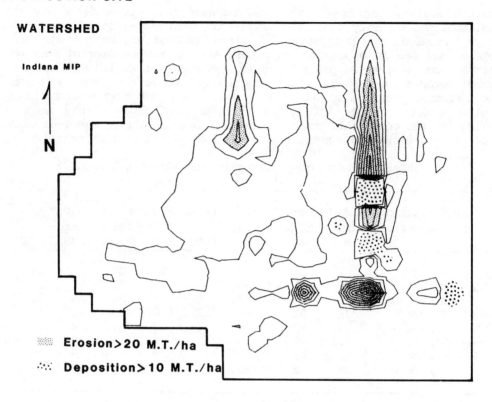

CONSTRUCTION SITE

WATERSHED

Indiana MIP

N

Erosion>20 M.T./ha

Deposition>10 M.T./ha

Figure 5 - Erosion/Deposition Contours for Construction Site -- Run 7 (SE,BR)

SUMMARY AND CONCLUSIONS

The merits of comprehensive watershed models in planning nonpoint source pollution control programs have been delineated. Some philosophical considerations of alternative modeling concepts applicable to simulating nonpoint source pollution on a watershed scale were also presented. Examples of the output from a distributed model have shown the special relevance of such models to national water quality planning programs. The discussion and examples are intended to support the following conclusions:

1. A distributed parameter modeling concept offers several advantages over lumped parameter approaches, including:

   a. potential increased accuracy,

b. increased information content in the output,

c. the capacity to utilize results of plot and small watershed research for development of component relationships.

2. The effectiveness of any water quality control program is entirely dependent upon the quality of those trained individuals charged with planning and implementing the program.

3. An effective watershed model is an essential tool for increasing the efficiency and effectiveness of those key individuals.

4. The detailed output of a distributed parameter model is essential for water quality planning programs because it allows planners to be site-specific in developing and evaluating the effectiveness of potential control measures. Such an approach is a prerequisite for recommending cost-effective measures for controlling nonpoint pollution from agricultural areas while still maintaining our productive capacity.

ACKNOWLEDGEMENTS

Development of the ANSWERS model and collection of much of the data used in that effort were sponsored by Region V, U.S. Environmental Protection Agency and the Agricultural Experiment Station, Purdue University in cooperation with the Allen County (IN) Soil and Water Conservation District and Indiana Heartlands Coordinating Commission. Additionally, the U.S. Department of Agriculture has supplied funds and personnel for data collection and practice implementation in the project areas where ANSWERS was developed and tested.

The author would like to thank Dr. Larry F. Huggins, Dr. Theo A. Dillaha, III, Dr. George R. Foster, and Dr. Edwin J. Monke who have contributed greatly to the development and utilization of the ANSWERS model.

REFERENCES

1. Amin-Sichani, S., "Modeling of Phosphorus Transport in Surface Runoff from Agricultural Watersheds", Ph.D. Thesis, Purdue University, W. Lafayette, Indiana, 1982, 157 pp.

2. Beasley, D. B., "ANSWERS: A Mathematical Model for Simulating the Effects of Land Use and Management on Water Quality", Ph.D. Thesis, Purdue University, W. Lafayette, Indiana, 1977, 266 pp.

3. Beasley, D. B., L. F. Huggins, and E. J. Monke, "ANSWERS: A Model for Watershed Planning", Transactions of the ASAE. St. Joseph, Michigan, Vol. 23, No. 4, 1980, pp. 938-944.

4. Beasley, D. B. and L. F. Huggins, "ANSWERS Users Manual", EPA-905/9-82-001. USEPA, Region V, Chicago, Illinois, 1982.

5. Dillaha, T. A., III, "Modeling the Particle Size Distribution of Eroded Sediments During Shallow Overland Flow", Ph.D. Thesis, Purdue University, W. Lafayette, Indiana, 1981, 189 pp.

6. Foster, G. R. and L. D. Meyer, "Transport of Soil Particles by Shallow Flow", Transactions of the ASAE. St. Joseph, Michigan, Vol. 15, No. 1, 1972, pp. 99-102.

7. Holtan, H. N., "A Concept for Infiltration Estimates in Watershed Engineering", ARS-41-51. Agricultural Research Service, USDA, Washington, D.C., 1961.

8. Huggins, L. F. and E. J. Monke, "The Mathematical Simulation of the Hydrology of Small Watersheds", Technical Report No. 1. Water Resources Research Center, Purdue University, W. Lafayette, Indiana, 1966.

9. Mannering, J. V., D. R. Griffith, and C. B. Johnson, "Tillage and Related Rainfall Studies", EPA-905/9-75-007. Proceedings of the Nonpoint Pollution Seminar, USEPA, Region V, Chicago, Illinois, 1975.

10. Mannering, J. V., "Crop Sequence and Fall Tillage Effects on Soil Erosion", EPA-905/9-76-005. Proceedings of the Best Management Practices for Nonpoint Source Pollution Control Seminar, USEPA, Region V, Chicago, Illinois, 1976.

11. Meyer, L. D. and W. H. Wischmeier, "Mathematical Simulation of the Processes of Soil Erosion by Water", Transactions of the ASAE, St. Joseph, Michigan, Vol. 12, No. 6, 1969, pp. 754-758.

12. Wischmeier, W. H. and D. D. Smith, "Predicting Rainfall Erosion Losses -- A Guide to Conservation Planning", Agriculture Handbook 537. Science and Education Administration, USDA, Washington, D.C., 1978.

13. Woolhiser, D. A., "Hydrologic and Watershed Modeling -- State of the Art", Transactions of the ASAE, St. Joseph, Michigan, Vol. 16, No. 3, 1973, pp. 553-559.

14. Yalin, Y. S., "An Expression for Bed-load Transportation", Proceedings of the Hydraulics Division of ASCE, New York, New York, Vol. 89, No. HY3, 1963, pp. 221-250.

MODELS OF STATISTICAL DISTRIBUTIONS FOR NPSP CONCENTRATIONS

A.MARANI, Universita' di Venezia, Facolta' di Chimica Industriale,
Dipartimento di Spettroscopia, Elettrochimica e Chimica Fisica,
Calle Larga S.Marta 2137 - 30123 VENEZIA, ITALY

G.BENDORICCHIO, Universita' di Padova, Facolta' di Ingegneria,
Istituto di Chimica Industriale, Via F. Marzolo 9 - 35131 PADOVA,
ITALY

ABSTRACT

Normal, Lognormal, Standard Gamma and Generalized Gamma
statistical distributions are analyzed for modeling experimental
distributions of total-N concentrations in the runoff from a test
agricultural catchment. Parameter estimation has been carried out
via the maximum likelihood method. Comparison among the various
distributions has been performed by several goodness-of-fit
statistics. A model is proposed for viable physical
interpretations of chemical concentrations gaged in agricultural
runoff. Model performance is compared with that of previous
approaches by means of two experimental data sets of total-N
concentrations. Among the results, the difficulty in estimation
of parameters via maximum likelihood method is noted. A critical
discussion is outlined with the statistical analysis of
numerical simulations of runoff and chemical losses. The
possibility of model calibration through data sets of simulated
distributions (and, hence via known parameter values) allows
thorough testing against experimental concentration distributions
of reliability of chemical and physical parameters characterizing
the setting of the watershed.

INTRODUCTION

Water pollution from nonpoint (NPS) and point (PS) sources holds
inherently random components tied to the fluctuations of a
variety of meteorological and source variables. Further
stochastic components are introduced by sampling techniques,
which affect the structure of experimental data sets. The
magnitude of random components is noteworthy as can be inferred
by the spreading of field data measurements, whereas such a
feature is common to all experiments performed on natural systems
in which data are gathered without perturbation of the system.

In the case of NPSP, research has dealt primarily with
deterministic approaches to simulation of the phenomena thus
confining statistical aspects to minor roles. Statistical
approaches have been used, however, to determine ranges of
variation or empirical relations [1] and to analyze Normal and
Lognormal models of nutrient concentration distributions [2,3].
Measured and simulated concentrations have also been compared
statistically [4,5].

Nevertheless, the state-of-the-art review of stochastic models of
NPSP illustrated the paucity of application, in contrast with the
wide development of stochastic hydrology; this is perhaps
justified by the limited sample sizes of water quality data.

Analogous trends have been reported in the field of air pollution
control where deterministic models have early success until
monitoring networks have yielded large data sets. Later on,
stochastic models have drawn more attention from the researchers,
and the modern trends combine stochastic and deterministic
components aiming at a rational comparison between simulations
and measurements.

The present work deals with the study of nutrient concentrations
in agricultural runoff and, in analogy to the air pollution
experience, analysis of models of concentration distributions.
Previously, a choice has been made among available distributions
by goodness-of-fit testing performed on the experimental data
sets. This fact is due to the lack of a physical rationale for
the models.

Early results of statistical analyses on NPSP data have shown a
tendency towards Lognormal models which nevertheless had been
compared only with Normal distributions.

Analogous patterns for air pollution have yielded the same
result, strengthening the belief that Lognormal distributions
were universal, until further models (in particular Weibull and
Gamma) were considered [6].

Recent development in air pollution research [7,8] seems to

indicate possible improvements induced by use of Weibull and Gamma models with respects to Lognormal models (and even to multi-parameter models such as the 4-parameter Beta), yet these developments also indicate the marked difficulty in identifying the best model for any given case. This shortcoming has allowed [9] the choice of a family of distributions (the Generalized Gamma) which embodies Weibull and Standard Gamma models. In the same pattern, air pollution experience has suggested comparison between Lognormal and Generalized Gamma models, in view of the simultaneous performance of comparisons with Exponential, Weibull and Standard Gamma models.

The present work also describes a computer code suitable for parameter identification. General considerations about the outcome of analyses performed on total-N data measured during two annual NPSP data collections carried out at the delivery location of runoff from the catchment of Zuccarello (Italy) are discussed, together with a conceptual model which substantiates the application of the Generalized Gamma distribution and its parameters.

The novel distribution is compared with the Generalized Gamma distribution by means of computed and measured total-N data, as from NPSP research [2,3,22].

THE PROBABILITY DISTRIBUTION FUNCTIONS

Probability distribution function (pdf) models, as from air quality literature, have been analyzed and listed in Table 1.

Such pdf's can be grouped into two classes: the first class embodies the various Lognormal forms and 4-parameter Beta distribution; the second class yields the distributions which can be regarded as special cases of four parameter Generalized Gamma (from applications in air pollution simulation).

The Generalized Gamma distribution ($\Gamma(\alpha)$ = gamma function of argument α), defined as

$$f(x; \alpha, \beta, \xi, \eta) = \frac{\beta}{\eta \, \Gamma(\alpha)} \left(\frac{x-\xi}{\eta}\right)^{\alpha\beta-1} \exp\left[-\left(\frac{x-\xi}{\eta}\right)^{\beta}\right] \qquad (1)$$

has been introduced by Amoroso [18] in the analysis of the distribution of economic income and discussed by Stacy [19]. Further work on the Generalized Gamma model has also been carried out recently [20,21] even through few applications have been made of the model.

The main feature of (1) consists of its capability to represent current models for special values of the parameters. This generality is particularly relevant to problems of identification of best models and optimal parameters for given experimental

Table 1 - Models of air quality data distributions.

TYPE	PROBABILITY DENSITY FUNCTION (pdf)	DOMAIN	REFERENCES
LOG NORMAL 2-PARAMETER	$\dfrac{1}{x\sigma\sqrt{2\pi}}\,\exp\{-(\ln x-\mu)^2/2\sigma^2\}$	$0<x<+\infty$ $-\infty<\mu<+\infty$ $0<\sigma^2<+\infty$	(10)(11)(12)(13)
LOG NORMAL 3-PARAMETER	$\dfrac{1}{(x-\xi)\sigma\sqrt{2\pi}}\,\exp\{-[\ln(x-\xi)-\mu]^2/2\sigma^2\}$	$\xi<x<+\infty$ $-\infty<\mu<+\infty$ $0<\sigma^2<+\infty$	(7)(8)(13)(14)(15)(16)
LOG NORMAL 4-PARAMETER	$\dfrac{\delta-\xi}{(\delta-x)^2\,\sigma\sqrt{2\pi}}\,\exp\left\{-\left[\ln\left(\dfrac{x-\xi}{\delta-x}\right)-\mu\right]^2/2\sigma^2\right\}$	$-\infty<\xi<x<\delta<+\infty$ $-\infty<\mu<+\infty$ $-\infty<\sigma^2<+\infty$	(14)
EXPONENTIAL 1-PARAMETER	$\dfrac{1}{\eta}\,\exp\{-x/\eta\}$	$0<x<+\infty$ $0<\eta<+\infty$	(17)
EXPONENTIAL 2-PARAMETER	$\dfrac{1}{\eta}\,\exp\{-(x-\xi)/\eta\}$	$\xi<x<+\infty$ $0<\eta<+\infty$	(12)
WEIBULL 2-PARAMETER	$\beta(x-\xi)^{\beta-1}\exp\{-(x-\xi)^\beta\}$	$\xi<x<+\infty$ $0<\beta<+\infty$	(11)
WEIBULL 3-PARAMETER	$\dfrac{\beta}{\eta}\left(\dfrac{x-\xi}{\eta}\right)^{\beta-1}\exp\left\{-\left(\dfrac{x-\xi}{\eta}\right)^\beta\right\}$	$\xi<x<+\infty$ $0<\beta<+\infty$ $0<\eta<+\infty$	(7)(8)(14)
GAMMA 2-PARAMETER	$\dfrac{(x/\eta)^{\alpha-1}}{\eta^\alpha\,\Gamma(\alpha)}\,\exp\{-(x/\eta)\}$	$0<x<+\infty$ $0<\alpha<+\infty$ $0<\eta<+\infty$	(11)(12)
GAMMA 3-PARAMETER	$\dfrac{1}{\eta^\alpha\,\Gamma(\alpha)}\left(\dfrac{x-\xi}{\eta}\right)^{\alpha-1}\exp\{-(x-\xi)/\eta\}$	$\xi<x<+\infty$ $0<\alpha<+\infty$ $0<\eta<+\infty$	(7)(8)(13)(14)
GENERALIZED GAMMA 4-PARAMETER	$\dfrac{\beta}{\eta\,\Gamma(\alpha)}\left(\dfrac{x-\xi}{\eta}\right)^{\alpha\beta-1}\exp\left\{-\left(\dfrac{x-\xi}{\eta}\right)^\beta\right\}$	$\xi<x<+\infty$ $0<\alpha<+\infty$ $0<\beta<+\infty$ $0<\eta<+\infty$	(9)
BETA 4-PARAMETER	$\dfrac{\Gamma(\alpha+\beta)}{(\delta-\xi)\,\Gamma(\alpha)\,\Gamma(\beta)}\left(\dfrac{x-\xi}{\delta-\xi}\right)^{\alpha-1}\left(\dfrac{\delta-x}{\delta-\xi}\right)^{\beta-1}$	$-\infty\leq\xi\leq x\leq\delta\leq+\infty$ $0<\alpha<+\infty$ $0<\beta<+\infty$ $0<\eta<+\infty$	(13)(14)

data.   In such   a case,   in fact,   best-fitting operators   define
models   and   parameters   without   any   further   need for   a   model
comparison based, for instance, on hypothesis tests which are seldom
selective.

The cumulative distribution function (cdf) derived from (1) is:

$$P(X \leqslant x) = F(x; \alpha, \beta, \xi, \eta) = \int_{\xi}^{x} f(x; \alpha, \beta, \xi, \eta) \, dx$$

$$= \gamma\left(\left[\frac{x-\xi}{\eta}\right]^{\beta}; \alpha\right) / \Gamma(\alpha)$$

(2)

where P(X≤x) is   the probability of an event X   less than or equal
to x; F(x;α,β,ξ,η)   is the   cdf of   x; γ(z,α)   is the   incomplete
gamma   function of argument z and parameter α.

The   software   needed   for   parameter   identification   of   Normal,
Lognormal and Generalized   Gamma distributions has been   run on an
IBM PC.   The software makes   use of maximum likelihood   method for
parameter   estimation.   The function to be maximized is   therefore
the likelihood function L, (joint pdf of the set of N experimental
values):

$$L = \prod_{i=1}^{N} f(x_i; \alpha, \beta, \xi, \eta)$$

(3)

The   algorithm   for   maximum   search   (which,   in   turn, for
computational reasons   has been transformed into   a minimum search
for   the o.f.   ln(1/L) was   Nelder and   Mead's Simplex.   The codes
allow for selective fixing of one   or more parameters in order for
the   effect of   each   factor   to be   pointed   out. In   particular,
Weibull (α =   1), Standard Gamma (β   = 1) or Exponential (α   = β =
1)   distributions can be studied.

The codes include the possibility   of interactive grouping of data
sets in   n classes and   of computing six auxiliary   statistics (of
general' use as   tests of   goodness of   fit): absolute   deviations
(AD);   weighted   absolute   daviations   (WAD);   chi-square   (CS);
Kolmogorov-Smirnov   (KS);Cramer-Von Mises-Smirnov   (CMS); and   sum
of   square   deviations (SSQ).   Such statistics, reported   in Table
2, may   be used either   for model   comparison or for   testing the
goodness of   fit with the maximum likelihood criterion.

The experimental data   analyzed in the present   work are concerned
with total-N measured   at the outlet of   an agricultural watershed
in the   area of Venice   (Italy) during two annual   campaigns (1976
and 1981).   The physical   setting of   the watershed,   the sampling
procedures   and the   data have   been described   in previous   works

Table 2 - Auxiliary statistics for goodness of fit.

T Y P E	FORMULATION	NOTATION
ABSOLUTE DEVIATIONS	$AD = N \sum_{i=1}^{n} |P_{oi} - P_{ei}|$	P_{oi} = observed proportion of data falling in the i-th interval;
WEIGHTED ABSOLUTE DEVIATIONS	$WAD = N \sum_{i=1}^{n} |P_{oi} - P_{ei}| \cdot P_{ei}$	P_{ei} = expected proportion of data falling in the i-th interval;
CHI - SQUARE STATISTIC	$CS = N \sum_{i=1}^{n} [(P_{oi} - P_{ei})^2 / P_{ei}]$	N = total number of observations in the data set;
KOLMOGOROV - SMIRNOV STATISTIC	$KS = \max |S_i - F_i|$	n = total number of intervals;
CRAMER - VON MISES - SMIRNOV STATISTIC	$CMS = N \sum_{i=1}^{n} (S_i - F_i)^2 \cdot P_{ei}$	S_i = observed cumulative probability of the right extreme of the i-th interval;
SUM OF SQUARED DEVIATIONS	$SSD = N \sum_{i=1}^{n} (S_i - F_i)^2$	F_i = expected cumulative probability at the right extreme of the i-th interval.

[2,3,22]. The 1976 campaign spanned 8 months (March-November) while the 1981 campaign spanned 12 months (the whole solar year). Table 3 illustrates, as an example, the comparison between Normal, Lognormal, Weibull, Standard Gamma and Generalized Gamma distributions performed via CS, KS, CMS, log(1/L) and SSD statistics.

The rank attributed to each distribution by the statistics is also shown in the parentheses. The CS statistics does not establish direct ranks because it depends upon the number of degrees of freedom (df). Furthermore it was deemed fit to identify confidence intervals of CS because the use of statistic as absolute criteria need be substantiated by critical analyses on quality and extent of the available data.

Table 3 – Comparison of total-N concentration models (1981 field data).

TYPE DISTRIBUTION	d.f.	CS	KS	CMS	ln(1/L)	SSD
NORMAL (2 - parameter)	9	44.7	18.8 (5)	.388 (5)	573.99 (5)	3.636 (5)
LOGNORMAL (3 - parameters)	8	29.2	14.8 (3)	.252 (4)	561.22 (4)	2.194 (4)
WEIBULL (3 - parameter)	8	22.1	11.6 (1)	.154 (2)	554.72 (2)	1.319 (1)
STANDARD GAMMA (3 - parameter)	8	25.5	15.5 (4)	.207 (3)	557.94 (3)	1.903 (3)
GENERALIZED GAMMA (4 - parameter)	7	22.2	12.9 (2)	.153 (1)	554.70 (1)	1.324 (2)

Table 3 suggests some observations suitable for extension to similar sets of data:
1) the Normal distribution has the lowest ability to describe asymmetric data (concentrations are positive quantities) and generally betrays the cause of the generating process;
2) the Lognormal distribution has limited capabilities to describe experimental data sets because it turns out to be more efficient

369

than Standard Gamma according to only one statistic (KS);
3) the function log(1/L) shows little sensitivity to model variations (about 4% ranging from the worst to the best distributions) and that holds particularly true whenever optimal parameter values are sought. This circumstance seems to suggest that parameter identification via minimum search for the o.f. log(1/L) is not to be pursued. Interestingly enough, this fact is known in the literature, although efficient alternatives are not proposed, other than the old-fashioned word of caution.
4) Weibull and Standard Gamma belong to the family of Generalized Gamma distributions. All results actually rank the Weibull one as better than Standard Gamma. Incidentally, Weibull outclasses even Generalized Gamma according to KS and SSD. This fact does not seem controversial because the optimal value of the α parameter of Generalized Gamma is $\alpha = 0.987$. Generalized Gamma is more than "fairly close" to Weibull according to the Table 3. They are practically identical. It would be interesting to see the significance of the differences. Nevertheless, in some instances, Generalized Gamma has turned out to be closer to Standard Gamma or even different from both.

Analogous comparisons performed on the 1976 total-N data set tend to substantiate the observations drawn from the 1981 data. The improved fit obtained by Generalized Gamma cannot be solely related to the presence of a further parameter: in fact, statistical analyses performed by four-parameter Lognormal or Beta models do not affect the above considerations.

PDFS OF CHEMICAL SPECIES CONCENTRATIONS IN THE RUNOFF

Probabilistic approaches to instantaneous unit hydrograph (IUH) have shown the equivalence of IUH and pdf of raindrop holding times within surface or subsurface soil layers. Raindrops are considered as non-interactiving water particles traveling through the catchment [23,24,25]. As far as water quality is concerned, a consistent approach should assume that the falling particles hold a known initial washout concentration and that, along the traveling through the catchment, they exchange with the soil available chemical species. With reference to a sample chemical species, each water particle holds an instantaneous concentration $c=c(t)$, dependent on the residence time in the soil and on the path traveled.

On assuming equivalent patterns from the chemical standpoint and that the concentration/time function possesses an analytical inverse, the concentration pdf at the control section is:

$$pdf(c) = IUH(t(c)) \cdot |dt/dc| \qquad (4)$$

Eq.4 is to be taken with caution whenever hydraulic effects of channeling are dominant.

IUH formulations consistent with (4) have been obtained by Lienhard [23] in terms of Generalized Gamma model at $\beta=2$ and $\alpha=3/2$. Lienhard's model may be modified, as suggested by the author himself [26], in order to yield a physical basis to the Generalized Gamma distributions.

Let M be the number of raindrops of instantaneous precipitation events reaching the gaging section of the watershed. Each water particle may travel along several paths to reach the gaging section, and each path is characterized by a residence time t. Such time t is a degenerate function of the path because it can be related to different paths. Upon discretization of the time into intervals (for simplicity sake, the time intervals Δt are equal), the paths can be grouped in classes. In this way every time level t_j, which characterizes the interval (t_{j-1}, t_j), pinpoints available paths (g is a degeneracy of the j-th level) for the raindrops having traveled through the catchment for a time t such that $t_{j-1} < t < t_j$.

Among M raindrops reaching the control section, let M_1 be the number of particles having traveled up to a time t_1 (through M_1 of the g_1 available paths), M_2 be the analogous up to t_2, and so on. Hence the reference scheme is:

time level $t_1, t_2, \ldots, t_j, \ldots$
time level degeneracy $g_1, g_2, \ldots, g_j, \ldots$
No. of time levels used $M_1, M_2, \ldots, M_j, \ldots$

Each set of "occupation" numbers $(M_1, M_2, \ldots, M_j, \ldots)$ defines a residence time distribution (RTD), of M water particles.

Under the assumption that each raindrop is distinguishable from and independent of the other raindrops, the probability W of realization for $(M_1, M_2, \ldots, M_j, \ldots)$ is given by, [27]:

$$W = M! \prod_j \frac{g_j^{M_j}}{M_j!} \tag{5}$$

The observed distribution holds necessarily the highest probability of the RTD that maximizes W. Nevertheless, a constraint is to be enforced, raindrop pdf's being subject to conservation laws.

The assumption that particles are singled out and not interacting yields particle number conservation, or:

$$M_1 + M_2 + \ldots + M_j + \ldots = M \tag{6}$$

while the intuitive assumption that any catchment holds a

characteristic time (supported also by the unimodal structure of RTD) requires conservation of expectation values of a time function.

A reasonable form for such a conservation is given by the parametric relation

$$M_1 t_1^\beta + M_2 t_2^\beta + \ldots + M_J t_J^\beta + \ldots = M T^\beta \qquad (7)$$

Eq.7 has been used by Lienhard [23] with $\beta=2$, whereas the watershed at hand had been characterized by the root-mean-square time of the physical hydrograph.

Constitutive laws for time levels need be assigned upon the hypotheses of a link between degeneracy and residence time t. Such laws can be expressed in parametric form as:

$$g_J = A t_J^{\alpha\beta - 1} \qquad (8)$$

Eq.8 has also been introduced by Lienhard [23] with $\alpha=3/2$ and $\beta=2$ considering that the time t required for a raindrop to arrive (after travelling any distance l) at the gaging station is roughly proportional to l. The author has assumed that the number of available paths depends upon the amount of area swept out by l. Such area should increase roughly as l-square, although for particularly long slender watersheds it might be closer to a linear function of l, i.e. $\alpha=1$.

Eq.8 is supported because it constitutes a simple model of the intuitive pattern which associates a larger number of available paths at the larger residence times within the catchment.

The search for the minimum of W (constrained by (6) and (7)), yields the pdf of residence time of raindrops (or the IUH) in the form:

$$IUH(t) = \frac{\beta}{\eta' \Gamma(\alpha)} (t/\eta')^{\alpha\beta - 1} \exp\left[-(t/\eta')^\beta\right] \qquad (9)$$

where : $\eta' = \alpha^{-1/\beta} \tau$

Water particles transfer mass of chemicals along their path to the gaging station. The law which rules this process may be written as:

$$\partial c/\partial t = h(C_e - c) \qquad (10)$$

where h is the mass-transfer coefficient (which embodies also contact area between water and soil) and C_e is a soil-water equilibrium concentration.

The time variable in (10) has a precise meaning of soil-water contact time and is radically different from current time t'.

It seems reasonable to assume that h and C_E depend on the contact time t, even though a dependence on the current time t' is foreseeable. This allows integration of (10) in the form:

$$c = C_E + (C_0 - C_E) \exp\left[-ht\right] \qquad (11)$$

which is carried out by considering h and C_E as constants of the time t and under initial conditions $C(0) = C_0$.

The parameters h, C_0 and C_E should be viewed as functions of the current time t'.

Generally h is affected by physical rather than chemical parameters (soil structure, specific surface area, etc.). It is slightly affected by the molecular nature of the compounds, while C_E depends upon the chemical nature of soil and molecules in addition to the amount of chemical spread onto the soil.

Eq.11 defines the concentration of a water particle which had previously been in contact with the soil for a time t. As a consequence, whenever (11) is made explicit with respect to t as:

$$t = (1/h) \ln[(C_E - C_0)/(C_E - c)] \qquad (12)$$

the contact time needed to reach concentration c is obtained.

Upon variable transformation (12) on (9), gaged concentration pdf's at the control station are obtained as:

$$f(c; \alpha, \beta, C_0, \eta'', C_E) = \frac{\beta}{\eta''(C_E - c)\Gamma(\alpha)} \left[\ln\frac{C_E - C_0}{C_E - c}\right]^{\alpha\beta - 1} \exp\left\{\left[\ln\frac{C_E - C_0}{C_E - c}\right]^{\beta}\right\} \qquad (13)$$

where:

$$\eta'' = \eta' h = \tau h \alpha^{-1/\beta} \qquad (14)$$

Eq.13 depends on 5 parameters $(\alpha, \beta, \tau, C_0, C_E)$ among which two purely hydrological parameters (α, β) are singled out. Furthermore C_0 and C_E can be regarded purely as chemical parameters while η'' is a mixed one, because it embodies α, β, τ and h (ref.eq.14).

Eq.13 is essentially a Generalized Gamma distribution. In particular, the cdf corresponding to (13) is given by the incomplete gamma function:

$$P(C \leqslant c) = F(c; \alpha, \beta, C_E, C_0, \eta'') = \frac{1}{\Gamma(\alpha)} \gamma \left\{ \left[\ln\left(\frac{C_E - C_0}{C_E - c}\right) / \eta'' \right]^\beta ; \alpha \right\}$$

(15)

It is worthwhile noting at low c values ($C_0 < c \ll C_E$) the function $\ln((C_E - C_0)/(C_E - c))$ can be expanded in series so that (13) becomes the Generalized Gamma distribution in the form (1) with:

$$\xi = C_0 \qquad \eta = \eta'' C_E = \eta' h C_E = \tau h C_E \alpha^{-1/\beta}$$

(16)

A physical rationale is therefore established for (1) and for the parameters α, β, ξ and η.

Eqs.9 and 13 may be extended to the case of several rainfall events. Nevertheless the quantities described cannot be measured directly. In particular, water concentrations at the gaging station at current time t' are built up through a particle mix generated by different residence times.

Nevertheless verification of the ability of (13) to describe measured concentration distributions for a long time span is deemed necessary.

DISCUSSION

The analysis of available experimental data has allowed a preliminary screening of (13) which has confirmed the expectations and raised some questions.

A preliminary check has been performed for comparison among the pdf of the families (1) and (13). Table 4 illustrates the results with total-N data gaged at Zuccarello (1981).

In particular, the results of Table 4 indicate that the maximum likelihood fit of (13) is top ranking; also according to all statistics shown in Table 2 (such a circumstance is not verified for eq.1). All the tests of fitting for eq.13 with five parameters rank definitely better than those for other distributions.

Figure 1 shows the plots of eq.1 and 13 together with the histogram of fitted experimental data. The plots of Figure 1, in particular, point out the differences in data interpretation for the two functions and the noteworthy variations corresponding to little changes in ln(1/L) (of the order of 1/1000).

The result seems to suggest the more flexible capability of (13)

Table 4 - Comparison of total-N concentration models (eqs.1,13 and 1981 field data).

TYPE–DISTRIBUTION		df	CS	KS	CMS	ln(1/L)	SSD
WEIBULL	eq. (1):3-P	8	22.1	11.6	.15	554.72	1.32
	eq.(13):4-P	7	21.6	11.3	.14	554.35	1.22
STANDARD GAMMA	eq. (1):3-P	8	25.5	15.5	.21	557.94	1.90
	eq.(13):4-P	7	24.6	14.8	.19	557.23	1.74
GENERALIZED GAMMA	eq. (1):3-P	7	22.2	12.9	.15	554.70	1.32
	eq.(13):4-P	6	20.4	7.7	.11	553.40	.96

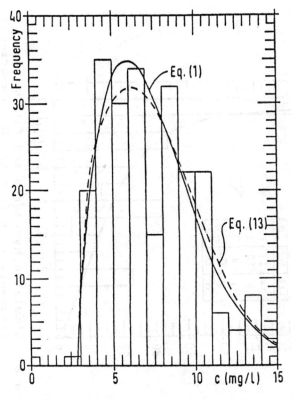

Figure 1 - Measured total-N hystograms gaged at Zuccarello (1981) and fitting by eqs.1 and 13.

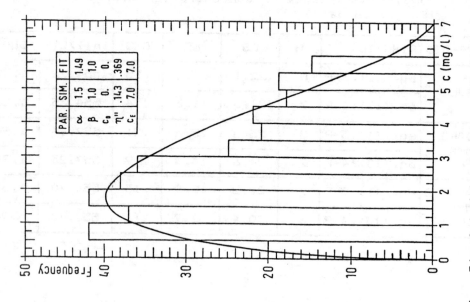

Figure 2 – Simulated hystogram and fitted frequencies as from the D.5 experiment.

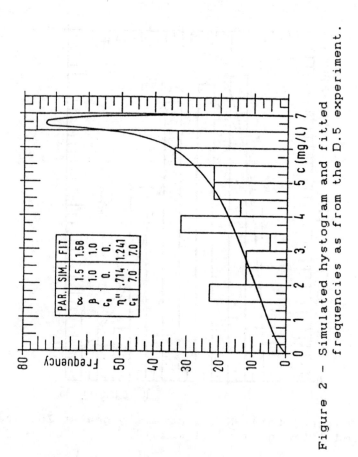

Figure 3 – Simulated hystogram and fitted frequencies as from the D.9 experiment.

376

to fit the experimental data even for long time spans, during which physical parameters cannot be assumed as constants and interactions of raindrops with particles from earlier events cannot be neglected.

It is to be noted that with respect to the other pdf's the function (1/L) of (13) holds a minimum value far better than could be explained by the presence of an additional parameter. In fact, Weibull and Standard Gamma do not appreciably benefit from the additional parameter.

The observations show that (13) may be considered as a "natural" distribution and that such a feature is maintained whenever extended field-data collections are analyzed.

The questions which have been raised are concerned with parameter identification and are largely tied to the fitting technique. In fact, the maximum likelihood method is a well-known parameter estimation procedure which leads asymptotically to an estimate with the greatest efficiency but not necessarily unbiasedness [28]. Furthermore, in the case of Generalized Gamma the parameters are not independent estimates. Very probably these facts help to foster the difficulties encountered in fitting experimental distributions by eq.13.

Numerical experiments have hence been performed aiming at analysis of such problems and long-terms effects on (13).

The experiments have consisted of simulation of nutrient losses using the computed data set to build hydrological and chemical parameters of the catchment by (13).

Simulations have been performed by choosing IUH as in (9), mass transport laws as in (10) and real gaged net precipitations.

Constant parameter estimation has been performed, and hence potential fitting problems were dependent on parameter variance or on validity of (13).

As an example two experiments (D.5 and D.9) have been considered whose main features are illustrated in Table 5.

In general, the maximum likelihood method yields an estimate of the parameters which does not optimize the other statistics and eq.15, computed with such values of the parameters, does not fit closely cumulative distributions of simulated data.

Only in the case of maximum likelihood restricted to the search of (α, η'') or (β, η''), does the maximum of L approximate the minimum of other statistics.

Table 5 - Basic statistics of D.5 and D.9 experiments.

STATISTICS	D.5	D.9
Min-Value	1.637	.373
Max-Value	6.937	6.090
Sample size	285	338
Mean	5.03	2.60
Standard dev	1.68	1.58
Skewness	- .65	.47
Kurtosis	2.18	2.10

Figures 2 and 3 illustrate data distributions for two experiments (D.5 and D.9) and the related pdf's (13). Figure 4 shows the related cdf of simulated and computed (via eq.15) data.

The sample experiments hold a statistical structure which is deemed sufficiently liable to characterize real data sets. In particular, the experiments emphasize some irregularities in the histograms which appear also for real data. Such discontinuities show the marked effects of raindrop mixing with earlier events which are amenable to modification of the scale factor η ".

The study of various statistics has tentatively allowed the setting of the parameter values which balance the various tendencies of the optimal searches.

The various distributions have been checked on total-N data collected at Zuccarello in 1976 and 1981. General statistics for such data are reported in Table 6.

Observed and computed (via eq.13) cdf's for the 1976 and 1981 campaigns are illustrated in Figure 5. In particular Figures 4 and 5 show that (13) is capable to fit the data sets even when mixing effects are likely to cause the failure of some assumptions underlying the model (13). Mixing, in fact, modifies the scale factor η " which defines an "apparent" time instead of the characteristic time. This fact is manifest also if a comparison of parameters in Figures 2 and 3 is carried out.

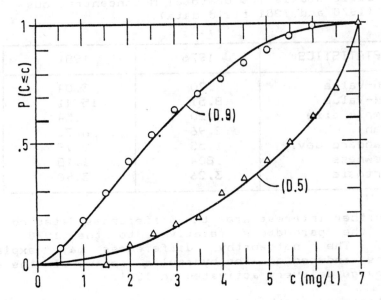

Figure 4 – Simulated and fitted cdfs as from D.5 and
D.9 experiments.

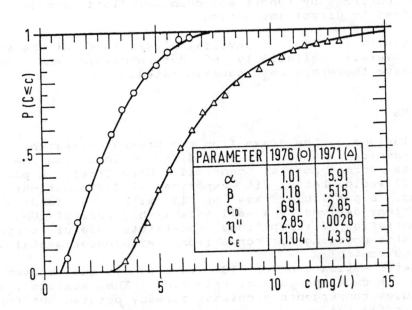

PARAMETER	1976 (○)	1971 (△)
α	1.01	5.91
β	1.18	.515
c_D	.691	2.85
η''	2.85	.0028
c_E	11.04	43.9

Figure 5 – Measured and fitted total-N concentration
cdfs (1976 and 1981 field experiment
collections).

Table 6 - Basic statistics of total-N concentrations
        (1976 and 1981 field data).

STATISTICS	1976	1981
Min-Value	.70	3.04
Max-Value	8.51	15.41
Sample size	120	154
Mean	2.96	6.71
Standard dev.	1.58	2.79
Skewness	.804	1.18
Kurtosis	3.26	3.90

Worthy of further interest are the differences detected among the values of the parameters relative to the 1976 and 1981 collections. The noteworthy differences may explain the modifications induced on the hydraulic regime of the catchment forced by sewage delivery activated in 1981.

General statistics of the data (Table 6) call for an increased use of fertilizers, while the analyses of pdf point clearly at substantial hydraulic regime alterations (α ranges from 1.01 to 5.91; β ranges from 1.18 to .515). Furthermore a modified chemical regime is also detected (C_o and C_∞ values). This fact, pointwise confirmed by runoff and chemical loss simulations, has been verified by direct inspection.

The sensitivity of (13) to physical alterations of the watershed and the overall reliability of deterministic and stochastic modeling have therefore been substantiated.

CONCLUSIONS

Some conclusions can be drawn from the present research:
1- NPS chemical concentration distributions can neither be regarded as Normal nor as Lognormal. More likely a pdf of the type (13) would better fit experimental distributions. On the other hand, a particular case of (13) with $\beta=1$ yields a pdf of concentrations derived by a well established form of IUH;
2- concentration distributions constitute useful supports to deterministic modeling for comparisons with experimental data and forecasting control;
3- parameter estimation of (13) is tied to the chosen fitting criterion of experimental distributions. The maximum likelihood method causes convergence problems already pointed out for models of the type (1) [21];
4- the choice of the fitting procedure for (13) cannot be carried out directly on experimental data fits because such models embody

mixing effects with earlier events. Suitable simulations for critical point evaluations of the various fitting procedures are being performed;
5- Eq.13 shows its usefulness even for long-term data evaluation if a modified physical meaning is attributed to the parameters.

Although this research is being pursued further, the preliminary results and their drawbacks appear noteworthy and encouraging for further applications.

ACKNOWLEDGEMENTS

The research was supported by funds provided by MAF, MPI and CNR.

REFERENCES

1- SMITH,R. V., STEWART, D. A., "Statistical models of river loadings of nitrogen and phosphorus in the Lough Neagh system", Water Research, 11, 631-635, 1977.
2- ZINGALES, F., ALESSANDRINI, S., BENDORICCHIO, G., COMIS, C., MARANI, A., PIANETTI, F., RINALDO, A., SARTORI BOROTTO, C., ZANIN, S., "Inquinamento delle acque da sorgenti diffuse: analisi statistiche", Inquinamento, 23, 1-4, 1981.
3- BENDORICCHIO, G., COMIS, C., ALESSANDRINI, S., "Evoluzione temporale dell'inquinamento da sorgenti diffuse di origine agricola", Inquinamento, in press.
4- HARTIGAN, J.P., QUASEBARTH, T.F., SOUTHERLAND, E., "Calibration of NPS model loading factors", Journal of Environmental Engineering, 109, 1259-1272, 1983.
5- HARTIGAN, J.P., FRIEDMAN, J.A., SOUTHERLAND, E., "Post-audit of lake model used for NPS management", Journal of Environmental Engineering,109, 1354-1370, 1983.
6- GEORGOPULOS, P.G., SEINFELD, J.H., "Statistical distributions of air pollutant concentrations", Environ. Sci. Technol., 16, 401A-416A, 1982.
7- BUTTAZZONI, C., LAVAGNINI, I., MARANI, A., ZILIO GRANDI, F., DEL TURCO, A., "Probability models for athmospheric sulfur dioxide concentrations in the area of Venice", J. Air Pollut. Control Ass., submitted for publication.
8- LAVAGNINI, I., BUTTAZZONI, C., MARANI, A., "Confronto di modelli di distribuzione statistica dei dati di qualita` dell'aria", Acqua Aria, in press.
9- MARANI, A., LAVAGNINI, I., BUTTAZZONI, C.,"The generalized gamma model of statistical distributions of air pollutant concentrations", submitted for publication.
10-LARSEN, R.I., "A new mathematical model of air pollutant concentration averaging time and frequency", J. Air Pollut. Control Ass., 19, 24-30, 1969.
11-BENCALA, R.E., SEINFELD, J.H., "On frequency distributions of air pollutant concentrations", Atmospheric Environment, 10, 941-950, 1976.
12-BERGER, A., MELICE, J.L., DEMUTH, C.L., "Statistical distribu-

tions of daily and high atmospheric SO_2-concentrations", Atmospheric Environment, 16, 2863-2877, 1982.

13-LYNN, D.A., "Fitting curves to urban suspended particulate data", Proc. Symp. Statistical Aspects Air Quality Data, Publ. EPA-650/4-74-038, U.S. Environmental Protection Agency, Research Triangle Park, N.C., 1974.

14-HOLLAND, D.M., FITZ-SIMONS, T., "Fitting statistical distributions to air quality data by tha maximum Likelihood method", Atmospheric Environment, 16, 1071-1076, 1982.

15-LARSEN, R.J., "An air quality data analysis system for interrelating effects, standards and need source reduction - Part 4. A three-parameter averaging time model", J. Air Pollut. Control Ass., 27, 454-459, 1977.

16-MAGE, D.T., OTT, W.R., "Refinements of the lognormal probability model for analysis of aerometric data", J. Air Pollut. Control Ass., 28, 796-798, 1978.

17-BARRY, P.J., "Use of Argon-41 to study the dispersion of stack effluents", Proceedings of the Symposium on Nuclear Techniques Environmental Pollution, I.A.E.A., STI/Publ./268, Austria, 1971.

18-AMOROSO, L., "Ricerche intorno alla curva dei redditi", Ann. Mat. Pura Appl., Ser.4, 21, 123-159, 1925.

19-STACY, E.W., "A generalization of the gamma distribution", Ann. Math. Statist., 33, 1187-1192, 1962.

20-JAMES, I.R., "Characterization of a family of distributions by the independence of size and shape variables", Ann. Statist., 7, 869-881, 1979.

21-LAWLESS, J.F., "Interference in the generalized gamma and log-gamma distributions", Technometrics, 22, 409-419, 1980.

22-ZINGALES, F., ALESSANDRINI, S., BENDORICCHIO, G., MARANI, A., PIANETTI, F., RINALDO, A., SARTORI BOROTTO, C., ZANIN, S., "Inquinamento dovuto alle acque di un bacino agricolo sversate nella laguna di Venezia", Inquinamento, 22, (12), 1-7, 1980.

23-LIENHARD, J.H., "A statistical mechanical prediction of the dimensionless unit hydrograph", J. Geophys. Res., 69, 5231-5238, 1964.

24-RODRIGUEZ-ITURBE, I., VALDES, J.B., "The geomorphologic structure of hydrologic response", Water Resour. Res., 15, 1409-1420, 1979.

25-GUPTA, V.K., WAYMIR, E., WANG, C.T., "A representation of an instantaneous unit hydrograph from geomorphology", Water Resour. Res., 16, 855-862, 1980.

26-LIENHARD, J.H., MEYER, P.L., "A physical basis for the generalized gamma distribution", Quart. Appl. Math., 25, 330-334, 1967.

27-DAVIDSON, M., "Statistical mechanics", McGraw Hill Book Company, New York, 1962.

28-HIMMELBLAU, D.M., "Process analysis by statistical methods", J.Wiley & Sons, Inc., New York, 1970.

DEVELOPMENT, TESTING, AND APPLICATIONS OF NONPOINT SOURCE MODELS FOR EVALUATION OF PESTICIDES RISK TO THE ENVIRONMENT

Lee A. Mulkey, Robert F. Carsel, and Charles N. Smith
U.S. Environmental Protection Agency
Environmental Research Laboratory
College Station Road
Athens, Georgia 30613

ABSTRACT

Nonpoint source loading models are currently being used in analyzing the risks posed to the environment and man from the agricultural use of pesticides. Pesticide pollution problems arise from widespread, diffuse chemical usage and are influenced by the management of soil and water in agricultural systems. Both surface water environments and groundwater systems can be impacted by pesticide use and nonpoint source models must accommodate these potential risks. Five operational pesticide runoff models have been evaluated and their similarities and differences discussed. The U.S. Environmental Protection Agency's pesticide leaching model, PRZM, is described and methodologies for model application in risk assessments are discussed.

INTRODUCTION

Pesticide use in agricultural production can pose risks to ecosystems and, in some cases, to humans. When pesticides move off-site via surface runoff and erosion, the traditional concern is toxicity to fish and wildlife (1). More recently, however, groundwater contamination by pesticides has raised concerns about drinking water hazards. In most cases (ignoring accidental or intentional spills), problems will arise if two conditions are met: the hydrologic cycle interacts with the soil-plant-pesticide system to transport chemicals in toxic amounts to rivers, lakes, or aquifers, and a sensitive or susceptible specie is present in or consumes the contaminated water.

Pesticide problems are easily placed in the context of nonpoint source pollution because they arise from widespread, diffuse chemical usage and because the management of soil and water in agricultural systems influences the use, fate, and transport of the chemicals. Indeed, the first systematic study of chemical nonpoint source pollution demonstrated that pesticides enter water bodies via runoff (21). In many cases, fish kills were attributed to off-site movement of agriculturally used pesticides (13). Similarly, the first simulation model designed to mathematically describe agricultural nonpoint source pollution was for pesticides (9).

Ideally the evaluation and management of pesticide nonpoint source pollution problems should be based on extensive monitoring of both the contamination levels (e.g., concentrations in surface or groundwaters)

383

and any subsequent damages (e.g., aquatic species reductions). Given a sufficient data base, the methods and management strategies to mitigate the problems could be developed and implemented where appropriate. Two problems with this approach render it impractical. First, the events leading to water quality problems are in many cases "rare" in the same sense as flooding events (2,3); thus, the need to sample.is random in time and often very short in duration. Second, the U.S. Environmental Protection Agency (EPA) is required to evaluate pesticides before they are released for widespread use.

A more prudent (and practical) approach than that of on-site monitoring is to combine an understanding of pesticide chemical properties and agronomic practices with appropriate descriptions of the hydrologic cycle to enable prediction of water quality impacts. Basic elements of this approach include development, testing, and application of mathematical models for evaluation of pesticide risk to the environment. The use of water quality modeling in general and pesticide nonpoint source modeling in particular in management decision making is discussed by Barnwell and Krenkel (3).

CONCEPT OF RISK

In a general sense risk implies the probability of damage. Kaplan and Garrick (17) have introduced the idea of risk as the set of "triplets:"

$$R = (< S_i, P_i, X_i >) \qquad (1)$$

where S_i = a scenario identification or description, i.e., What can happen?

           P_i = the probability of that scenario, i.e., How likely is it that S_i will happen?

and X_i = the consequence or measure of damage, i.e., What are the environmental effects?

It is convenient to adopt this concept of risk for pesticides and other nonpoint source pollutants. In brief, the scenairo of concern is the movement of pesticides from the use-site, say an agricultural field, to adjacent surface or ground waters. The probability of such movement for each of several concentrations is desired so that its effects or consequences may be estimated. The major role of nonpoint source modeling in risk assessment is to provide estimates of this probability. To complete the analysis, the probability that sensitive species are exposed to the pesticide must also be estimated before toxicity data can be used to infer damages. This paper is a description of the nonpoint source loading methods and models used to estimate the probability of pesticide exposure given sources defined as field-scale units.

MATHEMATICAL STATEMENT OF PROBLEM

Concise mathematical representation of the pesticide nonpoint source problem (as defined in this paper) is difficult because of the many processes acting simultaneously to produce temporally and spatially varying
concentrations in various environmental media. Indeed, this difficulty
has led most investigators to solve specific problems in specific media.
For example, models to predict runoff or models to predict volatilization
are available but rarely are these problems solved or modeled simultaneously. It is possible, however, to describe in general form the basic
equations governing the space-time concentration profile. A system of
mass-balance equations can be formulated as:

$$\sum_{i=1}^{n} \frac{\partial(C_i V_i)}{\partial t} = \sum_{i=1}^{n} \frac{\partial}{\partial x} D_i \frac{\partial(C_i V_i)}{\partial x} - \sum_{i=1}^{n} \frac{\partial(C_i V_i v_i)}{\partial x}$$

$$+ \sum_{i=1}^{n} APP_i(t) - \sum_{i=1}^{n} \sum_{j=1}^{m} k_j C_i V_i \tag{2}$$

where C_i = concentration of the pesticide in the ith environmental
conpartment, ML^{-3}

V_i = volume of the ith environmental compartment, L^3

D_i = coefficient of dispersion (includes diffusion) L^2T^{-1}

v_i = velocity vector for the media of the ith compartment, LT^{-1}

$APP_i(t)$ = pesticide application rate for the ith compartment on day,
t, MT^{-1}

k_j = pesticide transformation rate for the jth process, T^{-1}

n = number of environmental compartments or media

m = number of transformation processes acting on pesticide
within each compartment or media assuming first order
kinetics

x = space dimension, L

and   t = time, T

Solution of the equations represented above will yield the concentration
profile desired. Depending on the problem and the chemical properties,
the number of equations required may be limited. For example, a chemical
having a very low vapor pressure that has marked affinity for soils can
be evaluated for its potential to contaminate surface water by solving
only the mass balance equations for runoff and erosion. Thus, the various terms defined in equation 2 must be evaluated for soils, soil water,

runoff, and erosion and appropriate mass-balances for each medium or compartment written. Once boundary and initial conditions are specified and interactive terms are defined, the resulting equations can be solved numerically. The solution will be quite complex in some cases. For example the advective term, $\partial(C_i V_i v_i)/\partial x$, for runoff requires a solution to the rainfall-runoff process at a sufficient level of detail to adequately represent the temporal and spacial resolution desired. Given the wide range of environmental concerns and the equally wide range of chemical properties it is clearly desirable to have a general framework or model for pesticide exposure and risk assessment.

The need to evaluate probabilities in risk assessment imposes an additional and rather rigorous constraint on the solution to the basic equations. Namely, a dynamic solution that generates a time series of pesticide concentrations must be obtained. This is necessary to derive the frequency distributions required to estimate probabilities. The application of this concept to surface water concentrations is presented by Parkhurst et al. (23) and described in detail by Olsen and Wise (22). It is clear that prediction of water quality in this manner first requires that nonpoint source loads be expressed. Given the loads, a number of models are available to predict the in-stream concentrations as described by Mulkey and Falco (20), Ambrose et al. (2), and Burns et al. (4).

Deriving probability statements from time series is a straight forward process if one simply expresses the results as a cumulative distribution function. That is,

$$F(C_i) = \int_{-\infty}^{C_i} f(C_i)\, dC \tag{3}$$

where $F(C_i)$ = the probability that $C_i \leq$ some set value of C_i

and $f(C_i)$ = the frequency distribution of C_i

Proper selection of the equations of the form given by equation 2 and subsequent expression of their solution in the form of equation 3 formalizes the application of nonpoint source load models to environmental risk assessment. The problem remains to obtain solutions to the equations.

ALTERNATIVE SOLUTIONS AND MODELS

A general and rigorous solution of equations 2 and 3 is probably not possible and in any case would be impractical given the current data base and understanding of basic processes. Appropriate and approximate solutions, however, can be obtained for specific problems. Pesticide runoff and leaching from field or "small" land segments are two such problems for which nonpoint models have been successfully developed and applied.

Pesticide Runoff Models

A number of pesticide runoff loading models have been developed in the last decade (1,7,12,13,14,18,27,28). Most simulate the movement of pesticides and plant nutrients. All are based on the premise that processes influencing pesticide concentration (or mass) in the soil-plant profile must be combined with the runoff and erosion transport mechanisms that move the pesticide from its point of application to the boundaries of the area being analyzed. Lorber and Mulkey (19) have discussed the major differences among some of these models. Specific details of the hydrologic and erosion components for most will be discussed in other papers given at this conference. A summary of the pesticide process components of the major models is given in Table 1.

The importance of the differences outlined in Table 1 is difficult to assess in any general sense. In most cases the properties of the specific chemical, its formulation, and its mode of use will determine the relative importance of each process. For example, Smith and Carsel (25) have shown that ignoring foliar processes can lead to substantial errors in prediction of runoff losses. Only a relatively small amount of the registered pesticides, however, are intended for application to a mature or fully developed plant canopy. (Recent trends toward increased use of conservation tillage accelerate our need to more fully understand this process.)

Pesticide Leaching Models

Models that predict leaching of pesticides to groundwater from agricultural fields are less well developed than those for runoff summarized in Table 1. Model development has been recently accelerated, however, by revelations of positive groundwater contamination. The leaching processes have been investigated and specific models have been reported (10,15,24, 29), but in most cases these models were developed to elucidate processes and provide insights rather than provide a modeling framework for operational use. Recently, the EPA developed a generalized model for evaluating pesticide leaching--the Pesticide Root Zone Model (PRZM) (6).

The interactions of pesticides and soil-water-plant systems include surface runoff (in water and on eroded sediment), advection in percolating water, molecular diffusion, dispersion, uptake by plants, sorption to soil particles, volatilization, biological degradation, and chemical transformation. If a surface zone characterized by erosion and runoff losses is distinguished from subsurface zones where only vertical movement is allowed, mass balance equations of one dimension can be derived for both solution and sorbed phases of a pesticide. The further assumption of linear, reversible, and instantaneous equilibrium sorption yields equations 4 and 5 for the surface zone,

$$\frac{\partial(C_w\theta V)}{\partial t} = \frac{\partial}{\partial x} D \frac{\partial(C_w\theta V)}{\partial x} - \frac{\partial(vC_w\theta V)}{\partial x} - k_u TC_w - ROC_w$$

$$- C_w\theta V\Sigma k_w + F_d J_{ap} \quad, \text{ for } 0 \leq x \leq DPR \tag{4}$$

Table 1. Pesticide Process Representation in Current Pesticide Runoff Models

Model	Ref.	Transport Mechanisms	Degradation	Soil-Water Partitioning	Plant Processes	Other Options
PRS	7	Dissolved in surface runoff, sorbed to eroded soil	First order, lumped decay rate	Reversible, instantaneous, equilibrium, single-value linear isotherm	Foliar washoff and decay	Enrichment ratios
CREAMS	18	Dissolved in surface runoff, sorbed to eroded soil	First order, lumped decay rate	Reversible, instantaneous, equilibrium, single-value linear isotherm	Foliar washoff and decay	Enrichment ratios
CPS	27	Dissolved in surface runoff, sorbed to eroded soil	First order, lumped decay rate	Reversible, instantaneous, equilibrium, single-value linear isotherm	None	None
ARM II	12	Dissolved in surface runoff Dissolved in interflow Sorbed to eroded soil	Time-phased series of first order, lumped decay rates	Reversible and irreversible, single and non-single valued, Freundich isotherm	None	None
HSPF	16	Dissolved in surface runoff Dissolved in interflow Sorbed to eroded soil	Time-phased series of first order, lumped decay rates	Same as ARM II with added option for sorption/desorption kinetics	None	None

388

$$\frac{\partial(C_s\rho V)}{\partial t} = -EROC_s - C_s\rho V\Sigma k_s + (1-F_d)J_{ap} \qquad (5)$$

where: C_w = dissolved pesticide concentration in soil water, ML^{-3}

C_s = sorbed pesticide concentration, MM^{-1}

θ = soil volumetric water content, L^3L^{-3}

V = compartment volume, L^3

ρ = soil bulk density, ML^{-3}

D = coefficient of dispersion and diffusion, L^2T^{-1}

v = water pore velocity, LT^{-1}

k_u = pesticide plant uptake efficiency, $0 \le k_u \le 1.0$

T = volumetric plant transpiration rate, L^3T^{-1}

RO = volumetric rate of surface water runoff, L^3T^{-1}

Σk_w = summation of all first-order rate constants for transformation of dissolved pesticides, T^{-1}

F_d = fraction of applied pesticide in dissolved form

J_{ap} = pesticide application rate, MT^{-1}

t = time, T

x = vertical distance, L

DPR = depth of surface runoff zone, L

ERO = surface soil erosion rate, MT^{-1}

and Σk_s = summation of all first-order rate constants for transformation of sorbed pesticides, T^{-1}

For the lower subsurface zones,

$$\frac{\partial(C_w\theta V)}{\partial t} = \frac{\partial}{\partial x} D \frac{\partial(C_w\theta V)}{\partial x} - \frac{\partial(vC_w\theta V)}{\partial x} - k_uTC_w - \Sigma k_wC_w\theta V \qquad (6)$$

$$\frac{\partial(C_s\rho V)}{\partial t} = -\Sigma k_sC_s\rho V , \quad \text{for} \quad X \ge DPR \qquad (7)$$

For both cases the linear interaction term for sorbed and dissolved phases is

$$C_s = K_p C_w \qquad (8)$$

where K_p = partition coefficient, $L^3 M^{-1}$

It also is convenient, if not required by the experimental methods most often used to estimate transformation rate constants, to use the equality $\Sigma k_w = \Sigma k_s = \Sigma k$.

Equations 4-8 can be combined to yield basic equations for each zone written in terms of the dissolved pesticide concentration, C_w.

$$\frac{\partial(C_w(\theta+K_p\rho V))}{\partial t} = \frac{\partial}{\partial x} D \frac{\partial(C_w\theta V)}{\partial x} - \frac{\partial(vC_w\theta V)}{\partial x} - C_w[k_u T + kV(\theta+K_p\rho)$$

$$+ RO + EROK_p] + J_{ap} \qquad (9)$$

where   RO = RO for $0 \leq x \leq$ DPR

         ERO = ERO

and   RO = 0 for $X \leq$ DPR

         ERO = 0

Equation 9 is a variation of the advection-dispersion model most often derived as the basis for groundwater quality models. The plant uptake term, represented here as a simple linear function of plant transpiration, is not included in most representations. The runoff and erosion terms are rarely included. Equation 9 could be modified further to include the influence of a vapor phase pesticide component. For many very soluble pesticides, however, volatilization from within the soil profile is not a major mode of loss.

The PRZM model solves equation 9 by a finite difference scheme with daily time steps and variable depth dimensions in increments of 1.0 cm. Detailed descriptions of the model are provided by Carsel et al. (6) and specific use guidance is described in the user's manual (5).

MODEL TESTING RESULTS

Few systematic tests have been conducted for pesticide nonpoint source loading models. Most model development efforts include some testing, however, and results are often reported in the model documentation (12, 14,18). The lack of comprehensive data bases currently limit the extent to which models can be evaluated against field observations.

In lieu of rigorous testing against data, it is useful to compare the different models by sensitivity testing. A useful test that illustrates the combined influences of transport and partitioning is model sensitivity to the partition coefficient (19). Results of such a test are shown in Figure 1. The test illustrated in Figure 1 was obtained by first calibrating each model to the same watershed and then varying the partition coefficient. Thus, the results reflect differences in model structure and assumptions as well as variations in hydrologic and erosion predictions.

The models appear to follow the same trend. As the partition coefficient approaches zero, leaching dominates and runoff losses decrease. On the other extreme, as the partition coefficient becomes large, erosion dominates and the net loss decreases because of the relative differences in water and sediment volumes. A notable exception to this pattern is the influence of interflow transport simulated by the Agricultural Runoff Management (ARM) model. This phenomena has been evaluated (19) and when interflow is properly calibrated the divergence illustrated in Figure 1 does not occur. (In the absence of calibration data interflow should probably be ignored or set to near zero.)

An important conclusion arises from the results in Figure 1. Because all models respond similarly, the relative impacts for different chemicals will be the same. Similarly, these models used to evaluate best management practices should give consistent results regardless of which specific technique is used.

A useful performance test of environmental exposure or risk models is in the frequency domain. Because frequencies of events are desired for risk assessments, models should be tested for their ability to predict observed frequency distributions. A number of statistical issues immediately arise. Inadequate sample size and serial correlation may create problems in drawing appropriate statistical inferences. Nevertheless, Young and Alward (30) have tested the ARM model on 30 watershed-pesticide-years of data using nonparametric statistics to show that 22 out of 30 derived frequency distributions are acceptable at the 95% confidence level. Similar approaches to be reported elsewhere in this conference have been used to test the Hydrological Simulation Program--FORTRAN (HSPF) model, which has been applied to very large river basins with considerable success.

Pesticide leaching threats to groundwater can be assessed with the PRZM model. A major field project is now underway to provide the initial extensive testing of this new technique (8). Despite the current lack of data, some preliminary results are available. Carsel et al. (6) have shown that PRZM simulates concentration profiles of the pesticide aldicarb quite well for three locations in potato and citrus production. Testing also is ongoing for atrazine use on corn in the Southern Piedmont region of the United States based upon data collected by Smith et al. (26). Although not yet complete, typical results are shown in Figure 2.

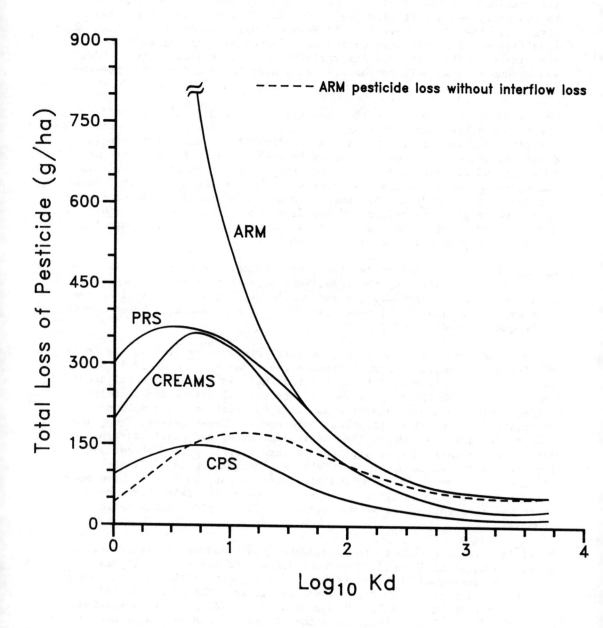

Figure 1. Sensitivity Analysis of Pesticide Runoff Models

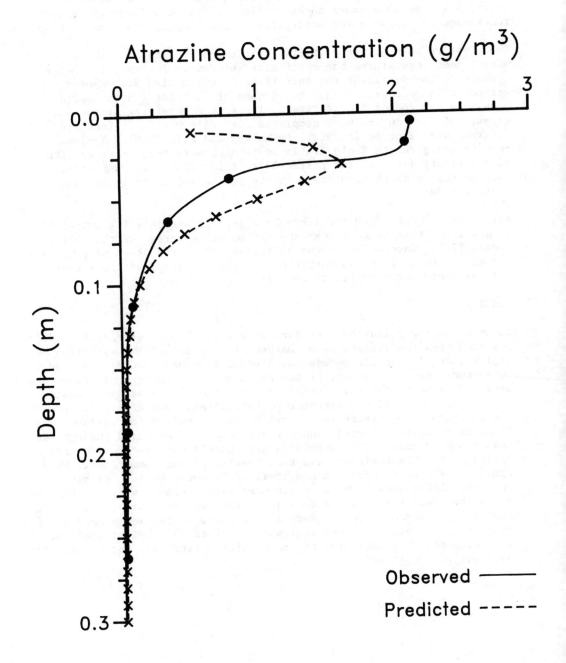

Figure 2. Performance Testing Results for the Pesticide Root Zone Model

MODEL APPLICATIONS FOR USE IN RISK ASSESSMENTS

The application of nonpoint source loading models to risk assessments is increasing. Notable among these applications is the investigation of runoff-aquatic impacts and prediction of leaching to groundwater supplies.

The exposure calculations for risk assessment are made for the environmental media containing the species of interest. For runoff a major concern is the pesticide concentrations to which fish are exposed over various time periods. Thus, the purpose of nonpoint source loading models is to produce time series of loads to a receiving stream. An example of such an analysis completed for EPA's Office of Pesticide Programs was reported in detail by Mulkey and Falco (20). Typical results are given in Table 2. These results were then combined with fish toxicity data for the pesticide to infer risk. The ARM model was used in this case to simulate loadings for inputs to an in-stream routing model.

More recently, the PRZM model has been used to evaluate the probability of pesticide threats to groundwater in major use areas of the United States (11). Cumulative frequency curves for a wide range of pesticide properties (half-life and partition coefficients) and soil-plant systems were produced for a number of locations.

SUMMARY

Nonpoint source pollution resulting from agricultural production can be analyzed with the assistance of mathematical models. The selection and application of specific models are dependent on many factors related to the nature and areal extent of the pollutant of interest. In cases where pesticide risk to the environment is the concern, the analysis must include estimates of the exposure probabilities. The magnitude, frequency, and duration of pesticide concentrations are required for surface waters draining agricultural areas and for groundwaters recharged through the soil-plant profile. The nonpoint source models used to provide mass loadings for such analyses have been identified and summarized. Five runoff models were described and their differences noted. Although specific differences in model structures exist, sensitivity testing has shown that relative response to major parameters is similar suggesting that all are appropriate for many applicatins. A new leaching model for estimating threats to groundwater was discussed and typical applications demonstrated. The extension of probability derivations for groundwater problems was described.

Table 2. Frequency of Occurrence of Pesticide for Given Durations at the Mouth of the Mississippi River Over a Five-Year Period

Pesticide Concentration	Duration (days)											
	1	2	3	4	5	6–10	11–15	16–20	21–30	31–40	41–60	61–90
$\geq$ 1 ug/l	7	6	5	5	4	2						
$\geq$ 0.1 ug/l	16	16	15	15	15	11	10	4	1			
$\geq$ 0.01 ug/l	14	14	13	13	13	12	11	9	6	5	2	

REFERENCES

1. Adams, R. T., and F. M. Kunisu. 1976. Simulation of Pesticide Movement on Small Agricultural Watersheds. U.S. Environmental Protection Agency, Athens, GA. EPA-600/3-76-066.

2. Ambrose, R. B., S. I. Hill, and L. A. Mulkey. 1983. User's Manual for the Chemical Transport and Fate Model TOXIWASP, Version I. U.S. Environmental Protection Agency, Athens, GA. EPA-600-3/83-005.

3. Barnwell, T. O., Jr., and P. A. Krenkel. 1982. The use of water quality models in management decision making. In: Water Pollution Research and Control, IAWPRC, S. A. Jenkins, Ed. Water Sci. and Technol. 14:1095-1107

4. Burns, L. A., D. M. Cline, and R. R. Lassiter. 1982. Exposure Analysis Modeling System (EXAMS): User Manual and System Documentation. U.S. Environmental Protection Agency, Athens, GA. EPA-600/3-82-023.

5. Carsel, R. F., C. N. Smith, L. A. Mulkey, J. D. Dean, and P. P. Jowise. 1984. User's Manual for the Pesticide Root Zone Model (PRZM) Release 1.0. U.S. Environmental Protection Agency, Athens, GA, EPA-600/3-84-109.

6. Carsel, R. F., L. A. Mulkey, M. N. Lorber, and L. B. Baskin. 1984. The pesticide root zone model (PRZM): a procedure for evaluating pesticide leaching threats to ground water. Accepted for publication in Ecological Modeling, 28:(pages unspecified).

7. Computer Sciences Corp. 1980. Pesticide runoff simulator (PRS) user's manual. Office of Pesticide Program, U.S. Environmental Protection Agency, Washington, DC.

8. Cooper, S. C., R. F. Carsel, C. N. Smith, and R. Parrish. 1984. Pesticide migration in the unsaturated and saturated soil zone for a field site within the Dougherty Plains area of southwest Georgia. Progress report for 1983. Draft report for EPA-USGS Interagency Agreement AD14930092.

9. Crawford, N. H., and A. S. Donigian, Jr. 1973. Pesticide Transport and Runoff Model for Agricultural Lands. U.S. Environmental Protection Agency, Athens, GA. EPA-660/2-78-013.

10. Davidson, J. M., G. H. Brusewitz, D. R. Baker, and A. L. Wood. 1975. Use of Soil Parameters for Describing Pesticide Movement Through Soils. U.S. Environmental Protection Agency, Athens, GA. EPA 600/2-75-009.

11. Dean, J. D., P. P. Jowise, and A. S. Donigian, Jr. 1984. Leaching Evaluation of Agricultural Chemicals (LEACH) Handbook. U.S. Environmental Protection Agency, Athens, GA, EPA-600/3-84-068.

12. Donigian, A. S., Jr., D. G. Beyerlein, H. H. Davis, Jr., and N. H. Crawford. 1977. Agricultural Runoff Management (ARM) Model - Version II. Testing and Refinement. U.S. Environmental Protection Agency, Athens, GA. EPA-600/3-77-098.

13. Duttweiler, D. W., and H. P. Nicholson. 1983. Environmental problems and issues of agricultural nonpoint source pollution. In: Agricultural Management and Water Quality. F. W. Schaller and G. W. Bailey, eds., Ames, Iowa State University Press.

14. Frere, M. H., C. A. Onstad, and H. N. Holtan. 1975. ACTMO: An Agricultural Chemical Transport Model. U.S. Department of Agriculture, Hyattsville, MD. Report No. ARS-H-3.

15. Gureghian, A. B., D. S. Ward, and R. W. Cleary. 1979. Simultaneous Transport of Water and Reacting Solutes Through Multilayered Soils Under Transient Unsaturated Flow Conditions. J. Hydrol. 41, 253-278.

16. Johanson, R. C., J. C. Imhoff, and H. H. Davis, Jr. 1980. User's Manual for Hydrological Simulation Program - FORTRAN (HSPF). U.S. Environmental Protection Agency, Athens, GA. EPA-600/9-80-015.

17. Kaplan, S., and B. J. Garrick. 1981. On the quantitative definition of risk. In: Risk Analysis vol. 1, no. 1. p. 11-27.

18. Knisel, W. G. (ed.). 1980. CREAMS - a field scale model for Chemicals, Runoff, and Erosion from Agricultural Management Systems. U.S. Department of Agriculture, Washington, DC. Conservation Research Report No. 26.

19. Lorber, M. N., and L. A. Mulkey. 1982. An evaluation of three pesticide runoff loading models. J. Environ. Qual. 11(3):519-529.

20. Mulkey, L. A., and J. W. Falco. 1983. Methodology for predicting exposure and fate of pesticides in aquatic environments. In: Agricultural Management and Water Quality. F. W. Schaller and G. W. Bailey (Eds.). Ames, Iowa State University Press.

21. Nicholson, H. P., A. R. Grzenda, G. J. Louer, W. S. Cox, and J. I. Teasley. 1964. Water pollution by insecticides in an agricultural river basin: occurrence of insecticides in river and treated municipal water. Limnol. and Oceanogr. $\underline{9}$:310-317.

22. Olsen, A. R., and S. E. Wise. 1982. Frequency Analysis of Pesticide Concentrations for Risk Assessment (FRANCO). U. S. Environmental Protection Agency, Athens, GA. EPA-600/3-82-044.

23. Parkhurst, M. A., Y. Onishi, and A. R. Olson. 1981. A risk assessment of toxicants to aquatic life using environmental exposure estimates and laboratory toxicity data. In: Aquatic Toxicology and Hazard Assessment: Fourth Conference. D. R. Branson and K. L. Dickson (Eds.). pp 59-71. Philadelphia, American Society for Testing and Materials. ASTM STP 737. p. 59-71.

24. Selim, H. M., J. M. Davidson, and P. S. C. Rao. 1977. Transport of Reactive Solutes Through Multilayered Soils. Soil Sci. Am. Pro., $\underline{41}$, 3-10.

25. Smith, C. N., and R. F. Carsel. 1984. Foliar Washoff of Pesticides (FWOP) Model: Development and Evaluation. J. Environ. Sci. Health. B19(3).

26. Smith, C. N., R. A. Leonard, G. W. Langdale, and G. W. Bailey. 1978. Transport of Agricultural Chemicals from Small Upland Piedmont Watersheds. U.S. Environmental Protection Agency, Athens, GA. EPA-600/3-78-056.

27. Steenhuis, T. S. 1979. Simulation of the action of soil and water conservation practices in controlling pesticides. Sect. 7. In: Effectiveness of Soil and Water Practices for Pollution Control. D. A. Haith and R. C. Loehr (Eds.) U.S. Environmental Protection Agency, Athens, GA. EPA-600/3-79-106. p. 106-146.

28. Steenhuis, T. S. and M. R. Walter. 1980. Closed form solution for pesticide loss in runoff water. Trans. ASAE. $\underline{23}$:615-620, 628.

29. Wood, A. L., and J. M. Davidson. 1975. Floumeturon and Water Content Distributions During Infiltration: Measured and Calculated. Soil Sci. Am. Pro., $\underline{39}$, 820-825.

30. Young, G. K., and C. L. Alward. 1983. Calibration and testing of nutrient and pesticide transport models. In: Agricultural Management and Water Quality, F. W. Schaller and G. W. Bailey (Eds.). Ames, Iowa State University Press.

FIELD DATA ACQUISITION SYSTEMS

L.F. Huggins
Professor and Head
Agricultural Engineering Department,
Purdue University, W. Lafayette, IN 47907, USA

ABSTRACT

The lack of reliable, comprehensive data on nonpoint water pollution and the environmental conditions that produce it is a primary factor limiting the design of cost effective controls. Unfortunately, collection of scientifically valid observations on the numerous processes that influence such pollution generation and transport is difficult and costly. Therefore, it is essential that water quantity and quality monitoring be improved to justify instrumentation costs and the long period of time required to collect useful information. Both philosophic considerations and practical guidelines for designing field data acquisition systems of increased value are presented.

INTRODUCTION

The single, fundamental recommendation that underlies all sections of this paper is that the scope and type of field data collection efforts undertaken as an integral part of Nonpoint Source (NPS) pollution research should be expanded. Monitoring efforts need to be substantially expanded because the current meager amount of comprehensive, reliable NPS data is a primary limitation to developing improved analytical tools to design control measures.

Nonpoint source water pollution is a broad term covering a variety of water quality problems characterized by spatially dispersed pollutant sources. Spatial variability means such pollution difficult to identify and expensive to control. While the physical characteristics of a catchment influence the rate of pollution generation, the originating processes are almost always hydrologic. Since spatially nonuniform watershed characteristics and hydrologic patterns determine the amount and type of NPS pollution, monitoring strategies capable of identifying cause-effect relationships must be sophisticated.

The diffuse nature of NPS pollution and the many physical processes and conditions that influence its generation make mathematical modeling the primary scientific methodology for devising control measures. While NPS pollution modeling is of relatively recent origin, numerous specific models have evolved. This plethora of models, while the bane of potential new users, is the inevitable result of an emerging science addressing a complex and multifaceted problem. Unfortunately, the dearth of comprehensive, reliable NPS field data is a serious handicap to the development of improved models or even a rational test of the suitability of a given model for a particular application.

Because monitoring NPS pollution is expensive and time-consuming, the value and availability of information collected should be maximized. This can be accomplished, but only if interdisciplinary needs are considered when new NPS data collection/management systems are implemented. Modern instrumentation technology and the cost of technical personnel are such that, for only modest incremental costs, the usefulness of hydrometeorological and water quality data collected could be greatly expanded. These benefits can be attained with proper design of data collection/management systems at a project's inception.

The first consideration in the development of a data acquisition system must be its effectiveness to meet individual project needs. Unfortunately, examination of current USA programs leads to the conclusion that only rarely are factors other than local project exigencies given any consideration. Broad NPS research needs and large scale acquisition system efficiencies are usually "beyond the scope" of narrowly defined, single-problem oriented projects. This philosophy has become pervasive because of pressures from funding agencies. However, researchers cannot be absolved from blame for their failure to more effectively resist this narrow, short-sighted use of public monies and their time. It is not clear whether this pattern has evolved because of poor communication skills or a lack of vision to perceive potential data base uses.

The goal of this paper is to suggest concepts and methodologies that are believed relevant to broadening the scope and utility of NPS data collection efforts.

FIELD MONITORING SYSTEM DESIGN

The availability of low cost, low power micro-processor technology and the continued rapid evolution of electronic components have revolutionized the feasible scope of field monitoring systems. However, if commercially available equipment is a reliable barometer, neither equipment manufacturers nor researchers who purchase and suggest new product configurations yet appreciate the true dimensions for improvement this technology offers. In short, with limited but noteworthy exceptions, the primary thrust has been to simply automate procedures developed for the strip-chart, single parameter recorder era.

Fundamental Concepts

Field observations of water pollution and environmental parameters that influence such pollution are essential to NPS modelers from two perspectives: to identify cause-effect phenomenon that form the foundation of deterministic watershed models and to enable comparisons between predictions and observed data when assessing the performance of particular deterministic models or developing statistically based models. While the same data can be usable for both purposes, cause-effect inferences require vastly greater detail and spatial resolution than model evaluation or statistical model building.

This paper is primarily concerned with development of field acquisition networks that will yield cause-effect data. It is the author's position that the limited utility of the more gross scale data sets often results in a poor benefit/cost ratio for fundamental research purposes.

Selecting monitoring sites within a given catchment is highly site specific and scale dependent; however, generalizations are possible. Since total catchment behavior is almost always desired by modelers, it is necessary to collect water yield and quality information at the outlet. Secondly, the spatial distribution of the primary forcing parameters throughout the watershed must be characterized. Finally, as the size of the monitored area increases, cause-effect data becomes economically feasible only from a few intensely instrumented sub-watersheds within the catchment. The inclusion of intensely monitored areas of nearly uniform physical character is essential to the cause-effect validity of the resulting data base. Selection of these small areas should be based on the extent to which that area's physical characteristics are present within the whole catchment and on convenience and serviceability with other stations.

A modern NPS field monitoring program must take into account recent instrumentation advancements as well as proven monitoring concepts of decades of hydrometeorological sciences research. In particular, new programs need to increase the number and areal distribution of parameters monitored, pay careful attention to sensor selection and their physical location within the target area, and be aware of operational economies afforded by new data storage modes developed to take advantage of the temporal characteristics of hydrologically controlled NPS data. Examples and elaboration of these three concepts seem appropriate.

What to Monitor? -- The recommendation to broaden the base of parameters monitored under each hydrometeorological project is straightforward. Modern, micro-processor based data acquisition systems have made this feasible because the primary cost of a field station is associated with its central control system and recording/communications equipment. Since basic equipment, housing and utilities are required for any size station, the per channel expansion costs are usually low for additional parameters.

The primary need is to anticipate that other NPS researchers as well as researchers in other disciplines, e.g. entomological and crop growth modeling, are likely to desire to use each data base for purposes other than outlined in the project which originally funds a monitoring effort. Monitoring a few otherwise non-critical parameters can usually be added at a minor cost while greatly increasing the scientific value of the data base.

Specific hydrometeorologic parameters required for a data base to be of general use include: instantaneous (1-minute) rainfall intensities, solar radiation, wind characteristics, soil temperatures and evaporation conditions. A wide variety of water quality parameters is also useful, although laboratory costs can make comprehensive determinations difficult to justify unless they are a prime project objective. As a minimum, determinations of nutrients and heavy metals, plus specifically suspected toxics should be routinely included.

Biological monitoring of stream and lake ecosystems is an often ignored and much needed element of any sound NPS monitoring program. In addition to the biological populations and the physical and chemical water conditions of a monitored catchment, the spatial distribution of and temporal changes to land use must be recorded.

Since NPS pollution is, by definition, generated from diffuse locations, the need to characterize the spatial distribution of physical conditions and hydrometeorological parameters over an entire catchment would seem self-evident. Unfortunately, this need substantially increases the cost and operational complexity of a monitoring system. The challenge for the system designer is to utilize advanced technology equipment, knowledge about the phenomena to be monitored (preferably enhanced with some preliminary modeling of the target catchment), and a familiarity with the physical character of the catchment to keep costs within budget while generating meaningful data.

Sensor Selection. -- Data continuity is vital to analyses of event driven phenomena typifying NPS pollution generation. Collection of continuous data is possible only with careful system design, selection of reliable equipment and meticulous attention to maintenance and recalibration.

Operational reliability of sensors and a system's recording equipment requires quality components. Even with quality components, redundancy must be designed into the monitoring scheme to achieve a high probability of continuous data from field stations.

The second major difficulty in obtaining a meaningful and continuous data base is collecting observations and samples that accurately indicate environmental conditions. The sensors selected must be stable and not distort the environmental parameter they purport to monitor. In this regard, collection and analysis of water and biological samples represent perhaps the greatest challenges.

Storage/Retrieval Techniques. -- Despite increasing storage capacities and speed of computer peripherals, automated data recording systems can generate such vast amounts of information that sophisticated storage/retrieval systems are essential to the data's utility. Langham (1971) eloquently developed the mathematical basis for a departure from the mode in which hydrometeorological data are usually stored, timed interval records. The traditional daily discrete observations (occasionally the frequency is increased to hourly intervals) of a parameter's magnitude are not only of limited value for many needs, they represent an inefficient means of storing information.

Hydrometeorological data are characterized by wide ranges in their rates of change; typically, long intervals of minor change followed by brief periods of rapid fluctuation. The classical method of recording all parameters at predetermined time intervals requires either the selection of a short interval with large volumes of redundant data or the acceptance of a high probability that rapid perturbations will not be recorded. Langham's solution to this dilemma is a recording mode called the incremental integral concept. It records the elapsed time required for each variable to change by a preselected increment. This mode, sometimes referred to as event recording, yields orders of magnitude improvements in storage efficiency, captures short duration perturbations of significant magnitude and still permits complete reconstruction of the temporal pattern of the variable. Information on acceleration rates, important to certain disciplines, is also preserved, Barrett, et al (1975).

Additional data storage efficiencies beyond the incremental integral mode can be implemented by storage formats. As an example, consider a single channel, all solid state event recorder manufactured by Omnidata International, Inc. Time-of-event information is stored in a programmable-read-only memory (PROM) erasable with ultra-violet light. Operating software for monitoring rainfall intensity with a tipping bucket rain gage achieves a one-minute time resolution for each bucket tip over a period of several months in a 2048-byte PROM. Instead of recording the number of minutes elapsed from when the recorder starts, necessitating 3 bytes of storage for each bucket tip, a unique "character" is recorded once every 4 hours. Every tip of the rain gauge records the number of minutes elapsed since the start of the last 4-hour data mark, a value that can also be stored in a single byte. This scheme allows the time of up to 1500 "events" to be stored over a 3-month period, a data capacity 2.2 times that obtainable in the same PROM with the more conventional format.

Equipment and Operational Considerations

Selection of a configuration of field transducers and data recording equipment to operate in an unattended location involves many considerations. An underlying assumption for the following recommendations is that most data will be recorded on a multi-channel device producing a computer compatible record. The older approach, an independent recorder for each parameter being monitored, is too costly when many parameters are required, often results in poor time synchronization between recorders and is incompatible with centralized communications.

While arranged in order of decreasing priority, the following list of system attributes is suggested as essential requirements, to be augmented with individual project considerations:
1. Unattended operational reliability over environmental extremes.
2. Compatibility with a wide variety of transducer outputs.
3. Battery powered operation.
4. An ability to communicate data and/or operational status of individual transducers to a central location.
5. Low unit cost for transducers and data transmission links.
Each of these items needs further elaboration to be viewed in proper perspective.

Field sensors and acquisition equipment must operate over a wide range of adverse environmental conditions with only infrequent site visits to provide maintenance. Until problems of keeping a sizable network of field instruments operational have been experienced first-hand it is difficult to appreciate the importance of this selection criteria. Central computer monitoring can detect only gross failures; it cannot detect deterioration of calibration accuracy nor effect the repair of a nonfunctioning sensor. Unfortunately, it is difficult to evaluate the operational reliability of prospective system components. Some factors that merit consideration in making an evaluation are: (1) quality of material and workmanship in the product; (2) design simplicity, especially for components utilizing electro-mechanical elements; (3) for sensors, the stability of the transducer, i.e., the principle by which the variable is sensed; (4) if available, field experience of other users of the equipment. Finally, it can hardly be overemphasized that the most effective means of assuring operational continuity and data integrity is to include as much redundancy as affordable.

The ability to accept "data" from transducers in more than one format can be a significant factor in the economic viability of a data acquisition system. This flexibility is important in three ways: incorporation of existing transducers into a new recording/communication system, ability to monitor a variety of parameters with low cost transducers, and delay of system obsolescence. While almost all modern systems require electronic signal inputs, this can take the form of a voltage, current or frequency proportional to the monitored variable or of a switch closure after a preselected increment of change. Compatibility with different signal forms will affect the feasibility of modifying an existing network of instruments to incorporate an automatic acquisition capability. Similarly, this flexibility lengthens the useful life of the acquisition/recording unit, normally an item of dominant expense, because the probability is increased of its being compatible with new transducers that might be added in the future or with monitoring a wide range of parameters.

Numerous examples could be given of locations that require environmental monitoring equipment to be battery powered because line power is not economically available. However, even when line power is available, battery operation is a very desirable, even essential, feature. Battery power can greatly increase the likelihood of an uninterrupted record of data in two ways. First, much hydrologic data is storm associated; it must be collected during periods when the reliability of line power is poorest. Second, the probability of lightning damage to field transducers and recording equipment is greatly increased when connected to line power, especially in remote regions. It is recommended that, if the use of line power is unavoidable for certain transducers, they be powered only intermittently for the shortest possible interval. This can be accomplished using battery powered relays that concurrently break the connection to all power lines to a transducer (hot, neutral and safety ground).

Costs for technically trained field personnel continue to increase rapidly relative to costs for computer-to-computer communications. Providing a mechanism for communication between a central computer and the remote field station(s) can simultaneously reduce the number of routine site visits required, improve the continuity of the data record by quickly detecting equipment malfunction and provide an enhanced operational capability via selective activation of field equipment on command from the central computer.

The latter can be particularly effective for closed-loop control. Multiple parameter condition tests of incoming data can readily be programmed into the central computer so that modified operational commands can be transmitted to the field monitoring equipment. A communications capability also provides an opportunity for redundant recording of data.

The ability of modern, micro-processor based data acquisition systems to accommodate many transducers at minimal incremental cost provides the opportunity to economically broaden the scope of data collected from field systems. Gage houses, acquisition/recording equipment and stream control sections together with wages for personnel to service a station usually dominate costs when transducer prices are reasonable. An increase in the number of parameters monitored can greatly increase the scientific value of the resulting data base and the economic justification for the entire station.

DATA MANAGEMENT CONSIDERATIONS

An implied purpose of collecting extensive field data is to make information available for subsequent scientific analysis and interpretation. If a monitoring system is well designed, the resulting data will be of value to more than a single study. The ultimate value of the collected data is often determined, especially for studies undertaken subsequent to the one generating the original data, by how well it is archived. Thus, data storage and retrieval considerations should be an integral part of the initial monitoring system design.

There are three major activities associated with any field data management system: (1) verification and editing of field data, (2) archival storage and retrieval, and (3) analysis and interpretation of data. Separation of activities (2) and (3) is especially important. Attempting to accommodate an analysis capability into the structure of the storage scheme can seriously compromise this vital element in a good data management system and is not recommended.

Verification/Editing

Data bases derived from field monitoring stations are subject to erroneous entries, missing information and periods when individual transducers are operating out of calibration. The task of eliminating these problems is substantial. Normally a combination of manual editing and automated range checking is desirable.

If the data acquisition network is connected to a central computer via a communications link as recommended, the existence of software to automatically examine all incoming data for out-of-range conditions can substantially reduce problems with the data base by early identification of field difficulties requiring on-site attention. Of course, the best technique is to include independent, redundant transducers for monitoring critical parameters. While more expensive, this approach has the advantage of detecting calibration drift as well as providing a continuous record during a transducer failure. Several specific recommendations in these areas have been presented by Wong, et al (1976).

Despite the best intentions, gaps will ultimately occur for some or all parameters. If a decision is made to fill these gaps by estimating values, e.g. data from the closest alternate station, an efficient means must be incorporated for flagging such information as estimated rather than observed.

Archival Storage and Retrieval

A most important and frequently overlooked requirements for a good data management system is a space efficient scheme for archiving the information. The most frequent mistake is selecting an archival storage scheme based on its convenience for accessing and analyzing data. Analysis functions usually occur infrequently and are short term activities in comparison with archiving requirements. Therefore, storage efficiency must be paramount when data are archived. This recommendation is in direct conflict with many general-purpose data base management structures. While such systems have much to offer for data analysis requirements and for interfacing separate simulation models, most are unsatisfactory for archival applications because of their copious storage demands.

To the extent practical, archived data should be stored in a "raw" form. That is, only clearly erroneous data should be modified before permanent storage. The use of alternate calibration coefficients, zero shift corrections, etc. should be accommodated, but stored in separate linked files or clearly identified in the original file.

Finally, it is crucial that documentation be developed about field equipment and laboratory analysis techniques employed. This should be summarized and reported as an uncertainty analysis for each monitored parameter. The importance of a quantitative estimate of inaccuracies or uncertainties associated with physical measurements cannot be over-emphasized, Kline and McClintock (1953). An explicit and comprehensive analysis of uncertainty is the only rational means by which potential users can determine the suitability of monitored data for a desired application. Similarly, provision should be made for machine-readable storage of all field notes.

Analysis and Interpretation

Data analysis should involve a distinct and separate process of transferring data from archival storage into a form convenient for the desired type of processing. This avoids compromising archival storage efficiency needs to accommodate short term or partial record analyses. The varied needs of this field make any detailed discussion beyond the scope of this paper. Only two recommendations are made: transferring data into a modern data base management system provides a flexible retrieval/analysis capability and considerable attention should be directed toward graphical data presentations.

LIMITATIONS TO NPS MONITORING

Partly as a result of demonstrated deficiencies of water quality models and partly because of naivete, the overwhelming majority of both the scientific and informed lay communities believe the only reliable way to evaluate NPS pollution and the effectiveness of control measures is by field monitoring programs. The perception, though often not explicitly stated, is that bottles of water must be collected and subjected to sophisticated laboratory analyses to establish "truth". Statistical correlations are all to often imputed to be cause-effect relationships. Modeling results are inferred to represent "theoretical or hypothetical" numbers changeable at the whim of a modeler to suit individual biases.

The primary problems of monitoring NPS pollution are not associated with the laboratory analysis of a collected sample, although significant difficulties persist for some constituents. Rather, they are with acquiring representative samples, determining the source of pollutants present in a sample, assessing the true significance of component levels, separating out influences of storm characteristics and assessing the impact of installed or proposed treatments on pollutant yields, i.e. determining cause-effect relationships.

There is a critical need for establishment of cause-effect relationships between alternative NPS control measures and the resulting pollution. Unfortunately, no economically feasible monitoring program can be devised that is capable of establishing cause-effect relationships on a watershed scale, even for areas as small as a few square kilometers, especially on a short-term basis. This situation prevails because of the storm-induced nature of NPS pollution, seasonal variations in weather patterns and the uncontrolled nature of the many factors which profoundly influence levels of such pollution. These same factors require that, to be of significant value, a NPS monitoring program must be of long duration; at least 7-10 years.

SUMMARY AND CONCLUSIONS

The lack of better quality hydrometeorologic and water quality data from various geographic regions is a primary impediment to more rapid progress in controlling NPS pollution. The scope and value of monitoring programs need to be broadened to meet the requirements for not only a range of water quality modelers, but concurrently for other disciplines. This can be accomplished at modest cost by increasing the number of parameters monitored and departing from the outdated mode of recording parameter levels at fixed time intervals.

Micro-processor based data acquisition systems offer the opportunity to obtain expanded data bases at reasonable costs; however, this will only come to fruition by properly designing monitoring programs to take advantage of this technology. The selection of data acquisition equipment should be based on its flexibility and the opportunity to increase data base integrity, via communications to a central computer system, while simultaneously reducing field labor requirements. The range of parameters monitored must be increased and recording modes that take advantage of computer technology must be adopted.

Three separate functional requirements should be incorporated into any NPS data management system: verification/editing, archival/retrieval and analysis/interpretation. In the trade-offs always involved between storage space and retrieval convenience, increasing emphasis needs to be given to efficient storage formats. Analysis requirements should be viewed independently from archival storage considerations.

Field monitoring of non-point source pollution needs to be expanded for small, single land use areas to quantify benefits of individual treatment systems. In this manner, information vital to the continued development of improved models can be obtained. Watershed scale monitoring is useful for establishing general levels of pollution and for testing the accuracy of comprehensive models or for developing statistical models, but not for establishing cause-effect relationships or component relationships for deterministic models. Monitoring efforts must be long term projects; the cost of establishing a gaging network and the year-to-year variability of storm induced data make 2-3 year efforts poor investments.

ACKNOWLEDGEMENTS

Many of the concepts and techniques discussed were developed and field tested under research concerned with non-point source pollution control from agricultural lands, sponsored by the U.S. Environmental Protection Agency and the Purdue Agricultural Experiment Station and coordinated by the Allen County Soil and Water Conservation District and the Indiana Heartland Coordinating Commission. Additional details of these studies may be found in Huggins and Mahler (1976) and in Preston (1982).

REFERENCES

1. Barrett, J.R., L.F. Huggins and W.L. Stirm. 1975. Environmental Data Acquisition. Envir. Entomology. 4:855-860.

2. Huggins, L.F. and S.J. Mahler. 1976. Environmental Data Acquisition and Real-time Computers. Proc. Best Management Practices for Non Point Source Pollution Control Seminar. EPA 905/9-76-005. pp. 164-170.

3. Kline, S.J. and F.A. McClintock. 1953. Describing Uncertainties in Single-sample Experiments. Mech. Eng. 75:3-8.

4. Langham, E.J. 1971. New Approach to Hydrologic Data Acquisition. Proc. ASCE, J. Hydro. Div. HY12:1965-78.

5. Preston, A.H., ed. 1982. Insights into Water Quality: Final Report. Chapter 5. Indiana Heartland Coordinating Commission.

6. Wong, G.A., S.J. Mahler, J.R. Barrett and L.F. Huggins. 1976. A Systematic Approach to Data Reduction using GASP IV. Proc. Winter Simulation Conf. pp 403-410.